CITRUS PROCESSING
Quality Control and Technology

Dan Kimball

Director of Quality Control, Research and Development

CALIFORNIA CITRUS PRODUCERS, INC.

 SPRINGER SCIENCE+BUSINESS MEDIA, LLC

An AVI Book
Copyright © 1991 by Springer Science+Business Media New York
Originally published by Van Nostrand Reinhold, New York, NY in 1991

International Thomson Publishing Asia
221 Henderson Road #05-10
Henderson Building
Singapore 0315

2 3 4 5 6 7 8 9 XXX 01 00 99 98 97

Library of Congress Cataloging-in-Publication Data

Kimball, Dan A.
    Citrus processing: quality control and technology / Dan A. Kimball.
        p.   cm.
    Includes bibliographical references and index.
    ISBN 978-0-442-00615-0   ISBN 978-94-011-3700-3 (eBook)
    DOI 10.1007/978-94-011-3700-3
    1. Citrus juices. 2. Citrus fruits--By-products. I. Title.
    TP562.K55   1991   90-25153
    663'.63--dc20   CIP

Contents

Preface

Citrus juices constitute the majority of the fruit juices consumed in the United States and around the world. Along with the rest of the fruit juice industry, they play a major role in the entire food industry as well. In spite of this prominence, few texts have been written on quality control technology; and most of the texts have been written by researchers who may possess great technical skill but generally are less familiar with daily routine quality control problems and concerns than quality control technologists are. On the other hand, quality control technologists and managers generally do not have the time and/or the talent to write books or communicate through scientific literature.

The author recognized the need for an updated, comprehensive, and easily understood text on citrus quality control. This text has been designed to be used not only by processors, bottlers, canners, and others involved in the citrus industry, but it can be of value to instructors and students of citrus technology. Researchers also can find value in the foundations laid down by the text, especially in regard to the needs and concerns of the processing industry. Also, consultants and marketing personnel will be greatly helped by understanding the concepts of this volume. Persons in related industries also will find many applications that can be easily adapted to their needs.

It may be impossible to foresee all problems and situations that can occur in various plants at various times. It has been said that the only thing that does not change is change itself. Constant revision of procedures, policies, and technical information should be an integral part of every quality control program. All quality control personnel should be receiving constant training, and procedures should be under continued scrutiny and development. Each company should have a quality control manual to supplement this text. The manual should outline procedures and policies specific to the individual plant's operation and facilities. Reference material should be orderly, readily available, and easy to understand.

A great lack of uniformity plagues a portion of the citrus industry, stemming

from a general lack of understanding of basic technical and managerial principles. The author hopes that this text will help unite the industry technologically,
in addition to functioning as a resource for persons who need up-to-date basic
information and techniques. The author welcomes comments and suggestions
about the text.

Chapter 1

Introduction

The commercialization of fruit juices has a long and colorful history. The first commercial fruit juice was produced in 1869 by Welch, who began bottling unfermented grape juice in Vineland, New Jersey. However, juice that could be preserved for long periods of time did not emerge until after Welch introduced the principles of heat sterilization. By the 1930s, flash pasteurization was developed, and it was during this time that fruit juices gained public notice as a significant source of vitamin C and began to increase in popularity. World War II made new demands on the juice industry, resulting in the development of dehydrated fruit juices and frozen concentrates. Prior to this time, California dominated the citrus industry with its fresh markets. With the advent of frozen concentrated orange juice and its increase in popularity, the citrus industry moved east to Florida.

Today, Florida grows approximately twice as much citrus as California, with about 90% of its fruit going to the processed markets. Florida's hot and humid climate induces thin peels that are easily scarred and develop poor color, and these cosmetic effects generally render Florida fruit inferior in the fresh markets. However, the juice flavor from such fruit is rated as superior. In California, the fresh markets still rule supreme, generating about ten times as much profit for citrus growers as processed fruit. The drier Mediterranean climate of California produces thick richly colored peels, which are more durable to scarring than Florida fruit, and deep-colored juices. Culls that do not make fresh fruit quality because of scarring, color, size, or freeze damage are sent to processing plants in California. Thus, in spite of the fact that one-third of the U.S. citrus is grown there, California is responsible for only about 10 to 15% of the juice market. Another factor that has affected the difference between California and Florida markets is the fact that in California a wider variety of agricultural products is produced over a much larger and more diverse geographical region than Florida. In Florida, citrus is king, a situation that led to the development of the Florida Department of Citrus, which has exerted great political influence in protecting a very important industry, second only to the tourist industry.

One Florida venture was to initiate a citrus industry in Brazil in order to avoid the economic damage of freezes in Florida. This budding industry was later sold to the Brazilians, who have, in recent times, emerged as a major contender in the international citrus juice market. With a few devastating freezes in Florida, it did not take long for the Brazilian citrus empire to establish markets in the United States and usurp Florida's previous position as king of the mountain in the citrus industry. Brazil has the largest juice processing plants in the world and is a leader in citrus technology in many areas.

Citrus processing consists mainly of the receiving and storage of fruit, fruit washing and grading, juice extraction, juice finishing or screening, heat treatment, packaging, and storing. Blending of previously prepared or purchased juices is very common, along with the manufacture of by-products such as pulp wash juices, oils and aromas, animal feeds, pectins, marmalades and jellies, drinks, flavors, wines, vinegars, fuels, and many more. Some areas grow citrus purely for oils or fragrances to be used in perfumes. All these products require a degree of quality control to ensure their marketability and consumer acceptance. However, even though there exists a wide diversity of citrus products, the majority of citrus products consist of 100% juices.

The chapters in this book treat the main parameters of citrus juices and some of the main by-products and other quality-control-related topics. New emerging technologies such as aseptic processing, debittering, reverse osmosis, and perhaps even supercritical extraction undoubtedly will improve the quality of citrus products but probably will have little impact on the basic principles of quality control outlined in this book. As new technologies and research advance, greater light is expected to be shed on citrus quality control and citrus processing. New concepts already are slipping over the horizon, such as those regarding the potential anticarcinogenic attributes of some components of citrus juices and the potential for debittering wastes to be used as antifeedants for some agricultural pests. Quality control personnel should be serious students of citrus research efforts and perhaps even participate in them when possible. On the other hand, researchers should be serious students of the industry and quality control technology and keep informed on what is happening in the industry because the industry is the prime benefactor of their research.

Even though approximately 70% of all juices consumed in the United States are citrus juices, it would be appropriate to mention that a significant portion of the juice industry is comprised of noncitrus juices. Citrus juices are preferred in an opaque form most closely resembling the natural juice. Other juices, such as apple, grape, and berry juices, are preferred in a clarified form, involving the use of filter presses, centrifuges, enzymatic treatments, and filters in various stages and engineering schemes. Except for this major difference between clarified and citrus juices, the quality control concepts of both are very similar, and the principles outlined in this text can be of great benefit to all. Even persons in the sugar industry can utilize much of this information in the crystallization of granular sugar as well as the manufacture of syrups and related products. Bottlers and canners of any type of juice will find much of this information valuable. Production, marketing, and other managers involved in fruit juice processing often need a reference source regarding related technology, a role that this text can fill.

The mission of a quality control department in a citrus juice processing facility can be divided into three main parts. The first is the measurement and control of juice characteristics. The procedures outlined in this text can serve as a complete guide to both the latest and traditionally accepted methods commonly used in the industry. Quality control goes beyond mere measurement of these parameters. It includes comparing these characteristics to customer or in-house specifications in determining compliance with the goals and policies of the company. Many operations await the decision of the quality control department before proceeding. This gives quality control the ability literally to control the quality of the products. Quality assurance is defined by some as being even more inclusive than this, with the added responsibility of ensuring that the products achieve a certain quality standard. This goes beyond mere measurement and reporting of discrepancies.

The second portion of the mission of a quality control department is to ensure the sanitation of the product. Good industrial hygiene involves a host of considerations, as detailed in Unit Two of the text. Modern lifestyles demand sanitary foods that are guaranteed to be free from any type of contaminants. Contamination can come from many places—the environment, a dirty plant, improper processing, inadequate sanitary procedures or enforcement, or even intentional contamination. Quality control personnel are responsible for ensuring a clean and healthy environment for food processing and safe, authentic, and palatable products. In fact, quality control personnel often are held criminally negligent in a court of law if sanitation problems reach litigation.

The third portion of the quality control mission involves management—management of people, procedures, policies, specifications, inventories, and information. Decisions have to be made on the spot, and problems must be quickly analyzed and corrected. Consequently, information must be readily available, and communication must be timely and thorough. The day will soon come when quality control decisions and functions will require the use of computers. Computers and programmable calculators already are widely used, and computer applications are an intrinsic part of this text. Citrus quality control software in GWBASIC for IBM-compatible personal computers entitled "CITQUIC" is available from the author, including not only those programs referred to in the text, but integration of these programs into a complete inventory management system and quality control program. One of the advantages of such software is that the GWBASIC programs can easily be modified to fit any commercial need.

Quality control personnel must always be prepared for any eventuality. If they wait until trouble hits, it then may be too late to gain the skills or knowledge needed to handle the situation properly. They should lay out concrete procedures and communication channels well in advance and follow those plans meticulously to avoid errors in judgment or conflicts within the company as well as with suppliers or customers. It should always be remembered, and it will always be true, that people are what make any quality control department or company successful and are a company's most valuable resource. The best thing an employee can do for the company is help it to make a profit, and the best thing a company can do for an employee is to make a profit. This mutual goal can be achieved only in an atmosphere of cooperation, integrity, and technical competence.

UNIT ONE

CITRUS JUICE CHARACTERISTICS

Chapter 2

Brix and Soluble Solids

Citrus juices contain a wide variety of chemical compounds, but none as prevalent as sugars or carbohydrates. Carbohydrates make up better than 80% of the soluble material in citrus juices, and of these soluble carbohydrates, half are in the form of sucrose. The sucrose molecule consists of one molecule of glucose and one molecule of fructose, as shown in Fig. 2-1. The other half of the carbohydrates in citrus juices consist of relatively even amounts of glucose and fructose, which result from natural enzymatic breakdown of the sucrose. Other carbohydrates play a minor role in overall juice composition. Because sucrose consists of one molecule of glucose and one molecule of fructose, the densities of aqueous solutions of sucrose mixed with equal portions of glucose and fructose are similar to the densities of 100% sucrose solutions.

Juice density is one of the most important quality control parameters in the juice industry. Juice densities are used in making conversions, back and forth, between volume and weight parameters in performing a host of important calculations in predicting juice blends and formulations and juice concentration, standardizing laboratory results, and managing inventories and marketing. Because carbohydrates occur in such high levels in citrus juices, juice density is determined by using methods and scales that apply to pure sugar solutions. The soluble material is not all in the pure form of the major carbohydrate components; so the soluble material, including some noncarbohydrates, is referred to as soluble solids rather than sugars. However, the soluble solids are treated as sugars in regard to density and other quality control measurements. Also, the soluble solids differ from the insoluble solids, such as cloud and pulp material, which contribute little to density measurements. Organic acids and their salts also occur in significant amounts in citrus juices and contribute to the soluble solids content. Corrections usually are applied to density measurements in order to account for these noncarbohydrate soluble solids as if they were carbohydrates, to facilitate the use of carbohydrate scales and tables.

Several scales have been developed to relate the measured specific gravity or density to the concentration of various solutions. The API scale, selected in

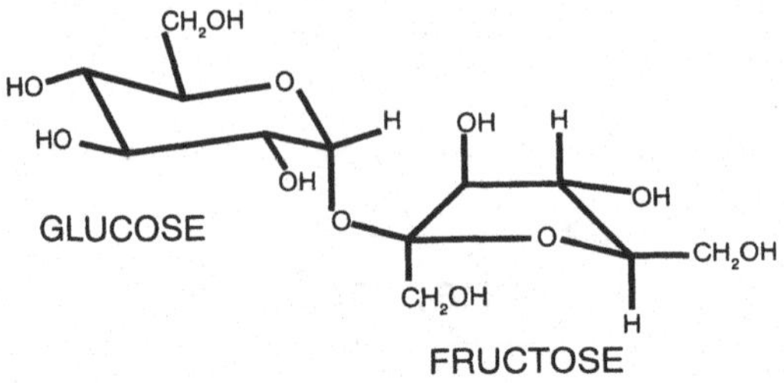

Fig. 2-1. A sucrose molecule, composed of a molecule of glucose and a molecule of fructose.

1921 by the American Petroleum Institute, the U.S. Bureau of Mines, and the National Bureau of Standards, is used for petroleum products. The Balling scale is used by the tanning and tanning extract industries. The Baumé scale was proposed in 1768 by Antoine Baumé, a French chemist, to measure the concentrations of acids and syrups that were lighter and heavier than water by using a different scale for each type of solution. The Quevenne scale is an abbreviated specific gravity scale used primarily by the milk industry. The Richter, Sikes, and Tralles scales were developed to measure the alcohol content in water via density, and the Twaddle scale is an attempt to simplify the density measurements of liquids heavier than water.

One of the best-known scales relating the concentration of sucrose solutions to solution density was published by Balling (Balling n.d.) and served as a basis for the development of the more complete and expanded table established by the German mathematician Adolf Ferdinand Wenceslaus Brix (1798–1870) in 1854. The original Brix density table, used exclusively by the sugar industry for many years, was based on the density of a sucrose solution at a standard temperature of 17.5°C. This scale based on 17.5°C still is used by the fresh citrus fruit industry for maturity tests. Domke (1912) proposed a table of densities of sucrose solutions according to concentration based on sucrose solution densities at 20°C, which has since become the standard for the processed juice industry. At least by 1941 the Brix scale was being used by the fruit juice industry in determining the sucrose equivalent of soluble solids; the term "Brix" or "degrees Brix" was being used interchangeably with the % sucrose or the % soluble solids by weight in fruit juices and was determined by using density measurements. This usage led to the Brix scale's becoming the standard for the measurement of juice concentration in the citrus and related industries.

BRIX BY HYDROMETER

The most economical commercial method of measuring the Brix of citrus juices uses a weighted spindle or hydrometer, illustrated in Fig. 2-2. The buoyancy of the hydrometer is directly proportional to the density of the solution, and the

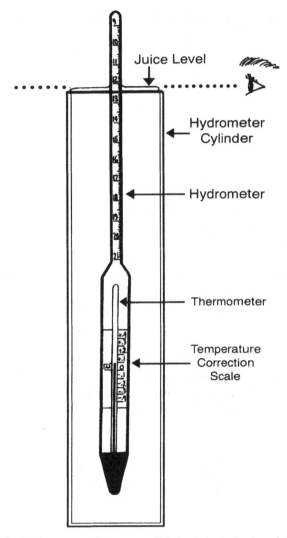

Fig. 2-2. Hydrometer used to measure Brix levels in single-strength juices.

degree that the hydrometer sinks into the juice can be calibrated to a Brix scale on the neck of the hydrometer, which can be read at the level where the juice surface strikes the hydrometer neck.

Dissolved gases affect this buoyancy, so deaeration of the juice aids in accurate Brix determinations by hydrometer. Air is incorporated into the juice during processing through agitation, mixing, and pumping.

Temperature also affects the density of the solution; therefore, many Brix hydrometers have a built-in thermometer accompanied by a temperature correction scale that can be used to correct the Brix reading. In the absence of such a scale, a thermometer and Tables 2-1 and 2-2 can be used for hydrometers based on 17.5°C and 20.0°C, respectively, to make the Brix correction for temperature. Hydrometers usually are used to measure the Brix of single-strength juices where acid corrections are small. Consequently, acid corrections generally are not used with hydrometers.

Measurement of Brix by Hydrometer

Equipment and Supplies

- Deaerating setup shown in Fig. 2-3.
- Brix hydrometer covering the Brix range of interest.
- Thermometer and Tables 2-1 or 2-2 if the hydrometer does not have a thermometer and correction scale built in.
- Hydrometer cylinder of sufficient size to allow the hydrometer to float freely in the juice.
- Adequate juicing apparatus if whole fruit is to be tested.

Table 2-1. Temperature corrections to the Brix using a hydrometer based on a standard temperature of 17.5°C (Florida Department of Citrus 1982).

Temp °C	Subtract from Brix	Temp °C	Add to Brix
5.0	0.45	18.0–18.5	0.05
6.0	0.40	19.0–19.5	0.10
7.0	0.40	20.0–20.5	0.15
8.0	0.35	21.0–21.5	0.20
9.0	0.30	22.0–22.5	0.25
10.0	0.30	23.0	0.30
11.0	0.25	23.5–24.0	0.35
12.0	0.25	24.5	0.40
13.0	0.20	25.0–25.5	0.45
14.0	0.15	26.0	0.50
15.0	0.10	26.5–27.0	0.55
15.5	0.10	27.5	0.60
16.0	0.05	28.0–28.5	0.65
16.5	0.05	29.0	0.70
17.0	0.00	29.5–30.0	0.75
17.5	0.00	30.5	0.80
		31.0	0.85
		31.5–32.0	0.90

Table 2-2. Temperature correction to the Brix using a hydrometer based on a standard temperature of 20.0°C (Charlottenberg 1900).

Temp. °C	0	5	10	15	20	25	30	35	40	45	50	55	60	70
						Observed Percentage of Sugar								
						Subtract from Brix								
0	.39	.49	.65	.77	.89	.99	1.08	1.16	1.24	1.31	1.37	1.41	1.44	1.49
5	.36	.47	.56	.65	.73	.80	.86	.91	.97	1.01	1.05	1.08	1.10	1.14
10	.32	.38	.43	.48	.52	.57	.60	.64	.67	.70	.72	.74	.75	.77
11	.31	.35	.40	.44	.48	.51	.55	.58	.60	.63	.65	.66	.68	.70
12	.29	.32	.36	.40	.43	.46	.50	.52	.54	.56	.58	.59	.60	.62
13	.26	.29	.32	.35	.38	.41	.44	.46	.48	.49	.51	.52	.53	.55
14	.24	.26	.29	.31	.34	.36	.38	.40	.41	.42	.44	.45	.46	.47
15	.20	.22	.24	.26	.28	.30	.32	.33	.34	.36	.36	.37	.38	.39
16	.17	.18	.20	.22	.23	.25	.26	.27	.28	.28	.29	.30	.31	.32
17	.13	.14	.15	.16	.18	.19	.20	.20	.21	.21	.22	.23	.23	.24
18	.09	.10	.10	.11	.12	.13	.13	.14	.14	.14	.15	.15	.15	.16
19	.05	.05	.05	.06	.06	.06	.07	.07	.07	.07	.08	.08	.08	.08
17.5	.11	.12	.12	.14	.15	.16	.16	.17	.17	.18	.18	.19	.19	.20
						Add to Brix								
21	.04	.05	.06	.06	.06	.07	.07	.07	.07	.08	.08	.08	.08	.09
22	.10	.10	.11	.12	.12	.13	.14	.14	.15	.15	.16	.16	.16	.16
23	.16	.16	.17	.17	.19	.20	.21	.21	.22	.23	.24	.24	.24	.24
24	.21	.22	.23	.24	.26	.27	.28	.29	.30	.31	.32	.32	.32	.32
25	.27	.28	.30	.31	.32	.34	.35	.36	.38	.38	.39	.39	.40	.39
26	.33	.34	.36	.37	.40	.40	.42	.44	.46	.47	.47	.48	.48	.48
27	.40	.41	.42	.44	.46	.48	.50	.52	.54	.54	.55	.56	.56	.56
28	.46	.47	.49	.51	.54	.56	.58	.60	.61	.62	.63	.64	.64	.64
29	.54	.55	.56	.59	.61	.63	.66	.68	.70	.70	.71	.72	.72	.72
30	.61	.62	.63	.66	.68	.71	.73	.76	.70	.78	.79	.80	.80	.81
35	.99	1.01	1.02	1.06	1.10	1.13	1.16	1.18	1.20	1.21	1.22	1.22	1.23	1.22
40	1.42	1.45	1.47	1.51	1.54	1.58	1.60	1.62	1.64	1.65	1.65	1.65	1.66	1.65
45	1.91	1.94	1.96	2.00	2.03	2.05	2.07	2.09	2.10	2.10	2.10	2.10	2.10	2.08
50	2.46	2.48	2.50	2.53	2.56	2.57	2.58	2.59	2.59	2.58	2.58	2.57	2.56	2.52
55	3.05	3.07	3.09	3.12	3.12	3.12	3.12	3.11	3.10	3.08	3.07	3.05	3.03	2.97
60	3.69	3.72	3.73	3.73	3.72	3.70	3.67	3.65	3.62	3.60	3.57	3.54	3.50	3.43

(Calculated from data on thermal expansion of sugar solutions by Plato and assumed that the hydrometer is of Jens 16 glass. Table should be used with caution and only for approximate results when the temperature differs much from the standard temperature or from the temperature of the surrounding air.)

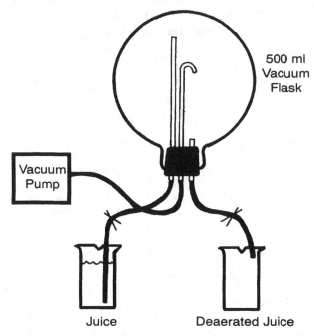

Fig. 2-3.   Deareation setup for hydrometer Brix measurements.

Procedure

1. Extract enough juice to fill the hydrometer cylinder, and place it in the juice beaker shown in Fig. 2-3.
2. Evacuate the vacuum flask with clamps closed. Open the clamp to the juice beaker, and draw the juice into the flask.
3. Turn off the vacuum source, and break the vacuum by leaving the juice beaker clamp open.
4. Open the second clamp, and allow juice to flow into the deaerated juice beaker.
5. Fill the hydrometer cylinder with the deaerated juice, and insert the hydrometer into the cylinder so that the juice overflows the cylinder. This is best done in a sink or outside.
6. Read the Brix where the level of the juice meets the Brix scale in the neck of the hydrometer, as shown in Fig. 2-2.
7. Using the thermometer, measure the temperature, and add or subtract the temperature correction to the Brix according to the scale in the hydrometer, or use Tables 2-1 or 2-2 if no correction scale is available.

BRIX BY REFRACTOMETER

Even though refractometers are more expensive than hydrometers, they require only 2 to 3 ml of sample, compared to about 200 ml or more required by the hydrometer, and deaeration generally is not necessary. Brix measurements can be performed much faster, and the Brix range of refractometers can be much broader, up to 0 to 70°Brix, which is important in processing plants that manufacture concentrated juices. These advantages quickly justify the use of refractometers in processing plants in spite of their cost. For an instrument that is used as often as a refractometer, with quality control, inventories, packaging, and marketing depending on the results, the cost of the instrument is trivial; the refractometer literally becomes the foundation of any citrus juice processing procedure. In fact, it is wise to have at least two refractometers so that if one needs to be repaired, the operations of the plant will not be hampered. Having two refractometers also provides a means of checking and comparing results to ensure accuracy.

The refractive principle that forms the basis of this type of measurement lies in the variability of the speed of light through two mediums of different density. To illustrate this principle, a simple analogy can be used. Suppose that a person is drowning offshore, as shown in Fig. 2-4, and a lifeguard is trying to determine which is the fastest way to reach her. Even though the shortest distance between two points is a straight line, it is not necessarily the fastest; in this case, the lifeguard probably can run faster than he can swim. Thus he would want to spend as much time as possible running and the least amount of time swimming. The speed that he can run and the speed that he can swim are determined by the density of the mediums through which he must pass, and determine the ideal angle that he should take to reach the drowning victim in the shortest possible time. It is a phenomenon of nature that light "knows" the ideal angle that will allow it to arrive at a certain point, going from one medium to another of different density, in the shortest possible time. As the densities of one or both of the mediums change, so does this angle. Thus a change in the angle of refraction permits the measurement of the density of citrus juices, as light passes through the juice at one density and then through a glass prism at another density, which can be translated into the concentration as degrees Brix.

The wavelength of light used with a refractometer has been shown to affect the angle of refraction, and, for this reason, a sodium vapor lamp has been suggested for use as a light source. However, any soft yellow lamp is sufficient for industrial purposes. The light first enters a fogged or ground glass lens that evenly scatters the light in all directions, as shown in Fig. 2-5. The scattered light then enters the sample and is refracted at the surface of the prism on which the sample lies. The critical ray is a ray that travels parallel to the surface of the prism and represents the minimum angle at which the scattered light coming

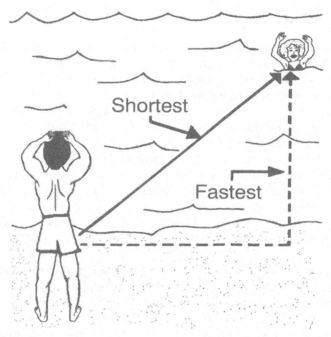

Fig. 2-4. Illustration suggesting why light refracts when going from one medium into another medium of greater density. Light has a "nose" for the fastest route and bends or refracts accordingly.

through the sample can strike the prism. Of course, a ray perfectly parallel to the surface of the prism never will strike the prism. If it changed its angle slightly, however, it would be refracted, defining a boundary for a dark zone or area, as shown in the figure. As perceived from the side, as illustrated, no scattered light coming from the surface of the prism can enter the dark zone. This dark area boundary can easily be calibrated to a Brix scale because both the Brix and the angle of refraction are determined by the sample density.

Because the standard error for Brix readings is ±0.1°Brix, the law of significant figures dictates that Brix values should be expressed only to the nearest tenth of a Brix (see Chapter 23). Brix values expressed to more decimal places than this imply deceptive accuracy. Acid corrections and temperature corrections often are expressed to two decimal places. When these corrections are applied, the final Brix value should be rounded to the nearest tenth of a Brix.

Several factors contribute to the errors in Brix readings. One of the greatest sources of errors is the fact that citrus juices, especially concentrates, produce an indistinct shadow or boundary between the dark and light zones of the refractometer due to insoluble matter in the juice. This fuzzy shadow engenders a wide range of opinions as to where the Brix value should be read. Whether

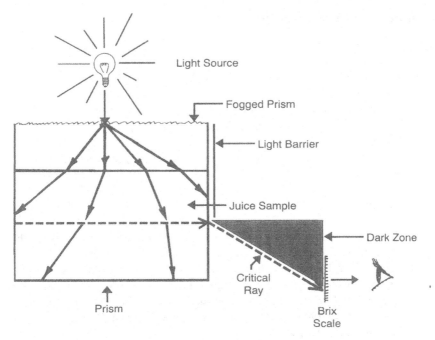

Light Source

Fogged Prism

Light Barrier

Juice Sample

Dark Zone

Critical Ray

Prism

Brix Scale

Fig. 2-5. Illustration of the formation of the shadow in a refractometer. No light can enter the dark zone because of refraction as the light passes from the sample into the prism at a different density.

the shadow is adjusted to the intersection of two cross hairs or used directly on a Brix scale itself, the Brix should be read in the middle of this fuzzy region, where it is estimated that a perfect shadow line would appear if the fuzzy line were clear and focused. Calibration of shadow readings can be achieved by having lab technicians measure the Brix of several samples and comparing the results of one lab technician to another. Those that are reading too high or too low will easily be detected and thus can be corrected. This type of comparison should be done as often as needed to ensure consistency and confidence in results. In fact, such multitechnician calibrations should be performed for all laboratory procedures, regularly or as often as needed.

Another source of Brix reading error is the refractometer itself. Most refractometers have an adjustment screw that permits calibration of the shadow. This can be done with standard sucrose solutions, but the use of distilled water to calibrate the refractometer according to the refractive index is sufficient in most cases. The refractive index is the ratio of the sine of the angle between the incident ray and the line perpendicular to the surface of the prism to the sine of the angle between the refracted ray and the same perpendicular line. The procedure for the calibration of a refractometer using distilled water is as follows.

Calibration of a Refractometer

Equipment and Supplies

- Refractometer with calibration screw.
- Distilled water.
- Dropper or plastic stirring rod.
- Wrench or screwdriver needed to adjust screw.
- Table 2-3.
- Thermometer if the refractometer does not provide for the measurement of the temperature.

Procedure

1. Measure the temperature of the water if the refractometer does not have a means of doing so.
2. Place a few drops of the distilled water onto the prism of the refractometer using a dropper or plastic stirring rod, being careful not to scratch the prism. If the temperature is to be measured by the refractometer, let the sample sit for a few minutes for the temperature to equilibrate. Otherwise, read the Brix right away.
3. Using Table 2-3, look up the refractive index for water at the temperature determined above.
4. Switch the refractometer to measure the refractive index instead of the Brix, and adjust the refractometer to read the refractive index determined from Table 2-3 regardless of where the shadow occurs.
5. Using the wrench or screwdriver, adjust the calibration screw until the

Table 2-3.   The refractive index of distilled water at various temperatures (Weast 1968).

°C	n_D	°C	n_D
10	1.3337	21	1.3329
11	1.3336	22	1.3328
12	1.3336	23	1.3327
13	1.3335	24	1.3326
14	1.3335	25	1.3325
15	1.3334	26	1.3324
16	1.3333	27	1.3323
17	1.3332	28	1.3322
18	1.3332	29	1.3321
19	1.3331	30	1.3320
20	1.3330		

shadow is aligned with the cross hairs or agrees with the refractometer reading from Table 2-3.
6. Switch the refractometer back to the Brix mode.

(The instructions may vary slightly from one instrument to another. The manufacturer's operational manual should be consulted for details.)

Temperature Corrections

Temperature fluctuations also affect the Brix reading for refractometers, and temperature corrections are required, as with the hydrometer. However, because the Brix is measured by using a different principle, the correction tables are different, as can be seen by comparing Tables 2-1 and 2-2 with Table 2-4.

Many modern refractometers make this correction automatically. By applying a double-nested least squares analysis to the data in Table 2-4, the following equation can be constructed, which can be used in computer applications instead of Table 2-4.

$$\begin{aligned}
\text{Cor}_T = {} & B^2(+1.425 \times 10^{-4} - 8.605 \times 10^{-6}T + 7.138 \times 10^{-8}T^2) \\
& + B(-2.009 \times 10^{-2} + 1.378 \times 10^{-3}T - 1.857 \times 10^{-5}T^2) \\
& + (-7.788 \times 10^{-1} + 1.700 \times 10^{-2}T + 1.100 \times 10^{-3}T^2) \quad (2\text{-}1)
\end{aligned}$$

The correction calculated above from the Brix (B) and the centigrade temperature (T) can be added to, if positive, or subtracted from, if negative, the Brix reading to give a temperature-corrected Brix.

Acid Corrections

As mentioned previously, citric acid corrections are necessary as well as temperature corrections. Appendix A gives the corrections that must be added to the Brix reading according to the % titratable acidity as citric acid, determined as described in Chapter 3. It must be remembered that other organic acids, such as malic acid, may comprise as much as 10% of the total organic acids present and may affect the Brix reading in a different manner from citric acid. Also, titration does not account for salts of the organic acids, which can be present up to about 20% of the total salts and acids and also affect the Brix reading (Shaw, Buslig, and Wilson 1983). The error in the Brix reading resulting from undetected salts and organic acids other than citric acid is usually insignificant on an industrial basis or, at least, is ignored by the industry. Again, by applying least squares regression analysis to Appendix A the following can be obtained:

$$\text{Cor}_A = {} = 0.014 + 0.192A - 0.00035A^2 \quad (2\text{-}2)$$

Table 2-4. Temperature corrections to the Brix reading using a refractometer based on a standard of temperature of 20°C (J. Intern. Sugar 1937).

Brix °C	0	5	10	15	20	25	30	35	40	45	50	55	60	65	70
						Subtract from the Brix									
10	.50	.54	.58	.61	.64	.66	.68	.70	.72	.73	.74	.75	.76	.78	.79
11	.46	.49	.53	.55	.58	.60	.62	.64	.65	.66	.67	.68	.69	.70	.71
12	.42	.45	.48	.50	.52	.54	.56	.57	.58	.59	.60	.61	.61	.63	.63
13	.37	.40	.42	.44	.46	.48	.49	.50	.51	.52	.53	.54	.54	.55	.55
14	.33	.35	.37	.39	.40	.41	.42	.43	.44	.45	.45	.46	.46	.47	.48
15	.27	.29	.31	.33	.34	.34	.35	.36	.37	.37	.38	.39	.39	.40	.40
16	.22	.24	.25	.26	.27	.28	.28	.29	.30	.30	.30	.31	.31	.32	.32
17	.17	.18	.19	.20	.21	.21	.21	.22	.22	.23	.23	.23	.23	.24	.24
18	.12	.13	.13	.14	.14	.14	.14	.15	.15	.15	.15	.16	.16	.16	.16
19	.06	.06	.06	.07	.07	.07	.07	.08	.08	.08	.08	.08	.08	.08	.08
						Add to the Brix									
21	.06	.07	.07	.07	.07	.08	.08	.08	.08	.08	.08	.08	.08	.08	.08
22	.13	.13	.14	.14	.15	.15	.15	.15	.15	.16	.16	.16	.16	.16	.15
23	.19	.20	.21	.22	.22	.23	.23	.23	.23	.24	.24	.24	.24	.24	.24
24	.26	.27	.28	.29	.30	.30	.31	.31	.31	.31	.31	.32	.32	.32	.32
25	.33	.35	.36	.37	.38	.38	.39	.40	.40	.40	.40	.40	.40	.40	.40
26	.40	.42	.43	.44	.45	.46	.47	.48	.48	.48	.48	.48	.48	.48	.48
27	.48	.50	.52	.53	.54	.55	.55	.56	.56	.56	.56	.56	.56	.56	.56
28	.56	.57	.60	.61	.62	.63	.63	.64	.64	.64	.64	.64	.64	.64	.64
29	.64	.66	.68	.69	.71	.72	.72	.73	.73	.73	.73	.73	.73	.73	.73
30	.72	.74	.77	.78	.80	.80	.81	.81	.81	.81	.81	.81	.81	.81	.81

where A is the % titratable acidity. The federal code utilizes the equation:

$$Cor_A = 0.012 + 0.193A - 0.0004A_2 \qquad (2\text{-}3)$$

which is slightly less accurate in the acid level ranges of commercial lemon and lime concentrations (CFR, Title 21, 146.132(a)(2) 1983).

The procedure for measuring the Brix level by refractometer is as follows.

Measurement of Brix by Refractometer

Equipment and Supplies

- Refractometer.
- Thermometer if refractometer does not provide for the measurement of the temperature.
- Table 2-4 if the refractometer does not automatically correct for the temperature.
- Appendix A and a % titratable acid value in order to make acid corrections to the Brix.
- Dropper or plastic stirring rod.
- Water and cleaning tissue for prism.

Procedure

1. Using the water and cleaning tissue, clean and dry the prism and the surface of the fogged glass used to scatter the light.
2. Prepare the sample by stirring with the plastic stirring rod and/or swirling, and, using the plastic stirring rod or sample applicator, place a few drops of the sample onto the prism. Care should be taken not to use metal applicators or other devices that may scratch the surface of the prism. Bubbles or foam also will distort the Brix reading.
3. Close the refractometer by lowering the fogged glass onto the sample, and make sure that it is securely in place.
4. Position the light source to shine through the fogged glass, adjusting the shadow to the cross hairs, and read the Brix, or read the Brix directly if the shadow falls directly on the Brix scale itself. Cold samples may need to sit a few minutes so that the temperatures can equilibrate.
5. Using Table 2-4 or Equation 2-1 and Appendix A or Equation 2-2, make the necessary acid and temperature corrections, and round off the corrected Brix to the nearest tenth of Brix. The standard Brix error is $\pm 0.1°$Brix. The Brix can be compared to the USDA grade standards depicted in Table 2-5.

Table 2-5. Minimum standards for Brix for USDA grades.

Orange Juice (47FR 12/10/82, 52.1551–52.1557)

Type of Juice	Grade A	Grade B
Pasteurized orange juice	11.0*	10.5*
Concentrated orange juice for manufacturing	11.8*	11.8*
Orange juice from concentrate	11.8*	11.8*
Reduced acid orange juice	41.8*	41.8*
Reconstituted reduced acid orange juice	11.8*	11.8*
Dehydrated orange juice	11.8*	11.8*

	unsw	sw	unsw	sw
Frozen concentrated orange juice	41.8	42.0	41.8	42.0
Reconstituted frozen concentrated orange juice	11.8*	11.8*		
Canned orange juice	10.5	10.5	10.0	10.5
Canned concentrated orange juice	41.8	42.0	41.8	42.0
Reconstituted canned concentrated orange juice	11.8*	11.8*	11.8*	11.8*

Grapefruit Juice (48FR 9/12/83, 52.1221–52.1230)

	Grade A		Grade B	
Type of Juice	unsw	sw	unsw	sw
Grapefruit juice	9.0	11.5	9.0	11.5
Grapefruit juice from concentrate	10.0	11.5	10.0	11.5
Frozen concentrated grapefruit juice	38.0	38.0	38.0	38.0
Reconstituted frozen concentrated grapefruit juice	10.6	10.6*	10.6	10.6*
Reconstituted dehydrated grapefruit juice	10.0	11.5	10.0	11.5
Reconstituted concentrated grapefruit juice for manufacturing	10.5		10.5	

Grapefruit and Orange Blend (6FR 11/1/72, 52.1281–52.1290)

	Grade A		Grade B	
Type of Juice	unsw	sw	unsw	sw
Single strength	10.0	11.5	9.5	11.5
Reconstituted	11.0	12.5	11.0	12.5

Tangerine Juice (2FR 7/1/69, 52.2931–52.2941)

	Grade A		Grade B	
Type of Juice	unsw	sw	unsw	sw
Concentrated tangerine juice for manufacturing	10.6		10.6	
Canned tangerine juice	10.5	12.5	10.0	12.5

*Juice solids only before or without sugar addition.

USE OF BRIX VALUES IN PROCESSING

Citrus juices are manufactured at a variety of concentrations or Brix levels, which can be adjusted up or down by using commercial evaporators, blending, or adding water. It is often more convenient to base marketing, packaging, and inventory systems on the weight of soluble solids without the water, rather than on the volume of the juice. Juice prices are based on so much per pound of soluble solids, whether the product is single-strength juice or concentrate. The weight of the soluble solids is determined by the weight or volume of the juice and the degrees Brix or concentration. The weight of the soluble solids is derived from the weight of the juice (W) and the Brix (B), by usng the following equation.

$$W_s = BW/100 \tag{2-4}$$

For example, if the net weight of a tanker of concentrate is 41,890 lb with a Brix of 60.2°Brix, the weight of the soluble solids, using Equation 2-4, is 25,218 lb. If the price of juice is \$1.30/lb soluble solids, the cost of such a tanker load of concentrate is $25,218 \times \$1.30 = \$32,783$.

Most juice products are packaged by weight, whether they be in cans, drums, or tankers. However, it is more convenient to use juice volumes in juice formulations, blends, and other in-house processing. In order to do this, the Brix, which is purely a weight/weight parameter, must employ a density conversion to convert the weight parameters to volume parameters. When blending is done, the easily measured volumes need to be converted to weight of soluble solids by using specific gravity tables or density equations according to Brix.

Many forms of density tables for aqueous sucrose solutions as a function of Brix have emerged in the past 140 years. The most accurate and up-to-date table was published by Chen (1983) and provides a basis for modern citrus quality control technology. Kimball (1986) used this table to generate regression equations that can be used instead of the table and are convenient for computer programming. The pounds of soluble solids as sucrose per gallon of solution (SPG) can be calculated from the Brix (B) and density (D in pounds of solution per gallon) as follows:

$$SPG = BD/100 \tag{2-5}$$

The density in grams of solution per milliliter (d) can be used to calculate D:

$$D = d(3785.306 \text{ ml/gal})/(453.59237 \text{ g/lb}) \tag{2-6}$$

Kimball's regression expression for d as a function of Brix is:

$$d = 0.524484e^{(B + 330.872)^2/170435} \tag{2-7}$$

which can be used in Equations 2-5 and 2-6 to give the following:

$$SPG = 0.0437691 B e^{(B + 330.872)^2/170435} \tag{2-8}$$

$$D = 4.37691 e^{(B + 330.872)^2/170435} \tag{2-9}$$

The above equations, expressed as a sole function of Brix, make it possible to use small computer subroutines in performing various quality control calculations. Some programming languages, such as RPG, cannot perform the exponential functions in Equations 2-8 and 2-9. The following Taylor series expansions of these equations can be used instead:

$$SPG = 0.0437691 B \left(1 + \sum_{n=1}^{n=6} (B + 330.872)^{2n}/(170435)^n n! \right) \tag{2-10}$$

$$D = 4.37691 \left(1 + \sum_{n=1}^{n=6} (B + 330.872)^{2n}/(170435)^n n! \right) \tag{2-11}$$

An example of a subroutine using RPG program language for Equation 2-10 can be found in Appendix B.

One application of the use of these equations is in calculating the amount of water (V_w) needed to reconstitute or dilute a given volume of citrus concentrate (V) at an excessively high Brix (B_H) to a desired Brix (B). First, the SPG_H and the SPG desired must be calculated by using Equation 2-8 (or 2-10) above, from the B_H and B values. Then the following equation can be used:

$$V_w = V(SPG_H - SPG)/SPG \tag{2-12}$$

For example, if we had 1000 gallons of 65.0°Brix concentrate and wanted to dilute it to 60.0°Brix, the water needed would be:

$$V_w = (1000 \text{ gal}) (7.135_{65°\text{Brix}} - 6.436_{60°\text{Brix}})/6.436_{60°\text{Brix}}$$

$$V_w = 109 \text{ gal}$$

A particularly irritating problem occurs when a tank is full, and the Brix is too high or too low. As there is no room for volume adjustments, some of the juice needs to be removed and replaced with concentrate of a higher Brix or water. The volumes involved can be calculated as explained in the following paragraphs.

Brix Is Too Low

If the Brix is too low in a tank filled to capacity, and we need to add a higher Brix concentrate to raise the Brix, we can use the following:

$$V_H = V(SPG - SPG_L)/(SPG_H - SPG_L) \tag{2-13}$$

where V_H is the volume of the juice that we want to replace in the tank, V is the volume desired or the tank capacity, SPG is calculated from the desired Brix using Equation 2-8, SPG_L is calculated from the original low Brix level in the tank, and SPG_H is calculated from the Brix of the concentrate that will be used to bring the Brix level up. For example, if we have a 1000-gallon tank full of 60.0°Brix concentrate that we want to raise to 62.0°Brix using 65.0°Brix concentrate, using Equation 2-13 we would get:

$$V_H = (1000 \text{ gal})(6.712 - 6.436)/(7.135 - 6.436) = 395 \text{ gal}$$

where 6.712 is the SPG desired (62.0°Brix) calculated from B, 6.436 is the SPG_L calculated from the existing low Brix (60.0°Brix), and 7.135 is the SPG_H calculated from the 65.0°Brix concentrate used for the adjustment. In other words, in this example, we would replace 395 gallons of 60°Brix concentrate with 65°Brix concentrate to get a full tank of exactly 62°Brix concentrate.

Brix Is Too High

If the Brix is too high, and we want to add water to reduce the concentration, we can use the following:

$$V_w = V_H(SPG_H - SPG)/SPG_H \qquad (2\text{-}14)$$

where V_w is the amount of water that is needed to replace a portion of the concentrate, V_H is the desired volume or the tank capacity, SPG_H is the pounds soluble solids per gallon calculated from the overly concentrated original juice Brix using Equation 2-8, and SPG is calculated from the desired Brix. For example, if we have a 1000-gallon tank full of 65.0°Brix concentrate, and we want to drop the Brix to 62.0°Brix, Equation 2-14 can be used as follows:

$$V_w = (1000 \text{ gal})(7.135 - 6.712)/7.135 = 59 \text{ gal}$$

with the SPG values calculated from the Brix values by using Equation 2-8 as before. In other words, we would need to replace 59 gallons of the 65°Brix with water to get a full tank of exactly 62°Brix concentrate.

We can easily see that involved computations are often required to account for one of the many parameters of citrus juice. Not only are these and the computations found hereafter complex, but they are often routine and must be performed over and over again. Such situations are ideally suited for computer applications. Computing devices range from hand-held programmable calculators to sophisticated multiterminal mainframe computers. Programming languages also differ considerably. However, the basic logic of how these equations can be used varies little and can be represented in the form of flow charts. These flow charts can be used as is or modified to fit any industrial need in the writing of programs for any computing device.

The flow charts in Fig. 2-6 through Fig. 2-9, and throughout the remainder of this text, illustrate some computer logic that can be used in routine citrus

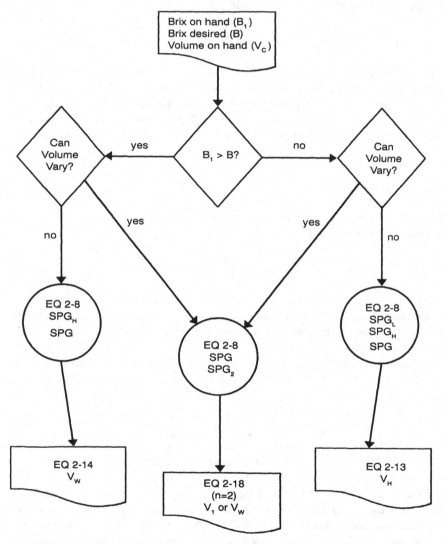

Fig. 2-6. Flow chart for a computer program that will determine the Brix adjustment in citrus juices.

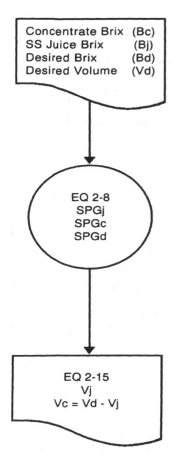

Fig. 2-7. Flow chart for programming a computer to calculate the volume of juice and high Brix concentrate to make a juice product.

quality control. There are four different box shapes, which imply specific functions. A rectangle with a curved or tear-sheet bottom represents input of data or the output of results. Within these boxes you will find the data that needs to be inputted or the variable that represents the final targeted output or answer. Diamond-shaped boxes represent decisions that have to be made by the computer or the operator. Circles represent the use of subroutines that have been found useful in more than one program or more than once in the same program. Plain rectangles represent computations that are unique to the program and do not justify the formation of a subroutine.

Figure 2-6 shows a flow chart that can be used to program a computer to calculate the amount of Brix adjustment needed in a tank with a Brix that is too high or too low, as discussed previously. The input data includes the Brix of the tank, the desired Brix, and the volume of the juice in the tank. The computer

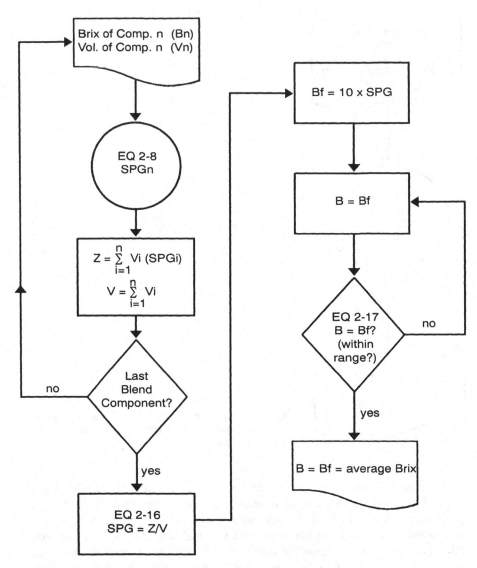

Fig. 2-8. Flow chart that can be used to program computers to calculate the average Brix of a blend.

then must determine if the Brix on hand is too high or too low compared to the desired Brix. It then must know if the volume of the juice can vary. If the tank is full, the volume cannot vary. If there is room in the tank to adjust the Brix level, then the volume can vary. Subroutines using Equation 2-8 can then be used to determine the *SPG* values needed, and Equation 2-13, 2-14, or 2-18

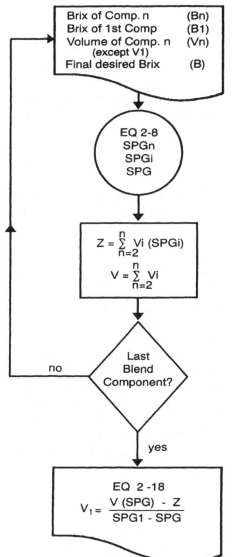

Fig. 2-9. Flow chart that can be used to program computers to calculate the volume of a single blend component needed in a blend to achieve a desired Brix.

can be used to determine the final volume of water or high Brix juice needed to achieve the desired Brix.

An example of the use of this flow chart using GWBASIC can be found in Appendix B. A similar program for the HP-41C programmable hand-held calculator can be found in Appendix C.

Fresh Cutback Juice

Because important volatile flavor components of the juice are lost during evaporation, a common, though obsolete, practice is to restore the lost flavor components by diluting high Brix concentrates with freshly extracted juice (cutback juice). This can be done by using the SPG_c calculated from the Brix of the concentrate in Equation 2-8, the SPG_j calculated from the Brix of the single-strength juice, the SPG calculated from the desired final Brix, and the corresponding volumes (V_c, V_j, and V):

$$V_j = V(SPG_c - SPG)/(SPG_c - SPG_j) \qquad (2\text{-}15)$$

For example, if we wanted to make 1000 gallons of 41.8°Brix concentrate from 12.0°Brix freshly extracted juice and 60.0°Brix concentrate, using Equation 2-15 we would get:

$$V_j = \frac{1000(6.436_{60.0°\text{Brix}} - 4.133_{41.8°\text{Brix}})}{(6.436_{60.0°\text{Brix}} - 1.047_{12.0°\text{Brix}})} = 427 \text{ gal } 12°\text{Brix juice}$$

To find the gallons of concentrate that would be needed, you can subtract 427 gallons of the single-strength juice from the needed 1000 gallons to get 573 gallons of concentrate (V_c). This method of using cutback juice has largely been replaced by the addition of aromas recovered from commercial evaporator condensates and single-strength or folded citrus oils recovered from citrus peel.

In Fig. 2-7, the volume of cutback juice and concentrate needed to make a concentrate blend can be determined. The four input values must be entered, the subroutine using Equation 2-8 must be used to calculate the SPG values, and Equation 2-15 can be used to calculate the final cutback juice and concentrate volumes. An example of a GWBASIC program using this flow chart can be found in Appendix B.

Brix Blending

The previous example of using cutback juice to blend with high Brix concentrates illustrates general juice blending using only two blend components. The sophisticated blending common in the industry today is the blending of multiple juice components in order to achieve a desired final product. One of the main specifications desired in the final product is the Brix value. To blend more than one juice component to achieve a desire Brix level or range, the following equation can be used:

$$SPG = \frac{V_1 SPG_1 + V_2 SPG_2 + V_3 SPG_3 + \cdots + V_n SPG_n}{V} \qquad (2\text{-}16)$$

where SPG and V are the final calculated values after blending n components. Once the final SPG has been calculated, the corresponding final Brix can be calculated by making successive approximations using the following equation:

$$B_f = SPG/(0.0437691\,e^{(B+330.872)^2/170435})\qquad(2\text{-}17)$$

Again, the exponential expression can be replaced with the Taylor series expansion in Equation 2-10 if it is necessary to facilitate computer programming. In using Equation 2-17, a first guess for the B value can be made by multiplying the SPG value by 10. The B_f value then can be calculated and compared to the original B value used. If they differ by more than $0.0001°$Brix, then the B_f value can become the new B value to calculate a new B_f value. The differences between the two Brix values should quickly converge to identical values, which then can be used as the final calculated exact Brix of the blend. Again, tables can be and have been used to convert back and forth between SPG and Brix values, but the use of these equations makes it possible to completely automate blend calculations.

The following example illustrates the use of these equations. Suppose that we want to calculate the final average Brix of a blend consisting of five lots (see tabulation below). The volumes of the given lots can be expressed in gallons or in 52-gallon drums, and the calculation is valid as long as the same units of volume are used throughout the calculation.

Lot #	Drums	Brix	SPG
1	3	58.8	6.273
2	2	61.4	6.629
3	1	65.3	7.178
4	11	61.4	6.491
5	8	60.1	6.450

The SPG values can be calculated from the given Brix values by using Equation 2-8 (or 2-10), or found from a table. They are given here for convenience, to illustrate the calculation procedure for the blend. Using Equation 2-16 we get:

$$SPG = \frac{3(6.273) + 2(6.629) + 1(7.178) + 11(6.491) + 8(6.450)}{25}$$

$$= 6.490$$

This SPG value can be used to determine the corresponding Brix by means of tables or Equation 2-17. A first guess for B in Equation 2-17 could be 10×6.490 or $64.9°$Brix, which converges to $60.4°$Brix quickly via successive approximations, which becomes the exact calculated Brix of the final blend.

In Fig. 2-8 the average Brix can be calculated for a blend. The Brix and the volume of the first component are entered, the subroutine using Equation 2-8 is used to obtain the *SPG* value, and summations are begun to obtain the intermediary values of Z and V. The computer must determine if this is the last record. If it is not, then the computer will prompt the operator is to input the data of the next component. This continues generating the summation values of Z and V until the data of the last blend component is entered. Equation 2-16 then can be used to calculate the average *SPG* value. The first guess in the successive approximations in Equation 2-17 is made by multiplying the *SPG* value by 10, and Equation 2-17 is used until the two Brix values are within a preset range. When the two Brix values are within range, they represent the output as the average Brix of the blend, the desired result. In Appendix B is a GWBASIC program example of the use of this flow chart. After entering the Brix and volume of the last component, you enter ''0'' when asked for the next Brix. The program will list the input data and calculate and display the average Brix and total volume. A similar program that can be used with an HP-41C programmable hand-held calculator can be found in Appendix C.

Many procedures require that the blend target a specific Brix or Brix range. In order to achieve the desired targeted Brix, the blend components need to be altered or rearranged. The final Brix can be fine-tuned to a specific target, usually by adjusting the volume or the Brix of a single component. Algebraic rearrangement of Equation 2-16 can be used to automate this procedure.

$$V_1 = \frac{SPG(V_2 + V_3 + \cdots + V_n) - V_2 SPG_2 - V_3 SPG_3 - \cdots - V_n SPG_n}{(SPG_1 - SPG)}$$

$$(2-18)$$

The *SPG* value can be determined from the desired or target Brix, and the exact amount of component 1 needed to achieve this Brix can be calculated. If a negative number is obtained, then you will know that you must vary another blend component in order to achieve the desired Brix. To illustrate the use of Equation 2-18, suppose that we wanted to know how many drums of blend component 1 in the previous example we would need to achieve exactly 60.0°Brix. Using the data given and Equation 2-18, we obtain:

$$V_1 = \frac{6.436(2 + 1 + 11 + 8) - 2(6.629) - 1(7.178) - 11(6.491) - 8(6.450)}{(6.273 - 6.436)}$$

$$V_1 = 11 \text{ drums}$$

The 3 drums of lot #1 given in the example would give a final Brix of 60.4, but it would take 11 drums to bring the Brix down to exactly 60.0. Equation

2-16 can be rearranged in several different ways to enable the varying of the Brix of a component rather than the volume, or the varying of serveral components. The more components you vary, the more complex the calculation becomes.

Figure 2-9 shows the flow chart that can be used to determine the volume needed of one blend component in order to achieve a targeted or desired Brix in a blend. The input data is, again, the Brix and volumes of the known blend components (all except blend component 1 in this case), as well as the final desired target Brix. The Brix value of the component whose volume will vary (component 1) needs to be entered as well. The Equation 2-8 subroutine is used to calculate the needed *SPG* values, and the intermediary summations are performed. If this blend component is not the last blend component, then the data from the next blend component is entered until all the blend components are entered and the proper summations performed. Equation 2-18 can then be used as shown in the flow chart to determine how much of blend component 1 is needed to achieve the desired Brix. An example of the use of this flow chart is illustrated in the GWBASIC program found in Appendix B.

Temperature Corrections to Density

The *SPG* values used above are based on the density of the juice at 20.0°C. These values are generally accurate enough for industrial purposes. Chen's table also lists densities as a function of temperature (0–50°C) at various Brix levels. Because most juices and concentrates are processed at much colder temperatures (-20 to 5°C), accounting for the effects of temperature on juice density may be in order, provided that sufficient computer resources are available. Making manual calculations to account for these temperature effects generally is not worth the trouble involved. By performing least squares regression analyses on Chen's data, the following equation was found to be accurate within the range of his data (0–50°C).

$$D = a + bB + cB^2 + T(d + e/B + f/B^2) + T^2(1/(g + hB))$$

$$a = 8.35100 \qquad e = 7.60258 \times 10^{-3}$$

$$b = 3.12158 \times 10^{-2} \qquad f = -5.24849 \times 10^{-2}$$

$$c = 1.47241 \times 10^{-4} \qquad g = -31481.1$$

$$d = -7.03313 \times 10^{-4} \qquad h = 98.7441 \qquad (2\text{-}19)$$

In the above equation, B is the Brix, T is the centigrade temperature, and D is the density in lb/gal and can be used in Equation 2-5 to calculate the *SPG* value.

QUESTIONS

1. Who developed the Brix scale and between what years?
2. Do both the fresh fruit industry and the processing industry use the same Brix scale?
3. Do Brix measurements, as described in this chapter, account for all the soluble solids in citrus juices?
4. List and compare the advantages and disadvantages of using hydrometers and refractometers in Brix measurements.
5. List six factors that affect Brix readings using a refractometer.
6. Why do hydrometers and refractometers use different temperature correction scales?
7. In determining the weight of soluble solids of citrus juices, can the Brix be used directly with weights of juices or with volumes of juices, and why?
8. Under what conditions would density tables be more convenient to use than the equations described in the chapter?
9. What is the basic inventory unit of citrus juices, and why?
10. List some reasons why you think a processor may want to blend juices.

PROBLEMS

1. If an uncorrected Brix of 10.6 were obtained from a hydrometer with a temperature reading of 15°C, what would the final corrected Brix be according to the fresh fruit industry? The processing industry?
2. One of the advantages of Equation 2-1 is that it can extrapolate temperatures at Brix values between the ones listed in Table 2-4. What is the temperature correction of a 12.0°Brix sample at 15°C according to Equation 2-1, and how does this compare to the estimated value from Table 2-4?
3. Using Equations 2-2 and 2-3, what are the acid corrections to the Brix of samples containing 0.76, 3.83, 24.76, and 37.86% acid? Compare the results to the values obtained from Appendix A.
4. You have a Brix reading from a refractometer that does not correct for temperature of 11.1°Brix. The temperature reading is 23°C with an acid level of 1.08% acid. What is the corrected Brix?
5. You have a Brix reading of 60.2°Brix from a refractometer that does not correct for temperature that gives a temperature reading of 16°C with an acid level of 5.23% acid. What is the corrected Brix?
6. How much water needs to be added to 457 gallons of 61.5°Brix concentrate in order to reconstitute it to 11.8°Brix juice? How much juice from concentrate can this make?
7. How much 64.5°Brix concentrate is needed to adjust a full 1150-gallon tank containing 39.6°Brix in order to obtain 41.8°Brix concentrate?

8. What would be the final Brix after blending the following blend components? Could this product be used to make Grade A Frozen Concentrated Orange Juice according to USDA standards?

Bulk Tank	Gallons	Brix
A	207	65.3
B	575	62.1
C	1102	12.3
D	162	39.6
E	572	59.3

9. How much 11.9°Brix cutback juice would be needed to blend with 267 gallons of 60.0°Brix juice and 472 gallons of 64.8°Brix juice in order to make FCOJ at 41.8°Brix?

10. Suppose you want to make a concentrate with a Brix value of 41.9°Brix using 8 drums of 59.7°Brix concentrate A, 6 drums of 64.6°Brix concentrate B, and cutback juice at 13.3°Brix. How many gallons of cutback juice do you need? Suppose that concentrate A is at 19°F, concentrate B is at 3°F, and the cutback juice is at 34°F. How many gallons of cutback juice do you now need in order to make the same FCOJ at 15°F? (52 gal/drum, °C = (°F − 32)5/9)

REFERENCES

Balling, C. J. N. *Gahrungschenie, 1*, pt. 1, 119.

Brix, A. 1854. *Z. Ver. deut. Zucker-Ind., 4*, 304.

Charlottenberg. 1900. Physikalisch-technische reichsanstalt, *Wiss. Abhandl. Kaiserliche Normal-Eichungs-Kommission, 2*, 140.

Chen, C. S. 1983. Accurate compilation of Brix, apparent specific gravity, apparent density, weight, and pounds solids of sucrose solutions, *Proc. Fla. State Hort. Soc., 96*, 313.

Domke. 1912. *Z. Ver. deut. Zucker-Ind., 62*, 302.

1967–1968. Weast, R. C., editor, *Handbook of Chemistry and Physics*, 48th edition, Chemical Rubber Co., Cleveland, OH, p. E-159, 1937.

J. Intern. Sugar, 39, 24s.

Kimball, D. A. 1986. Volumetric variations in sucrose solutions and equations that can be used to replace specific gravity tables, *J. Food Sci., 51*, 529.

Shaw, P. E., Buslig, B. S., and Wilson, C. W. 1983. Total citrate content of orange and grapefruit juices, *J. Agric. Food Chem., 31*, 182.

Chapter 3

Acids in Citrus Juices

Acids play a vital role in the quality of citrus juices, second only to the Brix in importance. They give the characteristic tartness or sourness of citrus products and have been acclaimed for their effectiveness as thirst quenchers. These acids and their salts replace many of the acids and salts lost by the body through vigorous exercise.

Acids in citrus fruits are formed from the energy-releasing citric acid cycle common to all life forms. This respiratory process, also known as the Krebs cycle or tricarboxylic acid cycle, breaks down stored carbohydrates to carbon dioxide and various organic acids, as shown in Fig. 3-1. This energy-generating process occurs in the mitochondria of the juice cell, as shown in Fig. 3-2. The juice cell is elongated and, when fitted together with other juice cells, forms juice vesicles that are visible in the fruit and can be manually separated from each other. Juice vesicles are organized into citrus sections in the fruit. As the juice cell grows, carbohydrate-laden fluid from the sap of the tree flows into the fruit and into the juice cell. Cell vacuoles, which have the function of storing food for the cell, absorb the aqueous carbohydrate fluid. As the fruit matures, the vacuoles grow so that they dominate the cell volume. The fluid in the vacuoles becomes the juice of the citrus fruit. As this accumulation is taking place, the mitochondria, the organelles next door to the vacuoles, are generating acids in the citric acid cycle. Carbohydrates are broken down into pyruvic acid in the membrane of the mitochondria. Once inside the mitochondria, the pyruvic acid enters into the citric acid cycle, generating the various acids.

As can be seen in Fig. 3-1, citric acid is the first acid formed in the cycle. Even though much of the way that citric acid gets into the vacuole is not understood, one of the most logical explanations appears to be that as soon as the citric acid is manufactured inside the mitochondria, it migrates through the membranes of the organelles into the juice vacuole before it can proceed with the rest of the citric acid cycle. There is evidence that mitochondria are very active throughout the lifetime of the fruit, indicating that not all of the citric acid escapes the mitochondria, but some proceeds in the cycle to provide energy

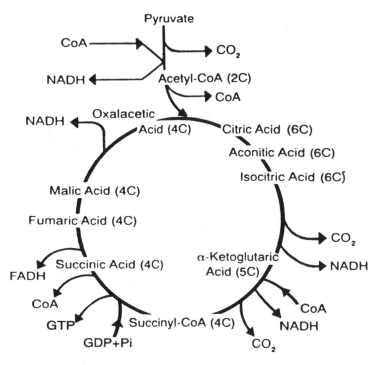

Fig. 3-1. Krebs or citric acid cycle that occurs in the mitochondria of the juice cell. Acids produced give citrus a characteristic tart flavor.

for fruit growth and maturation. Some observers have suggested that the aconitase enzyme that converts citric acid to aconitic acid may be dormant in the early stages of fruit maturity. It is postulated that this inactivity results in the early accumulation of citric acid in the fruit. As the fruit matures, and the aconitase enzyme becomes more active, the citric acid continues in the cycle, providing much-needed energy for fruit growth resulting in a slowing or end to citric acid accumulation in the juice cell vacuole. However, carbohydrates and water continue to accumulate in the vacuole, causing a dilution of acid concentration that gives a characteristic decrease of acid or tartness with fruit maturity.

The biological function of citric acid accumulation in citrus fruit is not only interesting to contemplate, but it may have a direct bearing on the prediction and understanding of citrus juice quality. There are two processes that govern citrus growth or the growth of any plant: photosynthesis, which generates carbohydrates from sunlight, and respiration in the mitochondria, as already mentioned. The first process forms the fuel, and the second burns it. The rates of these two processes, however, are not the same. At cooler temperatures, pho-

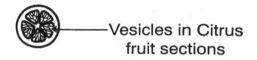

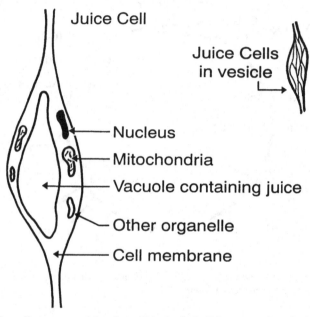

Fig. 3-2. Juice cell anatomy and location within the fruit. Juice accumulates in the vacuole and occupies the majority of the volume of the mature cell.

tosynthesis is more rapid than respiration, generating an excess of carbohydrates. At higher temperatures, respiration is more rapid than photosynthesis, resulting in a consumption of carbohydrate reserves, as illustrated in Fig. 3-3. The point where these two rates intersect and are equal is called the compensation point.

The compensation point temperature varies from fruit to fruit as well as with growing conditions. However, when this temperature is exceeded, it has been suggested that the fruit will draw on not only the carbohydrate reserve but the citric acid reserve found in the juice cell vacuoles, causing a drop in citric acid levels. This sudden drop of acid level is common with Valencia oranges in California during the heat of the summer when the fruit is being harvested, which seems to suggest that the compensation point for Valencia oranges in California is in the neighborhood of 100°F. Washington navels, which are harvested during winter months, never exhibit the same sharp drop in acid levels,

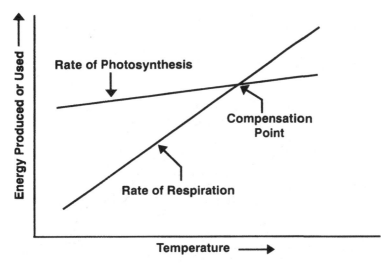

Fig. 3-3. Temperatures above the compensation point, where respiration exceeds photosynthesis, cause a high demand for citric acid, which is obtained from juice cell vacuoles. This causes a sudden drop in juice acidity.

but follow a smooth change in acid concentration characteristic of the dilution effect mentioned previously. Figures 3-4 and 3-5 illustrate seasonal changes in acid levels for California Washington navels and Valencia oranges where these trends can be observed.

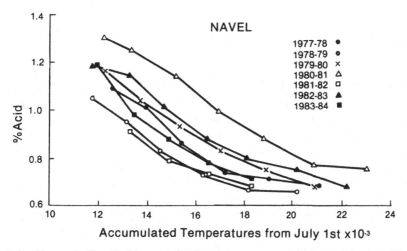

Fig. 3-4. Changes in % acid with accumulated average temperatures from July 1 just after bloom for navel orange juice. Smooth curves suggest a dilution effect in acid levels as the fruit matures (Kimball 1984). (Reprint from the *Proceedings of the Florida State Horticultural Society.*)

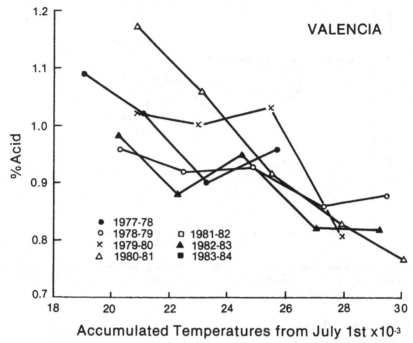

Fig. 3-5. Changes in % acid with accumulated average temperatures from July just after bloom for Valencia orange juice. Sudden drops in acid levels suggest temperatures exceeding the compensation point (Kimball 1984). (Reprint from the *Proceedings of the Florida State Horticultural Society.*)

ACID MEASUREMENTS

As mentioned in the preceding chapter, organic acids comprise a significant portion of the soluble solids in citrus juices. The primary acid found in these juices is the triprotic citric acid or tricarboxylic acid. Malic acid also is present, comprising about 10% of the total acid content. Unlike citric acid, malic acid levels remain fairly constant in the juice throughout fruit maturation. Also, as mentioned in Chapter 2, potassium and sodium salts of citric acid comprise about 20% of the total acid salt composition. These salts help buffer the acid, thus preventing sudden changes in pH during a harvesting season. These salts generally are ignored by the citrus industry, as are the differences between citric and malic acids. A single acid titration generally is performed, and the results are calculated as if all the acid were undissociated citric acid. In reality, all the acid is not citric. Because of the presence of the salts, the real structure of the citric acid in solution may be more that of a partially dissociated form, such as dihydrogen citric acid.

There are two basic ways that the acidity can be measured. Methods that determine the pH measure only the free hydrogen ions in the solution. Acid titrations with sodium hydroxide (NaOH) standards measure the total amount of acid hydrogen, whether free or undissociated. The flavor of citrus juices would be more closely associated with pH measurements because it is the free hydrogen ions that interact with the taste receptors on the tongue. However, maturity tests and Brix corrections preferably are determined by means of acid titrations, which better reflect the actual citric acid present—the reason being that pH values change very little in the acid range citrus juices and are difficult to quantify or relate to taste and maturity differences. Acid titration, the method used by the industry, provides a more stable and detailed scale for such measurements, although the results of pH and acid titrations are of course related.

Most organic acids are weak acids that change pH when titrated with a strong base, as illustrated in Fig. 3-6. The dashed curve represents the titration of a pure aqueous citrus acid solution identical in amount and concentration to its juice counterpart, which is illustrated with a solid line. The buffering action of the weak acid can be seen by the small change of pH during the early stages of the titrations. As the number of equivalents of base added approaches the number of equivalents of acid in the sample, or when neutralization occurs, a sharp inflection point emerges that signals the endpoint of the titration. Because citric

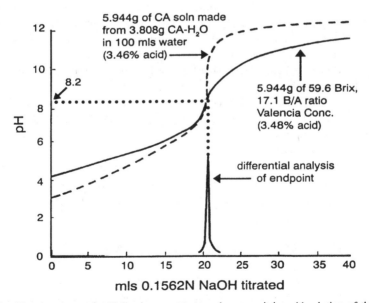

Fig. 3-6. Titration curves for Valencia concentrate and a pure citric acid solution of the same weight and concentration, illustrating the proper pH endpoint in acid titration of citrus juices.

acid is a triprotic acid, three separate inflection points would be expected, one for each hydrogen as it comes off the molecule. However, the equilibrium constants for these three dissociations are all within an order of magnitude; so a single inflection point is observed, as seen in the figure. The equilibrium constants for both of the dissociations of the diprotic malic acid are also close to those of citric acid. Thus, malic acid becomes indistinguishable from citric acid in these acid titrations.

There has been some controversy about what pH should be taken as an endpoint for an acid titration. The Association of Official Analytical Chemists suggests a pH endpoint of 8.1 (AOAC 1980). The inflection point is somewhat broad and slightly skewed, so that slight variations in what is taken as the endpoint can cause a significant error in the acid determination. However, as shown in Fig. 3-6, the pure citric acid solution gave a standard uniform inflection point that intersected the juice titration curve at a pH of 8.2. This value is used by the USDA in determining grade standards for citrus juices and concentrates, and its use was supported by Sinclair and Bartholomew (1945). Accordingly, this pH endpoint should be and generally is used in citrus juice acid determinations. Phenolphthalein indicator endpoints generally occur in this pH range and can be used if a pH meter is not available.

Even though the concentration of the sodium hydroxide used in the titration does not affect the final results, it does affect the way that the results are calculated. The Association of Official Analytical Chemists (AOAC) recommends a concentration of $0.1N$ (N = normality = number of moles of H^+ or OH^- ions per liter of solution) (AOAC 1980). The fresh fruit industry uses $0.1562N$ NaOH because all one must do is divide the milliliters titrated by the weight of the sample in order to obtain the % acid as citric acid. Another popular concentration is twice $0.1562N$ or $0.3125N$ NaOH, which is convenient in titrating high acid juices such as lemon or lime and is used in Florida for orange juice as well. The method used in Florida involves multiplying the normality (0.3125) by a factor of 0.064, followed by multiplying by 100% and dividing by the volume of the sample. The result is a w/v percentage of acid as citric acid. The full calculation is:

$$(0.3125N) \left(\frac{1 \text{ mole CA}}{3 \text{ moles OH}^-} \right) \left(\frac{192.12 \text{ g CA}}{\text{mole}} \right) \left(\frac{1 \text{ liter}}{1000 \text{ ml}} \right) \left(\frac{100\%}{? \text{ ml sample}} \right)$$

or:

$$(0.3125)(0.064)(100\%)/(\text{ml sample}) = \text{w/v \% citric acid}$$

Such a w/v % cannot be used to calculate the Brix/acid ratio, however. This value must be divided by the juice density to get the w/w % citric acid needed to calculate the Brix/acid ratio. The density of single-strength juice remains essentially unchanged between 11.0°Brix and 13.0°Brix at 1.04 to 1.05 g/ml, and the w/v % can be divided by 1.04 to get the w/w % citric acid. When one is using the California method to determine the acid in a large number of single-strength juice samples using a 10 ml pipette, the standard weight of 10.4 or 10.5 grams of sample can be used in calculating the acid level. If the NaOH solution is at a concentration other than the convenient 0.1562N, the following equations can be used to calculate the % acid:

$$\% \ acid_{based \ on \ sample \ weight} = \frac{(Normality \ used)(ml \ titrated)}{(0.1562N)(grams \ of \ sample)} \qquad (3\text{-}1)$$

$$\% \ acid_{based \ on \ sample \ volume} = \frac{(Normality \ used)(ml \ titrated)}{(0.1562N)(1.04 \ g/ml)(ml \ sample)} \qquad (3\text{-}2)$$

In routine quality control operations, large amounts of NaOH solution are consumed, as they are during the beginning of the harvest season when fruit maturation tests are performed. The following procedure can be used to make a 50-gallon lot of 0.1562N NaOH solution. The procedure involves the use of one-gallon distilled water containers that not only provide the water, but also the containers for repackaging the NaOH solution after it is made. These one-gallon containers of NaOH solution can then be sold to growers or packing houses for maturation tests, in addition to being used by the processing plant. These commercial solutions generally need to be certified by the regulatory agency when it certifies fruit maturation tests, or when fruit is mature enough to harvest. In Florida, maturity tests are performed by a regulatory agency, and twice the amount of sodium hydroxide can be used to provide 0.3125N solutions.

Preparation of NaOH Solution

Equipment and Supplies

- 50-gallon container with drain spigot at bottom.
- NaOH pellets (about 1200 g).
- 50 to 55 one-gallon containers of distilled water.
- 50 labels identifying the lot, including date and other information required by local regulatory agencies as well as the customers.
- High-speed agitator for the 50-gallon container.
- 50-ml buret and stand.
- Phenolphthalein indicator (1% in 1:1 methanol and water) or pH meter.
- Balance to weigh 1200 g of NaOH pellets.
- 25 ml volumetric pipette.

- 150 ml beaker or flask.
- About 1 g benzoic acid in 50 ml of ethanol *or* 25 ml of standard 0.1562*N* HCl solution. (Acids other than benzoic acid can be used if the proper equivalent weight is used in the caculations in the procedure.)
- 50 snap-on plastic caps that can be used to reseal one-gallon plastic containers.
- Magnetic stirrer and stirring bar (optional).

Procedure

1. Clean and dry the 50-gallon container. Empty the 50 one-gallon containers of distilled water into the 50-gallon container, discarding the used caps.
2. Add about 1180 g of the NaOH pellets to the water and agitate until the pellets are completely dissolved. This amount of NaOH should actually produce a concentration of about 0.1590*N*. However, the excess NaOH is needed to compensate for atmospheric carbon dioxide neutralization during agitation.
3. Weigh between 0.4 and 0.5 g of benzoic acid into a beaker and record the exact weight, *or* pipette 25 ml of the standard 0.1562*N* HCl solution into the beaker. If an acid other than benzoic acid is used, the 121.13 g/mole molecular weight for benzoic acid should be replaced by the molecular weight of the acid used, divided by the number of acid hydrogens per molecule of acid.
4. If benzoic acid is used, add about 50 ms of ethanol to solubilize the acid.
5. Rinse the buret with the NaOH solution from the tank, discard the rinse, fill the buret with the NaOH solution, and zero the buret or adjust the level to 0 ml.
6. Add 5 drops of the phenolphthalein indicator to the acid in the beaker and/or rinse the electrode from the pH meter, carefully dry the electrode, and insert it carefully into the acid solution so that the solution level is above the reference wire in the electrode. These electrodes should be kept in solution at all times when not in use, and the pH meter should be calibrated from time to time according to the manufacturer's instructions.
7. Titrate the acid solution with the NaOH solution while stirring or swirling it until a persistent pink color of the indicator is observed, or to a pH of 8.2 if using a pH meter.
8. Using Fig. 3-7, determine the milliliters of titrant from the buret, and calculate the normality as follows: The Agricultural Commissioner's Office in California uses standard 0.1562*N* HCl to standardize the basic solution and allows a deviation of ± 0.10 ml from the exact 25.00 ml of

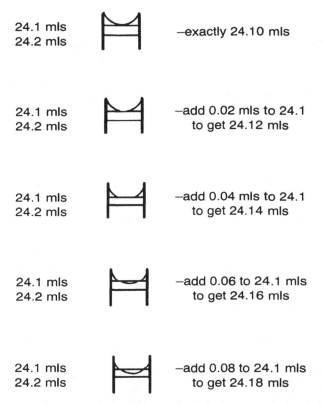

24.1 mls
24.2 mls   —exactly 24.10 mls

24.1 mls   —add 0.02 mls to 24.1
24.2 mls   to get 24.12 mls

24.1 mls   —add 0.04 mls to 24.1
24.2 mls   to get 24.14 mls

24.1 mls   —add 0.06 to 24.1 mls
24.2 mls   to get 24.16 mls

24.1 mls   —add 0.08 to 24.1 mls
24.2 mls   to get 24.18 mls

Fig. 3-7. How to read a buret properly: Place a white card with an even heavy black line behind the buret and about $\frac{1}{16}$ inch below the meniscus. The dark line of the meniscus then can be compared to the figure.

the NaOH solution needed in the titration in order to obtain exactly $0.1562N$ NaOH to certify the solution. This corresponds to an error of about $\pm 0.01\%$ acid when titrating juice.

$$\text{Normality} = \frac{(\text{grams}_{\text{acid}})\,(1\ \text{mole}/122.13\ \text{g})\,(10^3\ \text{ml}/\text{liter})}{(\text{ml titrated})} \quad (3\text{-}3)$$

or (using HCl solution):

$$\text{Normality} = (0.1562N)\,(25\ \text{ml HCl}/\text{ml NaOH titrated}) \quad (3\text{-}4)$$

For example:

$$(0.477\ \text{g benzoic acid})\,(8.188)/25.00\ \text{ml} = 0.1562N\ \text{NaOH}$$

or:

$$(0.1562N)(25.00 \text{ ml HCl}/24.65 \text{ ml NaOH}) = 0.1584N \text{ NaOH}$$

9. If the normality is too high, use the following to determine the amount of water to add where N is the high normality:

$$\text{ml water} = (50 \text{ gal})(3785 \text{ ml/gal})(N - 0.1562)/0.1562 \quad (3\text{-}5)$$

For example:

$$(50)(3785)(0.1584 - 0.1562)/0.1562 = 2665 \text{ ml water}$$

10. If the normality is too low, use the following to determine the grams of NaOH pellets needed to add to the solution where N is the low normality:

$$\text{g Na OH} = (50 \text{ gal})(3.785 \text{ l/gal})(40 \text{ g}_{\text{NaOH}}/\text{mole})(0.1562 - N)$$

$$(3\text{-}6)$$

For example:

$$(50)(3.785)(40)(0.1562 - 0.1544) = 13.626 \text{ g NaOH}$$

12. With the solution at the proper concentration, using the spigot at the bottom of the 50-gallon container, refill the one-gallon containers, snap on new sealing lids, and apply a label to each container according to the instructions of the inspector.

Handling NaOH Solutions

Sodium hydroxide solutions are very susceptible to two forms of contamination from the air. One is the absorption of moisture, and the other is the absorption of carbon dioxide. Carbon dioxide comprises about 1 to 2% of the atmosphere at sea level, and when dissolved in aqueous solutions undergoes the following reaction:

$$CO_2 + H_2O \rightarrow H_2CO_3 \text{ (carbonic acid)} \quad (3\text{-}7)$$

The carbonic acid then will proceed to neutralize the NaOH solution, changing the concentration. For this reason it is important to isolate NaOH solutions from the atmosphere. When one is using one-gallon containers, excessive opening and closing of the container as well as excessive agitation in the presence of air will result in a decrease in the NaOH normality. Syphon dispensing systems thus are recommended, as shown in Fig. 3-8. As air is drawn into the NaOH reservoir during dispensing, it passes through ascarite CO_2 absorbent and Drierite moisture absorbent. Ascarite turns white upon saturation, and Drierite turns from blue to rose red and should be changed when needed.

Both manual titration and automated titration are used in citrus processing. The advantages of auto-titration, which uses a motor-driven buret that stops at a preset pH, include faster analyses and the need for less effort by lab technicians. Manual titrations, as depicted in Fig. 3-8, are nearly as rapid as auto-

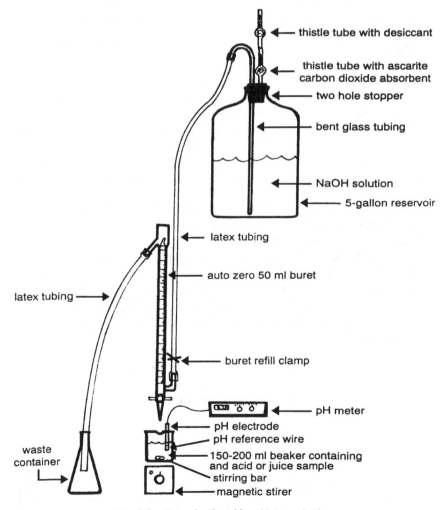

Fig. 3-8. Setup for titratable acid determination.

mated methods and cost much less. Either method is suitable for industrial purposes. The following titration procedure assumes a manual titration.

Acid Titration

Equipment and Supplies

- Setup shown in Fig. 3-8.
- 0.1562N NaOH.
- Phenolphthalein indicator (1% in 1:1 methanol and water) or pH meter.
- Lab balance for concentrate or 10 ml pipette for single-strength juice.

Procedure

1. Pipette 10 ml of single-strength juice or weight 5 to 10 g of concentrate into a 200 ml beaker. Record the exact weight of the concentrate.
2. Add 5 drops of phenolphthalein indicator, or insert a clean dry pH electrode into the solution so that the reference wire is in contact with the solution.
3. Insert a magnetic stirring bar, and place the beaker on the magnetic stirrer, or, if a magnetic stirrer is not available, use a 200 ml Erlenmeyer flask and swirl it during titration.
4. Fill the buret with the 0.1562N NaOH solution (0.3125N for lemons or limes) by opening the clamp at the base of the buret as shown in Fig. 3-8 until the buret is automatically zeroed.
5. Titrate with the NaOH solution until a slight darkening of the juice persists. The combination of the orange color of the juice and the pink color of the indicator is brown. If a pH meter is used, titrate to a pH of 8.20.
6. Read the buret to the nearest 0.02 ml as shown in Fig. 3-7 and calculate the % acid as citric acid using one of the following equations:
   - *Single-strength juice*:

$$\% \text{ acid} = (\text{ml titrated})/10.4 \text{ g} \qquad (3\text{-}8)$$

   - *Orange, tangerine, or grapefruit concentrate*:

$$\% \text{ acid} = (\text{ml}_{titrated})/(g_{conc.}) \qquad (3\text{-}9)$$

   - *Lemon or lime concentrate*:

$$\% \text{ acid} = (\text{ml}_{titrated})(0.3125N)/((g_{conc.})(0.1562N)) \qquad (3\text{-}10)$$

The USDA grade standards for acid content are generally incorporated in the Brix/acid ratios, discussed in Chapter 4. Only canned orange juice (47FR 12/10/82, 52.1557) and lemon juice products have acid standards (CFR, Title 21, 146.114, 146.120, 146.121.). In California and Arizona sweetened or unsweetened canned orange juice must have an acid level of 0.70 to 1.40% acid for Grade A, and 0.60 to 1.40% acid for Grade A in other areas. Grade B sweetened canned orange juice requires 0.60 to 1.60% acid in any location, and unsweetened canned orange juice requires 0.55 to 1.55% acid for the same grade. Lemon juice requires at least 0.70% acid in reconstituted lemonade formulations in order to meet the Standards of Identity for Frozen Concentrate for Lemonade and Frozen Concentrate for Artificially Sweetened Lemonade. The Standards of Identity for Lemon Juice require at least 4.5% acid.

PROCESSING LEMON OR LIME JUICES

Concentrating Lemon (or Lime) Juice

Because lemon and lime juices contain a higher percentage of organic acids by weight than carbohydrates, lemon and lime juice concentrations are determined as the weight of acid as anhydrous citric acid (ACA) per unit volume of juice—grams ACA/liter or *GPL* (grams per liter). Instead of the sucrose equivalent weight of soluble solids being the standard for inventories and marketing, the weight of acid is the standard unit of measure. Even so, Brix measurements are necessary in order to obtain densities in quality control calculations and in the making of concentrates, as in-line refractometers are convenient to use. There is no accurate in-line citric acid detector available for lemon evaporators, and there are no accurate density scales available based on citric acid. Also, there are few evaporators dedicated solely to lemon processing. The use of refractometers and the Brix scale makes it possible to process other types of citrus fruit. The lemon or lime *GPL* values can be calculated from Brix data and acid measurements.

Most lemon concentrates are evaporated to 400 or 500 $\pm$ 5 *GPL*. To accomplish this, the uncorrected observed Brix that is desired must be determined so that the evaporator operator will know at what Brix level to operate the evaporator. Hence, the acid and the temperature-corrected Brix of the inbound single-strength lemon juice must be known along with the % acid. First, the SPG_j of the single-strength juice must be found from the Brix by using Equation 2-8 (or 2-10) and the density d from Equation 2-7. The following equation then can be used to calculate the SPG_c of the final desired lemon concentrate:

$$SPG_c = (GPL_{desired})(1 \text{ liter}/1000 \text{ ml})(100\%)(SPG_j)/Ad \quad (3\text{-}11)$$

or:

$$SPG_c = (GPL)(SPG_j)/10Ad \qquad (3\text{-}12)$$

where A is the % acid of the single-strength lemon juice. The Brix corresponding to SPG_c can be found by using Equation 2-17 or equivalent density tables. This is the acid-corrected Brix, B_c. Because the evaporator operator needs to know the uncorrected observable Brix, B_o, the acid correction for this Brix must be determined and subtracted from B_c. This can be done by finding the density, d_c from B_c by using Equation 2-7, which enables the calculation of the final acid concentration, A_c, according to:

$$A_c = (GPL_{\text{desired}})(1 \text{ liter}/1000 \text{ ml})(100\%)/d_c \qquad (3\text{-}13)$$

or:

$$A_c = GPL/10d_c \qquad (3\text{-}14)$$

Next, by using Equation 2-2 or Appendix A, the acid correction to B_c according to A_c can be determined and subtracted from B_c to give the needed uncorrected Brix. If the temperature of the concentrate can be determined, the temperature correction to the Brix also can be subtracted from B_c by using Equation 2-1 or Table 2-4.

For example, if we wanted to concentrate 9.7°Brix single-strength lemon juice with 5.37% acidity to 400 *GPL*, using Equation 3-12, we would get:

$$SPC_c = \frac{(400 \; GPL)(0.838_{SPG@9.7°\text{Brix}})}{10(5.37\% \text{ acid})(1.0359 \text{ g}/\text{ml}_{@9.7°\text{Brix}})} = 6.026 \text{ lb sol}/\text{gal}$$

which gives 57.0°Brix by using Equation 2-17. If we use Equation 2-7, 57.0°Brix corresponds to a density of 1.2680 g/ml, which in turn is used in Equation 3-14 to give:

$$A_c = (400 \; GPL)/(10(1.2680 \text{ g}/\text{ml})) = 31.55\% \text{ acid}$$

Using this value in Equation 2-2 or Appendix A gives an acid correction to the Brix of 5.7°Brix. The corrected Brix (57.0°Brix) minus the correction (5.7°Brix) is 51.3°Brix, which the evaporator operator should adjust the evaporator to produce according to an in-line refractometer. If the temperature of the concentrate is determined to be 28°C, the temperature correction to the Brix of 0.64°Brix also should be subtracted from the 51.3°Brix, to give 50.7°Brix, which should be run on the evaporator according to Equation 2-1 or Table 2-4.

As with the other quality control calculations discussed thus far, the above calculation lends itself to computer application. A flow chart outlining the steps in such a program is shown in Fig. 3-9. A suggested GWBASIC program using Fig. 3-9 can be found in Appendix B. A similar program, which can be used with an HP-41C programmable hand-held calculator, can be found in Appendix C.

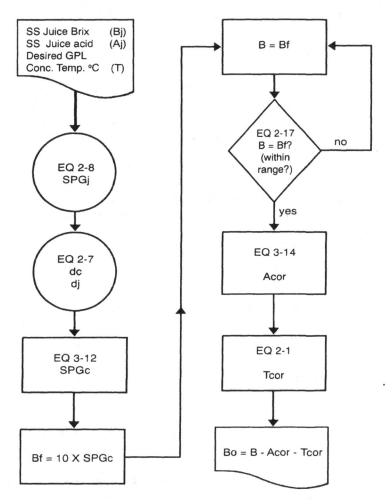

Fig. 3-9. Flow chart that can be used to program computers to calculate the uncorrected Brix needed from an evaporator when processing lemon concentrate.

Determining the Concentration of Lemon (or Lime) Concentrate

The actual concentration of lemon or lime concentrates can be determined by using the following equation:

$$GPL = Ad(1000 \text{ ml/liter})/100\% = 10Ad \qquad (3\text{-}15)$$

where A is the % acid, and d is the density calculated from the corrected Brix by using Equation 2-7. For example, if we were to make a 56.2°Brix lemon concentrate with an acid content of 31.50%, Equation 3-15 would become:

$$(10)(31.50\% \text{ acid})(1.2634 \text{ g/ml@56.2 Brix}) = 398 \text{ } GPL$$

which is within the standard 400 ± 5 GPL specification. Figure 3-10 shows a flow chart that can be used to write a computer program for such a calculation, such as the sample GWBASIC program found in Appendix B. A similar program that can be used with a hand-held HP-41C programmable calculator can be found in Appendix C.

Since GPL and SPG are similar parameters, the blend equations (2-12 through 2-16) for other citrus juices can be used for lemon juice blending simply by replacing all SPG values with GPL values. For example, suppose we want to dilute 1000 gallons of 420 GPL concentrate to 400 GPL. The water needed can be calculated from Equation 2-12:

$$V_w = (1000 \text{ gal})(420_{GPL} - 400_{GPL})/400_{GPL} = 50.0 \text{ gal}$$

If we have a full 1000-gallon tank of 380 GPL concentrate, and we want to raise the concentration to 400 GPL using 500 GPL concentrate, Equation 2-13 can be used:

$$V_H = (1000 \text{ gal})(400_{GPL} - 380_{GPL})/(500_{GPL} - 380_{GPL})$$
$$= 167 \text{ gal } 500 \text{ } GPL \text{ conc.}$$

Thus 167 gallons of the 500 GPL concentrate would be needed to replace 167 gallons of the 380 GPL concentrate in order to bring the concentration up to the needed 400 GPL.

Similarly, if we have a full 1000-gallon tank of 418 GPL lemon concentrate and want to dilute it to 400 GPL, using Equation 2-14 we obtain:

$$V_w = (1000 \text{ gal})(418_{GPL} - 400_{GPL})/(418_{GPL}) = 43 \text{ gal}$$

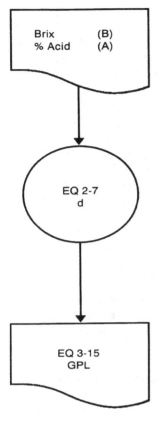

Fig. 3-10. Flow chart that can be used to program computers to calculate the GPL of lemon concentrates.

This is the amount of the 418 *GPL* concentrate that would have to be replaced with water to get 1000 gallons of exactly 400 *GPL* lemon concentrate.

In blending several lemon concentrates, the final average GPL_f can be calculated as follows, using the data given in the tabulation and Equation 2-16:

Lot #	# of Drums	GPL
1	9	398
2	4	409
3	3	500
4	16	404
5	5	395

$$GPL_f = (9(398) + 4(409) + 3(500) + 16(404) + 5(395))/ (9 + 4 + 3 + 16 + 5) = 410 \; GPL$$

REDUCED ACID JUICES

A recent advance of citrus processing is the use of weak anionic exchange resins to remove organic acids from citrus juices (Varsel 1980). This technique is advantageous in situations where blending may not be so practical in reducing acid levels. However, the use of ion exchange may be permitted only for juices that allow such in the standards of identity, such as Reduced Acid Orange Juice. Nineteen different types of ion exchange resins have been approved by the Food and Drug Administration (FDA) to date for use in food processing (CFR, Title 21, 173.5–173.25.), and many more are expected to emerge. Weakly basic resins favor the removal of stronger organic acids, such as citric acid, which is important in citrus processing where retention of the weaker nutritional ascorbic and folic acids is preferred. The presence of larger amounts of citric acid also favors its removal during ion exchange. The reaction that takes place during this process is:

$$3R^+ \cdot OH^- + C_6H_5O_7^{3-} \rightarrow (R^+)_3 \cdot C_6H_5O_7^{3-} + 3OH^- \qquad (3\text{-}16)$$

The positive unit of the resin (R^+) exchanges its hydroxide ion to neutralize the juice acidity. Overall, citrate is removed, and water is formed. As the resin becomes saturated with citrate, the resin efficiency declines, and regeneration with a basic solution is required.

Two types of columns containing these resins have been considered. An up-flow fluidized bed approach has the advantages of treating juices with normal pulp levels and of enhancing efficiency by more vigorous and thorough contact between the juice and the resin. The main disadvantage is the difficulty in making resins of sufficient particle size or density that they will not be carried out with the juice. A down-flow packed column approach has the advantage of increased surface availability of the smaller resin, which enhances contact efficiency as well as the efficiency of regeneration. The main disadvantage is that the juice needed to have the pulp removed to a level of at least 2% or less in order to avoid plugging the column. The high cost of depulping centrifuges or UF membranes adds to the capital costs. Backwashing the column to remove entrapped pulp may determine how often regeneration needs to be done, rather than saturation of the resin with citrates. Overall, the down-flow process has proved to be the preferred industrial technique. The larger the particle size is, the less the depulping required. Insoluble solids removed in depulping can be restored after column treatment to replenish the original character of the juice.

As explained in Chapter 15, the acids in citrus juices serve to guard against pathogenic organisms that flourish in less acid foods. For this reason, care should be taken in manufacturing reduced acid juices so that the pH does not remain above 4.6 for any length of time. Timely blending of reduced acid juices with other juices is a convenient way to achieve this. The standards of identity specify that reduced acid concentrate must have a Brix/acid ratio of 21–26 (CFR, Title 21, 146.148.).

Ion exchange and its cousin, adsorption techniques, are being used to remove bitter principles in citrus juice, and often are used in conjunction with acid removal. A single resin can be used for both. If only debittering is desired, saturation of the resin with commercial acids as a last step in regeneration will prevent significant citric acid loss during resin treatment.

QUESTIONS

1. What contributions do acids make to the overall quality of citrus juices?
2. Where exactly is citric acid manufactured in citrus fruit?
3. Why does citric acid accumulate in fruit, and how and why does it change as the fruit matures?
4. Why are pH measurements not used to measure the acid content of citrus juices?
5. What two compounds are found in the air that can affect the concentration of NaOH solutions, and how do they?
6. In measuring the citric acid levels in lemon juices, why are solutions of NaOH used that have a higher normality?
7. Are Brix measurements necessary for lemon juices, since concentrations, inventories, and marketing are based on the amount of acid only?
8. What is the standard *GPL* error range?
9. What are the standard *GPL* values for lemon or lime concentrates?
10. What is the difference between *SPG* values and *GPL* values when it comes to blend calculations?

PROBLEMS

1. What would be the % acid of a lemon concentrate sample weighing 1.684 g using 10.62 ml of a $0.7810N$ NaOH solution?
2. In the standardization of a 5-gallon NaOH solution, the meniscus on the buret is flat on the 25.00 ml mark with just the tip of the meniscus below the mark, and 25 ml of a $0.1562N$ HCl solution was being titrated. How much NaOH or water must be added to get exactly $0.1562N$ NaOH?
3. Exactly 28.12 ml of $0.3125N$ NaOH solution was used to titrate 10 ml of single-strength lemon juice. What is the % acid?
4. A 9.462 g sample of orange concentrate was titrated with a $0.1562N$ NaOH solution with the meniscus on the buret just touching the 45.60 mark. What is the % acid?
5. How much water do you need to add to a 20-gallon NaOH solution standardized with 0.436 g benzoic acid using 22.42 ml of NaOH solution in the titration?

6. What is the *GPL* of a lemon concentrate containing 32.46% acid with a Brix of 51.2?
7. At what Brix would the evaporator operator run the evaporator when concentrating an 8.9°Brix single-strength lemon juice with an acid content of 4.98% in order to get a 500 *GPL* concentrate at 18°C?
8. How many gallons of water do you need for 1216 gallons of 496 *GPL* lemon concentrate in order to make 400 *GPL* concentrate?
9. If you have a full 3264-gallon tank of 408 *GPL* lime concentrate and want to change it to a full tank of 400 *GPL* lime concentrate, what do you do?
10. How many gallons of 398 *GPL* lime concentrate do you need to blend with the following in order to get a final 500 *GPL* product? How many gallons do you have total? (*Hint*: See Equation 2-18.)

Lot #	Gallons	GPL
1	?	398
2	520	520
3	1344	494
4	672	512
5	365	540

REFERENCES

AOAC. 1980. *Official Methods of Analysis*, 13th edition, 22.061.

Kimball, D. A. 1984. Factors affecting the rate of maturation of citrus fruits, *Proc. Fla. State Hort. Soc.*, 97, 40.

Sinclair, W. B., and Bartholomew, E. T. 1945. Analysis of organic acids of orange juice, *Plant Physiol.*, 20, 3.

Varsel, C. 1980. Citrus juice processing. *Citrus Nutrition and Quality*, S. Nagy and J. Attaway, eds. ACS Symposium Series 143, Washington, D.C., 237.

Chapter 4

The Brix/Acid Ratio

The empirical Brix/acid ratio, found by dividing the acid-corrected and temperature-corrected Brix by the % titratable acidity w/w as citric acid (B/A ratio), is one of the most commonly used indicators of juice quality as well as fruit maturity. In California, the fruit harvested for the fresh fruit markets needs a B/A ratio of at least 8 : 1 or 8, whereas the fruit harvested for juice in Florida must have a B/A ratio of at least 10 : 1 or 10. Even through fruit destined for juice in Florida requires a B/A ratio of 10, commercial Florida juices must have a B/A ratio of at least 13, which can be achieved through blending. Consumers of citrus juices generally prefer a B/A ratio of 15 to 18, depending on the product and individual tastes.

It is generally agreed that taste is a four-dimensional phenomenon consisting of sweet, sour, salty, and bitter. Taste is sensed by receptors about 60 microns in diameter, which are budlike structures having numerous microvilli about 2 microns long that project into the taste pores on the tongue. Taste is detected by a disturbance of the electrical charges on the receptor due to contact with a food substance. These receptors are very sensitive to abuse, and are replaced every 12 to 17 days. The receptors respond to more than one taste, in various combinations and intensities.

In citrus products, the sourness of the organic acids and the sweetness of the sugars compete for the same receptor sites on the tongue. Thus the actual amount of sugar or acid is of less importance in the taste of citrus products than the ratio of the two; and this is why the B/A ratio (or simply just the ratio) is so important as a flavor quality indicator in citrus juices. It also means that an overabundance of acid sourness or sugar sweetness can be blended out by using various combinations of the high- and low-ratio juices. This is done by using the volume (V_n), the pounds soluble solids/gallon (SPG_n) found from the Brix and Equation 2-8 and the B/A ratio (R_n) for each component of the blend of n components, in the following equation:

$$R = \frac{V_1 SPG_1 + V_2 SPG_2 + V_3 SPG_3 + \cdots + V_n SPG_n}{V_1 SPG_1/R_1 + V_2 SPG_2/R_2 + V_3 SPG_3/R_3 + \cdots + V_n SPG_n/R_n}$$

$$(4\text{-}1)$$

If the Brix values of all the components are essentially the same, the more compact equation below can be used instead of Equation 4-1:

$$R = \frac{V_1 + V_2 + V_3 + \cdots + V_n}{V_1/R_1 + V_2/R_2 + V_3/R_3 + \cdots + V_n/R_n} \qquad (4\text{-}2)$$

For example, if we wanted to calculate the final ratio of the following components, Equation 4-1 would be used as shown below.

Lot #	# Drums	B/A Ratio	Brix
1	12	13.0	65.3
2	9	16.7	60.3
3	18	12.3	62.4
4	6	21.8	59.6
5	4	18.9	60.3

$$R = \frac{12(7.178) + 9(6.477) + 18(6.768) + 6(6.382) + 4(6.477)}{12(7.178)/13.0 + 9(6.477)/16.7 + 18(6.768)/12.3 + 6(6.382)/21.8 + 4(6.477)/18.9} = 14.3$$

Using equation 4-2, we obtain:

$$R = \frac{12 + 9 + 18 + 6 + 4}{12/13.0 + 9/16.7 + 18/12.3 + 6/21.8 + 4/18.9} = 14.4$$

Each equation gives a slightly different answer because the Brix values are slightly different. The wider the range of Brix values, the greater the difference will be in the results of the two equations. Figure 4-1 shows a flow chart that can be used to program a computer to calcuate the average B/A ratio. Appendix B contains a sample GWBASIC program that utilizes the flow chart in Fig. 4-1.

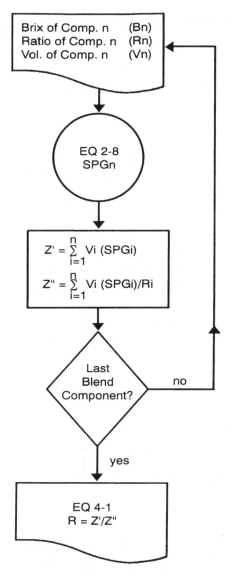

Fig. 4-1. Flow chart that can be used to program a computer to calculate the average Brix/acid ratio in a blend.

Suppose that you want to know how much of a certain component you will need to in order to achieve a certain desired ratio. Equations 4-1 and 4-2 can be rearranged to the following:

$$V_1 = \frac{V_2 SPG_2(1/R - 1/R_2) + V_3 SPG_3(1/R - 1/R_3) + \cdots + V_n SPG_n(1/R - 1/R_n)}{SPG_1(1/R_1 - 1/R)}$$

(4-3)

and:

$$V_1 = \frac{V_2(1/R - 1/R_2) + V_3(1/R - 1/R_3) + \cdots + V_n(1/R - 1/R_n)}{(1/R_1 - 1/R)}$$

(4-4)

For example, if we wanted to calculate how much of Lot #1 we would need to achieve the 14.3 ratio calculated in the previous example, using Equation 4-5 and 4-6 we would obtain with $R = 14.3$:

$$V_1 = \frac{\begin{array}{c}9(6.477)(1/R - 1/16.7) + 18(6.768)(1/R - 1/12.3) + 6(6.382) \\ (1/R - 1/21.8) + 4(6.477)(1/R - 1/18.9)\end{array}}{7.178(1/13.0 - 1/14.3)}$$

$V_1 = 11$ drums

or:

$$V_1 = \frac{\begin{array}{c}9(1/R - 1/16.7) + 18(1/R - 1/12.3) \\ + 6(1/R - 21.8) + 4(1/R - 1/18.9)\end{array}}{1/13.0 - 1/R}$$

$V_1 = 14$ drums

Here, the two equations again give different answers because the latter equation does not take into account the higher Brix of the first component and thus is less accurate than the first equation. Also, 11 instead of the original 12 drums were calculated because of rounding-off errors. Apparently 11 or 12 drums will give us essentially the desired 14.3 overall B/A ratio. Figure 4-2 gives a flow chart that can be used in programming the above calculations. A sample of a GWBASIC program using this flow chart can be found in Appendix B.

In reviewing the last three chapters and the given examples of computer ap-

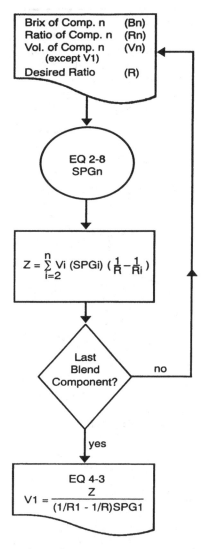

Fig. 4-2. Flow chart that can be used to program a computer to calculate the volume needed of one blend component to achieve a desired B/A ratio.

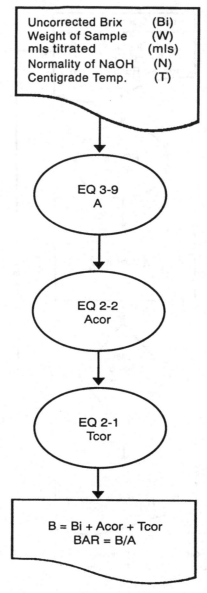

Fig. 4-3. Flow chart that can be used to program a computer to calculate the B/A ratio from laboratory data.

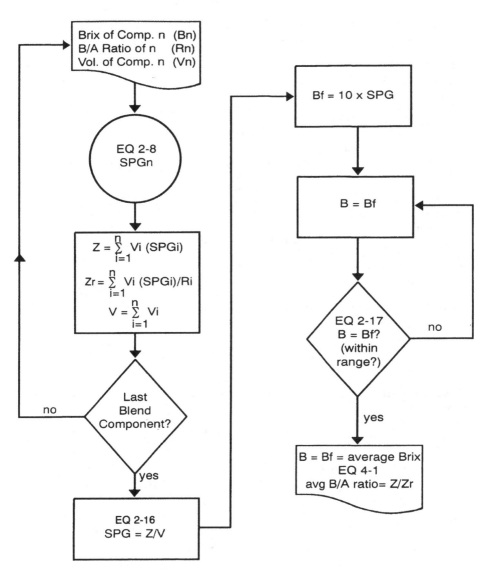

Fig. 4-4. Flow chart that can be used to program computers to calculate the average Brix and the Brix/acid ratio of a blend simultaneously.

Table 4-1. USDA standards for Brix/acid ratios.

Type of Juice	Grade A unsw	Grade A sw	Grade B unsw	Grade B sw
Orange Juice (47FR 12/10/82, 52.1551–52.1557)				
Pasteurized orange juice				
California/Arizona	11.5–18.0	12.5–20.5	10.5–23.0	
Other areas	12.5–20.5		10.5–23.0	
Concentrated OJ for				
manufacturing	8.0–24.0		8.0–24.0	
Orange juice from concentrate				
California/Arizona	11.5–18.0	12.5–20.5	11.0–23.0	
Other areas	12.5–20.5		11.0–23.0	
Reduced acid orange juice	21.0–26.0		21.0–26.0	
Dehydrated orange juice	12.0–18.0		10.5–19.0	
Frozen concentrated OJ				
California/Arizona	11.5–19.5	12.0–19.5	10.1+	
Other areas	12.5–19.5	13.0–19.5	10.1+	
Canned orange juice				
Less than 11.5°Brix	10.5–20.5		9.5–20.5	
Greater than				
or equal to 11.5°Brix	9.5–20.5		9.5–20.5	
Canned concentrated OJ	11.5–20.0	12.0–20.0	9.5–20.0	10.0–20.0

Grapefruit (48FR 9/12/83 52.1221–52.1230)

Type of Juice	Grade A unsw	Grade A sw	Grade B unsw	Grade B sw
GfJ and GfJ from conc.	8.0–14.0	9.0–14.0	7.0+	9.0+
Frozen concentrated GfJ	9.0–14.0	10.0–13.0	7.0–16.0	8.0–13.0
Conc. GfJ for				
manufacturing	6.0+		5.5+	
Dehydrated grapefruit	8.0–14.0	9.0–14.0	7.0+	9.0+
juice				

Grapefruit and Orange Blend (6FR 11/1/72, 52.1281–52.1290)

Type of Juice	Grade A unsw	Grade A sw	Grade B unsw	Grade B sw
Single strength				
Less than 11.5°Brix	9.5–18.0		8.0+	10.5+
Greater than 11.5°Brix	8.5–18.0	10.5–18.0	8.0+	10.5+
Reconstituted				
Less than 11.5°Brix	9.5–18.0		9.0+	10.5+
Greater than 11.5°Brix	9.5–18.0	10.5–18.0	9.0+	10.5+

Tangerine (2FR 7/1/69, 52.2931–52.2941)

Type of Juice	Grade A unsw	Grade A sw	Grade B unsw	Grade B sw
Concentrated TJ for				9.0–21.0
maunfacturing	9.0–18.0			
Canned tangerine juice	10.5–19.0	11.5–19.0*		11.5*

*If the Brix is above 16.0, the B/A ratio can be below 11.5.

plications, one may notice that the programs can be consolidated and used to calculate the Brix, % acid, B/A ratio, and water needed to adjust the Brix if necessary. Figure 4-3 is a flow chart that can be used to construct such a program along with the sample GWBASIC program found in Appendix B.

In blending, the average Brix and the B/A ratio can be calculated by combining the flow charts of Fig. 2-8 and Fig. 4-1, to create the flow chart shown in Fig. 4-4. This flow chart can be used to generate a program such as the one found in Appendix B using GWBASIC language. In this program, after adding the data of the components, you merely need to press "enter" and the results will appear. You also can add more data after a calculation is made. A similar program for HP-41C programmable calculators can be found in Appendix C.

Because the standard error for B/A ratios is ±0.1, these ratios should be expressed or rounded off to the nearest tenth of a ratio point. Table 4-1 gives the USDA B/A ratio grade standards for citrus products.

In the fresh fruit industry, once the fruit is mature enough for harvesting, little attention is paid to the B/A ratio. However, juice processors are concerned about the B/A ratio throughout the season. The rate of increase is linked to the rate of changes in the Brix and acid levels in the fruit, and of these two parameters, the change in the acid levels generally has the greater effect on the B/A ratio. Brix levels range from about 9 to 15, whereas acid levels may range from 0.5% to 1.5%, a threefold increase. Smooth acid changes, as depicted in the last chapter for Washington navel oranges, have been shown to generate linear relationships with heat accumulation, as shown in Fig. 4-5. These linear relationships have been empirically correlated with such parameters as crop size (C in tons/acre), tree growth or age (Y), maximum accumulated temperatures between May and August just after bloom (M), and accumulated heat during the growing season (H), according to the following equation:

$$R = H(0.0629Y - 0.03076C + 1.237)/1000 + 6.607M/1000 - 78.42$$

$$(4-5)$$

Equation 4-5 gave a confidence level of 98% in a statistical t-test comparison between calculated and measured B/A ratios for the seven seasons illustrated in Fig. 4-5. Even though Equation 4-5 may not be accurate with citrus growth in other geographical areas or under different conditions, it does illustrate some of the main influences on the B/A ratio. Further research is needed in this area.

QUESTIONS

1. What are the minimum B/A ratios required for harvesting oranges in the United States?

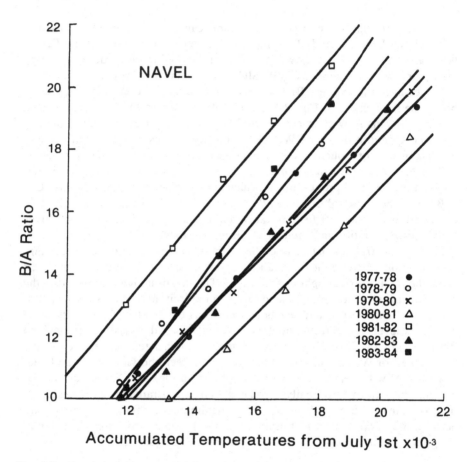

Fig. 4-5. Correlation of heat versus B/A ratio development for navel oranges from fruit samples from three California packing houses (Kimball 1984). (Reprint from the *Proceedings of the Florida State Horticultural Society.*)

2. Which of the four dimensions of taste is affected by citrus juices?
3. What actually causes taste, and what type of compounds that are found in citrus juices are most likely to create taste sensations?
4. Is it true that if you add enough sugar to the juice it will become sweet regardless of how much acid is present?
5. What is the major factor that influences the development of the B/A ratio throughout a growing season?

PROBLEMS

	Blend A				Blend B		
Lot #	# Drums	Brix	B/A Ratio	Lot #	# Drums	Brix	B/A Ratio
1	4	60.1	14.6	1	16	66.4	10.2
2	10	59.5	21.6	2	3	58.7	14.6
3	7	60.0	18.7	3	11	62.8	19.1
4	24	60.4	13.9	4	8	60.5	16.3
5	9	59.9	15.1	5	5	64.0	9.9

1. What is the average B/A ratio of blend A, and which equation would be best to use?
2. What is the average B/A ratio of blend B, and which equation would be best to use?
3. How many drums of 59.8°Brix 20.4 B/A ratio are needed to bring up the ratio in blend B to 14.0?
4. How many drums of 60.1°Brix 22.3 B/A ratio juice are needed to bring up the B/A ratio in blend A to 16.4?
5. Suppose that it is September and you want to know if navel oranges will be mature enough for harvesting in California by November 1. You know that the summation of the maximum summer temperatures from May through August gave 10,879 degree days, that the estimated crop load is about 13.5 tons/acre, and the estimated summation of daily temperatures from July 1 up until November 1 is 10,702 degree days for the upcoming season ($Y = 10$). Based on this information and that found in the chapter, is it likely that navel oranges will be mature enough to harvest by November 1?

Chapter 5

Testing of Fruit Samples

When fruit is brought into a processing facility, proper procedures must be followed in order to fairly credit the owner of the fruit for that which is processed. The fruit owner may be a member of a cooperative or part owner of the plant, or the fruit may have been purchased by the plant itself. In some areas, this process is regulated by governmental agencies in order to ensure impartiality, such as is done in Florida. Because it is most efficient for processors to operate continuously, batch processing of the fruit belonging to an individual grower generally has been replaced by the collection of a random fruit sample from the fruit prior to processing. An analysis made from tests on this fruit sample is used to determine the value of the load of fruit brought into the plant. These tests include juicing the fruit sample to determine the theoretical juice yield, as well as measuring the Brix and acid levels to determine the theoretical weight of soluble solids that will be produced from the corresponding load of fruit. There are many ways that this can be done. The following suggested procedure involves the use of a weigh ticket that originates at the weigh station, a copy of which accompanies the fruit sample so that the results can be associated with the proper load of fruit.

Analysis of Fruit Samples

Equipment and Supplies

- Random fruit collector (preferably mechanical).
- Sample containers (you should have enough containers to hold one day's worth of samples).
- Fruit sample conveyer (belt, rollers, hand truck, pallet, etc.).
- Scales to weigh samples (preferably washable scales).
- Fruit receipt or weigh ticket for sample identification.
- Brix measuring equipment (see Chapter 2).
- Acid measuring equipment (see Chapter 3).

- Juice container or bucket (2–3-gallon).
- Ladle and juice sample containers (numbered paper cups work well).
- Juice extractor, preferably one that is similar to or simulates commercial extraction in the plant.

Procedure

1. Obtain a random representative sample weighing approximately 30 to 40 lb, using a mechanical sampler and adequate sample containers. Be sure that a fruit receipt or sample identifier is attached to each sample. Convey the samples to the sample extracting area.
2. Weigh the sample and note the gross weight on the receipt or on a fruit record form.
3. Extract the fruit according to the design of the extractor, preferably in a manner similar to the commercial method used in the plant. Collect the juice in a 2–3-gallon container or bucket.
4. Reweigh the empty fruit sample container, and subtract its weight from the sample gross weight to get the net weight of the fruit sample.
5. Weigh the bucket of extracted juice to get the gross juice weight. Using the ladle, stir the juice in the bucket, and fill the hydrometer tube if you are using a hydrometer, or the juice sample container if you are using a refractometer.
6. Discard the remaining juice in the bucket, and reweigh the empty bucket. Subtract this weight from the gross weight to get the wet weight of the juice.
7. If a hydrometer is used, measure the Brix as explained in Chapter 2. Once the corrected Brix is determined, use a portion of the juice in the hydrometer tube for the acid titration.
8. The juice sample can be used for Brix and acid determinations if a refractometer is used.
9. The acid titration can be performed as explained in Chapter 3, and the Brix, % acid, and Brix/acid ratio should be recorded on the fruit sample receipt or record.

The yield can be reported as gallons of juice per ton of fruit, but the most popular yield expression is pounds of soluble solids per ton of fruit. This can be calculated from the fruit sample data by use of the following equation:

$$sol/ton = (2000 \ lb/ton)(W_j)(F)(B)/(W_f)(100\%) = 20W_jFB/W_f \quad (5\text{-}1)$$

where W_j is the net weight of the juice, W_f is the net weight of the fruit, B is the corrected Brix, and F is a factor applied to account for fruit loss during grade-out in normal processing. The generally accepted value of this factor for oranges, grapefruit, and tangerines is 0.85, allowing for a 15% loss of fruit, whereas for lemons and limes the accepted factor is 0.78.

The gallons of juice per ton of fruit can be found by first obtaining the density D (lb/gal) from the Brix by using Equation 2-9 or 2-11, and then using the following:

$$\text{gal/ton} = (2000 \text{ lb/ton})(W_j)F/W_f D \qquad (5\text{-}2)$$

This value is helpful in estimating the size and uses of processing equipment.

Also of value in estimating the size and uses of the equipment employed in peel processing, as well as in crediting fruit growers for the resulting profits in the sale of peel products, is calculation of the pounds of peel per ton of fruit. This can be done by using fruit sample data, as follows:

$$\text{peel/ton} = (2000 \text{ lb/ton})(W_f - W_j)/W_f \qquad (5\text{-}3)$$

Each of the three above parameters can be used with the total net weight of the fruit to give the total weight of solids, gallons, and weight of peel from a given load of fruit. For example, if 20,172 lb of fruit were brought into a processing plant with a random fruit sample weighing 34.1 lb (including a 3.8-lb container) that yielded 14.2 lb of juice (including a 1.1-lb bucket) with a corrected Brix of 12.6, Equations 5-1, 5-2, and 5-3 would give the following results:

$$\text{sol/ton} = 20(13.1)(0.85)(12.6)/30.3 = 92.6 \text{ sol/ton}$$

$$\text{gal/ton} = 2000(13.1)(0.85)/(30.3)(8.745) = 84.0 \text{ gal/ton}$$

$$\text{peel/ton} = 2000(30.3 - 13.1)/30.3 = 1135 \text{ lb peel/ton}$$

As with other quality control calculations, fruit sample data processing can be performed most efficiently on a computer. Figure 5-1 is a flow chart that can be used to write the appropriate programs. A suggested GWBASIC program using such a flow chart can be found in Appendix B, with a similar program for HP-41C programmable calculators in Appendix C.

Even though fruit sample analysis is used to credit fruit owners for the fruit processed, processors generally are responsible for ensuring that the results of such tests are sufficiently accurate. Also, the processors need to monitor actual yields to ensure that they compare to the theoretical yields of the fruit samples, so that they will not shortchange the fruit owners or themselves. For example, if, in a given week, a processor recovers 100,000 lb of soluble solids but credits the fruit owners with 110,000 lb of soluble solids, he will have shortchanged himself by 10,000 lb of soluble solids, which he will have to pay for from some

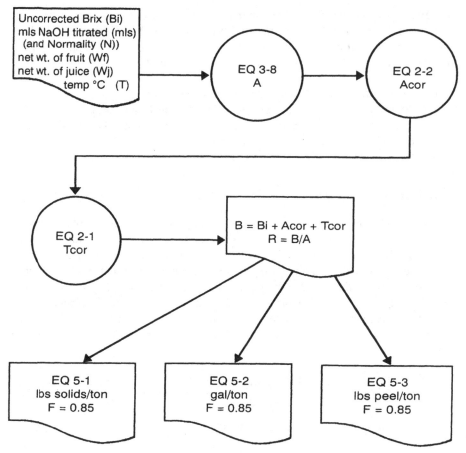

Fig. 5-1. Flow chart that can be used to program a computer to calculate the fruit sample data needed to credit a fruit owner for his or her fruit that is processed.

other funds. Conversely, if he credits the fruit owners with only 60,000 lb of soluble solids, he will have cheated the fruit owners out of 40,000 lb of soluble solids. This is why it is important to obtain as representative fruit samples as possible, with results that match actual industrial yields as closely as possible. Production efficiency can easily be measured by using:

$$\% \text{ efficiency} = S_a(100\%)/S_e \qquad (5\text{-}4)$$

where S_a is the lb soluble solids of *actual* recovery, and S_e is the lb soluble solids *estimated* from the fruit sample analyses. The ideal % efficiency is 100 to 110%, which gives the processor the freedom and the economic means to make slight adjustments in returns to fruit owners and to remain financially solvent. If the % efficiency is outside the ideal range and/or fruit sample results appear out of line, the following checks should be made:

1. Check fruit sample data and calculations for errors.
2. If possible, collect another sample and redo the fruit sample test.
3. Check the sample extractor to see if it is simulating actual production, and make legal adjustments if necessary.
4. Check the commercial extractors to see if they are producing proper yields.

LEMONS AND LIMES

Lemons (and limes) require a slightly different approach to fruit sample analysis because their juice concentration is based on the weight of acid per unit volume. The procedure for extraction of the juice from lemon fruit samples is the same as with oranges, and the juice is tested in the same manner except that the NaOH solutions used in the acid titration may be more concentrated here. Once the fruit has been tested, the calculations of the results differ from those performed for other citrus fruit. First, the density, D, needs to be calculated from the Brix by means of Equation 2-9 (or density tables), and used in the yield equations below:

$$\text{lb acid/ton} = \frac{2000 \text{ lb/ton}(W_j)(\% \text{ acid})F}{W_f(100\%)} = \frac{20(\% \text{ acid})W_jF}{W_f} \quad (5\text{-}5)$$

$$\text{gal/ton} = 2000 \text{ lbs/ton }(W_j)F/W_fD \quad (5\text{-}6)$$

where W_j is the net weight of the lemon juice, W_f is the net weight of the lemon fruit, and F is the production loss factor of 0.78 for lemons and limes. For example, a lemon fruit sample weighing 31.2 net lb yielding 11.6 lb net weight of juice with a corrected Brix of 9.7 and a % acidity of 5.37% would give:

$$20(11.6)(5.37)(0.78)/(31.2) = 31.1 \text{ lb acid/ton}$$

$$2000(11.6)(0.78)/(31.2)(8.644) = 67.1 \text{ gal/ton}$$

The lb peel/ton is calculated the same way as for orange fruit samples, mentioned previously. Figure 5-2 shows a flow chart that can be used in programming computers to perform these calculations, and a sample GWBASIC program can be found in Appendix B. A similar program for the HP-41C programmable calculator can be found in Appendix C.

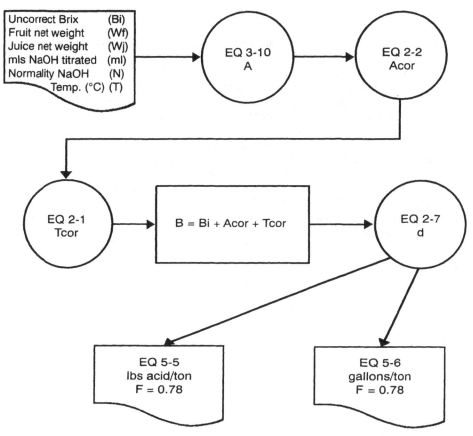

Fig. 5-2. Flow chart that can be used to program computers to calculate data from lemon or lime fruit sample testing.

QUESTIONS

1. Why are actual yields not used to credit fruit owners for juice and by-product yields?
2. What is the purpose of the fruit receipt or ticket?
3. Why and how often should actual yields be determined?
4. Why is a different factor, F, used with lemon fruit samples compared to other citrus fruit?
5. What would probably be the cause of an 80% level of efficiency in processing tangerines?

PROBLEMS

1. What would be the yield in sol/ton, gal/ton, and peel/ton of a fruit sample with a weight of 27.6 lb (including a 3.4-lb container) that yielded 13.4 lb (including a 1.1-lb container) of juice that had a corrected Brix of 13.4?
2. If grower A sent 19.6 tons of fruit into a processing plant that gave a yield of 106.2 lb sol/ton, and grower B sent in 22.4 tons of fruit with a yield of 92.2 lb sol/ton, who would receive more money for the fruit?
3. Is the efficiency of a plant that obtains 315.6 tons of oranges in one week with a total solids of 31,150 lb credited to fruit growers or owners, and which produces 5073 gallons of 60.0°Brix concentrate from that fruit, within an acceptable range?
4. What would be the yield in lb acid/ton, gal/ton, and peel/ton of a lemon fruit sample with a net weight of 28.6 lb yielding a net weight of 9.8 lb of juice with a corrected Brix of 9.2 and 6.12% acid?
5. If the fruit sample in problem 4 were grapefruit instead of lemon, what would the fruit sample yields be?

Chapter 6

Citrus Oils, Aromas, and Essences

The sweetness of carbohydrates and the sourness of organic acids impart about the same flavor characteristics to all citrus juices. However, the distinctive flavor of an individual variety can be attributed to varying components in the oils, aromas, and essences found within the juice. Although some compounds contribute more to flavor than others, exact citrus flavor duplication would require a delicate balance of a host of contributors. About 87% of the fresh squeezed flavor of orange juice has been restored to freshly evaporated concentrate by using d-limonene, ethyl butyrate, citral, and acetaldehyde (Ahmed, Dennison, and Shaw 1978). It is well known that oxygenated hydrocarbons are major contributors to citrus flavor, with d-limonene acting more as a carrier of these flavors than an actual contributor itself. The excess d-limonene levels, however, produce the characteristic ''oil burn'' taste that can irritate the skin or eyes. The d-limonene in the peel will burn the edges of the mouth when one is eating fresh fruit, and d-limonene fumes will burn the eyes during juice extraction in unventilated areas. Natural variations of the delicate balance of flavor-producing compounds are primarily responsible for vintage and off years. Processing also has an effect on these compounds and can make or break the quality of a particular citrus product.

Not only do citrus oils, aromas, and essences contribute to juice quality, but they can be commercially recovered and used to enhance juices and other food products as well as be used in other industries. Because these oil products were extracted from natural sources, the citrus fruit itself, they can be added back to juices without violating federal standards of identity. This method of restoring citrus flavor to concentrates generally has replaced the older method of adding freshly squeezed cutback juice to high-Brix concentrates. Citrus oils also have been used in the manufacture of textiles, chemicals, and other commodities. Their toxicity to certain insects, such as houseflies, fire ants, and fleas, as well as their herbicidal and flammable nature, suggest other promising uses.

OIL COMPOSITION

Citrus oils are found primarily in oval-shaped sacs in the flavedo or colored portion of the peel, and they act as a natural toxic barrier to many microorganisms and insects. Citrus oils are composed primarily (over 90%) of d-limonene, a sesquiterpene, with other monoterperenes and sesquiterpenes being found in trace amounts.

In orange oils, 111 volatile constituents have been found (Shaw 1977), including 5 acids, 26 alcohols, 25 aldehydes, 16 esters, 6 ketones, and 31 hydrocarbons. Nonvolatile constituents comprise about 1.5% of the orange oils, including waxes, coumarins, flavonoids, carotenoids, tocopherols, fatty acids, and sterols.

Grapefruit oil is characterized by the presence of nootkatone, which gives a grapefruit flavor and aroma. It consists of 20 alcohols, 14 aldehydes, 13 esters, 3 ketones, and 14 monoterpenes and sesquiterpenes. The nootkatone increases with fruit maturity. The nonvolatile portion of grapefruit oil comprises about 7% of the oil, including coumarins, flavonoids, tocopherols, and waxes.

Tangerine oils contain the distinctive nonvolatile tangeretin in significant amounts (4%), along with 24 alcohols, 11 aldehydes, 4 esters, 2 ketones, 7 acids, 24 hydrocarbons, and 2 ethers.

Lemon oils are believed to contain about 2% nonvolatiles, primarily in the form of coumarins, along with 18 alcohols, 16 aldehydes, 11 esters, 3 ketones, 4 acids, and 23 hydrocarbons.

Lime oils have been known to contain 12 alcohols, 7 aldehydes, 4 esters, 1 ketone, and 22 hydrocarbons, with 7% nonvolatiles, mostly in the form of coumarins.

OILS IN CITRUS JUICES

The hydrocabrons are the major constituents of citrus oils, primarily as d-limonene (over 90%). In lemon and lime oils, significant amounts of α-pinene occur (about 5%). As mentioned above, d-limonene acts as a carrier of flavor for other oxygenated compounds and contributes little to flavor itself. However, excess d-limonene can be detected as an oil burn on the tongue and thus should be kept in check. For these reasons, oil levels in citrus products have been considered important quality parameters and generally merit routine monitoring.

Scott and Veldhuis (1966) developed a method of measuring the oil in citrus juices (or any product containing citrus oils) that has been accepted by the industry, and their method (known as the Scott method) generally has replaced the Clevenger method, which involved steam distillation and collection of the oil from the product. The Scott method is much easier and faster, with greater accuracy and precision, than the Clevenger method. Scott oil values are higher then Clevenger levels, suggesting a more efficient analysis. The Scott method involves bromine addition to the two double bonds in d-limonene, the measurement of d-limonene as a representation of the oil level having become standard in the industry. Citral, the *cis–trans* mixture of neral and geranial that occurs in lemon and lime oils in quantities up to 4%, reacts with half as much bromine as d-limonene. Also α-pinene reacts with the same amount of bromine as d-limonene, even though α-pinene has only one double bond. Scott attributed this phenomenon to some sort of secondary reactions or molecular rearrangement of the α-pinene.

Because bromine is very volatile in pure form or in solution, straight standard bromine solutions are unreliable in concentration. Scott used a bromide–bromate salt solution to titrate the acidified alcoholic distillate from a juice sample, as this salt solution was found to have a very stable concentration. As soon as the bromide–bromate solution comes into contact with the acidified distillate, an acid-catalyzed oxidation–reduction reaction takes place between the bromide and the bromate to form bromine, which can react immediately with the double bonds of d-limonene, according to the following equation:

$$BrO_3^- + 6H^+ + 5Br^- \rightarrow 3Br_2 + 3H_2O$$

$$\begin{array}{cccc} \text{bromate} & \text{acid} & \text{bromide} & \text{bromine} \\ \text{ion} & & \text{ion} & \end{array}$$

$$(6\text{-}1)$$

The $E°$ value for the above reaction is $+0.43$ V, which is indicative of the thermodynamic favorability of the reaction. Once the bromine has been generated, it proceeds to react with d-limonene, and the other minor constituents mentioned above, according to the following reaction:

$$(6\text{-}2)$$

2Br₂ bromine + d-limonene → brominated d-limonene

What makes this basic chemistry effective as a tool in routine industrial quality control is detection of the endpoint by means of methyl orange, a common acid–base indicator. Bromine reacts preferentially with d-limonene until the unbrominated terpene is exhausted. It then reacts with the pink or violet methyl orange according to the following reaction:

Methyl Orange

Br_2 + NaO_3S—⟨ ⟩—N=N—⟨ ⟩—N $(CH_3)_2$
Bromine

NaO_3S—⟨ ⟩—N—N—⟨ ⟩—N $(CH_3)_2$

Brominated Methyl Orange

(6-3)

The conjugated resonance structure of the double bonds in methyl orange is responsible for its color. Bromination disrupts this conjugation and results in a loss of color, with the color loss signaling the end of the titration. Titration beyond the endpoint causes a slight yellow color to appear, which resembles the color of dilute unreacted bromine. Scott reported that the use of a blank in the titration produced a value of 0.2 ml or 0.0008% oil, which generally can be disregarded. Experience in working with blanks for specific citrus products will facilitate their proper use.

Scott Method for % Oil

Equipment and Supplies

- Setup shown in Figs. 6-1 and 6-2.
- 10 ml auto-dispenser or graduated cylinder.
- 25 ml pipette or graduated cylinder.
- Hot pad to enable handling of hot glassware.
- Isopropyl alcohol.
- Dilute HCL solution made from equal parts concentrated HCl and water. (Care should be taken to add the concentrated acid to water under a fume hood or in a well-ventilated area. Never add water to concentrated acid.)
- 0.1% methyl orange indicator (0.1 g dry methyl orange to 100 ml of distilled water).
- $0.0247N$ Br^-–BrO_3^- solution. (Dissolve 0.688 g of $KBrO_3$ and 3 g (excess) KBr in one liter of distilled water. This solution should be standardized.)
- Two 100 ml volumetric flasks for standardization.
- 1 ml pipette for standardization.

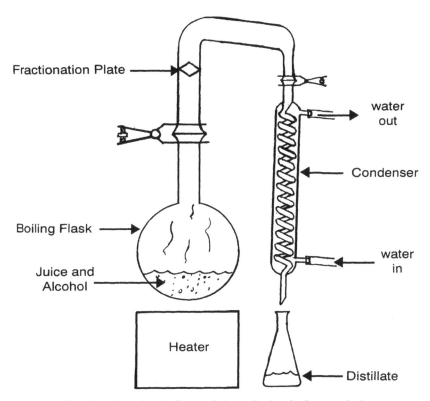

Fig. 6-1. Setup for distillation of citrus oil using the Scott method.

- 10 ml pipette for standardization.
- About 10 ml of pure citrus oil for standardization.

Standardization of Bromide–Bromate Solution

1. Add 10 ml of citrus oil to a 100 ml volumetric flask using a 10 ml pipette. Fill the flask to the mark with isopropyl alcohol. Mix the solution well.
2. Withdraw 1 ml of this solution and add it to a second 100 ml volumetric flask, and fill it to the mark with isopropyl alcohol. Mix well.
3. Pipette 10 ml of this solution into the boiling flask shown in Fig. 6-1 along with about 25 ml of water and another 15 ml of isopropyl alcohol.
4. Proceed with the oil determination below, beginning with step number 3.
5. Divide the dilution (0.010) by the milliliters titrated (should be about 10 ml) to get the factor (about 0.0010).

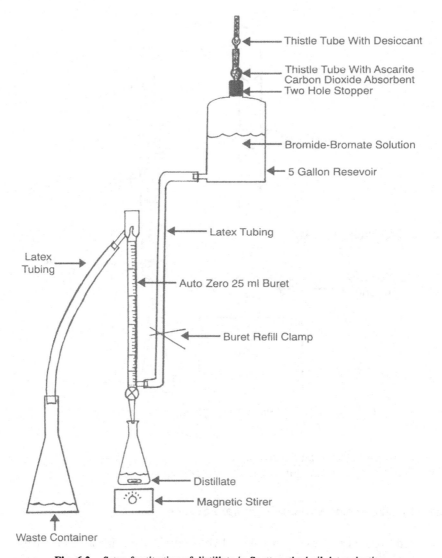

Fig. 6-2. Setup for titration of distillate in Scott method oil determination.

Procedure

1. Using a pipette or graduated cylinder, add 25 ml of 11.8°Brix juice or sample into the boiling flask in Fig. 6-1. Boiling chips may be used to facilitate boiling.
2. Using a graduated cylinder, add 25 ml of isopropyl alcohol and 25 ml

distilled water to the boiling flask, and heat the mixture to boiling using the setup in Fig. 6-1.

3. Distill until moisture can be seen inside the distillation apparatus in the form of water beads. Even though 25 ml of alcohol was added to the sample, more than 25 ml should be collected in the distillation because of the azeotropic nature of water/alcohol mixtures. Premature termination of the distillation will result in artificially low oil levels.

4. Using an auto-dispenser or a graduated cylinder, add about 10 ml of the diluted acid and 1 drop of methyl orange indicator to the distillate. An additional 25 ml of isopropyl alcohol can be added to the distillate to provide more volume for mechanical stirring but is not necessary for the titration.

5. Titrate the distillate, while stirring, with the bromide–bromate to a colorless endpoint.

6. Oil levels are reported as the % by volume in 11.8°Brix juice. This is found as follows:

$$\% \text{ oil } v/v = \frac{(\text{ml titrated})(\text{factor})(100\%)}{(\text{ml of sample})} \qquad (6\text{-}4)$$

where the factor is found from the above standardization. For example, if 1.5 ml of bormide–bromate were titrated, Equation 6-4 would become:

$$(1.5 \text{ ml})(0.0010)(100\%)/(25 \text{ ml}) = 0.006\% \text{ v/v oil}$$

The USDA grade standards for citrus products can be found in Table 6-1. Even through grade A orange juice is allowed up to 0.035% oil, oil burn can be detected below 0.020%, depending on the juice. Commercial orange juices generally contain about 0.015 to 0.025% oil.

OIL LEVELS IN BLENDS

It is sometimes of interest to determine the final oil in a blend so that one can predetermine the amount of additional oil or oil-based flavors without exceeding any maximum oil level specifications. For example, if an oil specification requires the oil level to be 0.020% or lower, and you want to add as much oil-based flavor enhancer as possible, the average oil content of the components in a blend must be determined prior to any further oil addition. Once the oil level of the blended components has been calculated, the amount of flavor enhancer that can be added without exceeding oil juice specifications can be calculated. It is important to determine these parameters prior to blending because of the difficulty of mixing the immiscible oil with the aqueous juice. Oil addition is

Table 6-1. Maximum oil levels (% v/v) allowed under USDA grade standards

	Grade A	Grade B
Orange (47FR 12/10/82, 52.1551–52.1557)		
Frozen concentrated orange juice	0.035	0.040
Orange concentrate for manufacturing	none	one
Orange juice from concentrate	0.035	0.045
Canned orange juice	0.035	0.055
Reduced acid orange juice	0.035	0.040
Canned concentrated orange juice	0.035	0.040
Dehydrated orange juice	0.035	0.045
Pasteurized orange juice	0.035	0.045
Grapefruit (48FR 9/12/83, 52.1225–52.1227)		
Grapefruit juice	0.020	0.025
Grapefruit juice from concentrate	0.020	0.025
Frozen concentrated grapefruit juice	0.020	0.025
Concentrated grapefruit juice for manufacturing	none	none
Dehydrated grapefruit juice	0.020	0.025
Grapefruit and orange (6FR 11/1/72, 52.1288–52.1290)		
Grapefruit and orange juice	0.035*	0.055
Tangerine (2FR 7/1/69, 52.2931–52.2941)		
Canned tangerine juice	0.025	0.035
Concentrated tangerine juice for manufacturing	none	none

*Grade standards are the same for sweetened and unsweetened.

best done as the components are being added. The average oil of the blend components can be found from the following equation:

$$O_a = \frac{V_1 O_1 S_1 + V_2 O_2 S_2 + V_3 O_3 S_3 + \cdots + V_n O_n S_n}{V_1 S_1 + V_2 S_2 + V_3 S_3 + \cdots + V_n S_n} \qquad (6\text{-}5)$$

where volumes (V), % oils (O), and the pounds of solids per gallon (S) (calculated from the Brix using Equations 2-8 or 2-10) of the blend components are used to calculate the final average oil of the blend (O_a). If the Brix values of the blend components are similar, the S values may be omitted. The amount of oil needed to raise the oil level to a certain level can be calculated by using the following equation:

$$\text{ml oil} = (O_f - O_i) \frac{(3785.306 \text{ ml/gal}) SV}{100\% (1.046 \text{ sol/gal})} = (O_f - O_i)(36.19)(SV) \qquad (6\text{-}6)$$

where the final desired oil level (O_f), the initial oil level (O_i), the pounds of solids per gallon (S) (found from the Brix using Equation 2-8 or 2-10), and the juice volume (V) can be used to calculate the milliliters of oil that need to be added. In a blend, the three latter values should be the calculated average values for the final blend. If the calculated average oil level of a blend is too high to allow you to add as much oil or flavor enhancer as you would like, one of the blend components can be exchanged for one with less oil to maintain the same overall volume, or a different volume of one of the blend components can be used. In the former case, the following equation can be used:

$$O_1 = O_f + (S_2 V_2 (O_f - O_2) + S_3 V_3 (O_f - O_3) + \cdots$$
$$+ S_n V_n (O_f - O_n)) / S_1 V_1 \qquad (6\text{-}7)$$

In the latter case, the following equation can be used:

$$V_1 = \frac{S_2 V_2 (O_f - O_2) + S_2 V_3 (O_f - O_3) + \cdots + S_n V_n (O_f - O_n)}{O_1 S_1 (1 - O_f / O_1)} \qquad (6\text{-}8)$$

Both of the above equations are rearrangements of Equation 6-5. Again, if the Brix values of the components are similar, the S values can be omitted.

Let us take an example to illustrate the use of the above equations. Suppose that we had the following blend components:

Lot #	# Drums	Brix	Sol/Gal	% Oil
1	18	65.0	7.135	0.006
2	12	62.1	6.726	0.011
3	16	59.8	6.409	0.023
4	7	60.6	6.518	0.009
5	7	64.8	7.107	0.013

Also, suppose that we want to add as much flavor enhancer as possible, not to exceed 0.020% oil. First, Equation 6-5 would be used to determine the average oil in the blend as follows:

$$\frac{\begin{array}{c}18(0.006)7.135 + 12(0.011)6.726 + 16(0.023)6.409 \\ + 7(0.009)6.518 + 7(0.013)7.107\end{array}}{18(7.135) + 12(6.726) + 16(6.409) + 7(6.518) + 7(7.107)}$$
$$= 0.012\% \text{ oil}$$

As you can see, the units of volume can be anything as long as the same units are used throughout the calculation. In order to raise the oil level from 0.012%

to the desired 0.020%, Equation 2-16 can be used to find the average pounds of solids/gallon:

$$SPG = \frac{18(7.135) + 12(6.726) + 16(6.409) + 7(6.518) + 7(7.107)}{18 + 12 + 16 + 7 + 7}$$

$$= 6.784$$

Next, Equation 6-6 can be used to calculate the milliliters of oil needed to raise the oil level from the average of 0.012% to 0.020%:

$$(0.020 - 0.012)(6.784)(36.19)(60 \text{ drums} \times 52 \text{ gal/drum}) = 6128 \text{ ml}$$

Because the "36.19" constant is based on gallons as a volume unit, the 60 drums had to be converted into gallons assuming 52 gallons per drum.

Now suppose that your product formulation requires at least 2.5 ml of oil-based flavor enhancer per gallon of concentrate. For the 3120 gallons (60 drums × 52 gal/drum) of concentrate in the example, you would need to add 7800 ml of flavor enhancer or oil, an amount greater than the 6128 ml of oil permitted by the specification. Rearrangement of Equation 6-6 will allow us to determine the maximum allowable oil level that will permit 7800 ml of enhancer to be added and still maintain a final oil level of 0.020%:

$$O_i = O_f - (\text{ml oil})/VS(36.19) \tag{6-9}$$

In the example this gives:

$$0.020 - 7800/3120(6.784)(36.19) = 0.010\%$$

This means that we need to reduce the 0.012% average oil level in the blend to 0.010% by changing the blend components. Suppose that we wanted to exchange lot 3 for a lot with a lower oil content with a Brix of 60.1 ($S = 6.450$). Equation 6-7 would become:

$$0.010 + \frac{\begin{pmatrix} 18(0.010 - 0.006)7.135 + 12(0.010 - 0.011)6.726 \\ + 7(0.010 - 0.009)6.518 + 7(0.010 - 0.013)7.107 \end{pmatrix}}{16(6.450)}$$

$$= 0.013\% \text{ oil}$$

which represents the maximum oil level that the replacement lot could contain. If a lot with such an oil level is difficult to find, the number of drums of lot 3 can be changed instead of the oil level, using Equation 6-8 as follows:

$$\frac{18(0.010 - 0.006)7.135 + 12(0.010 - 0.011)6.726 + 7(0.010 - 0.009)6.518 + 7(0.010 - 0.013)7.107}{0.023(6.409)(1 - 0.010/0.023)}$$

$$= 4 \text{ drums of lot 3 instead of 16}$$

This maze of calculations is best performed by computers. A flow chart illustrating the programming logic is shown in Fig. 6-3. A GWBASIC program that utilizes the flow chart can be found in Appendix B, along with a related program for the HP-41C programmable calculator in Appendix C. After adding the data for the blend components using the HP-41C program, enter "0" for the Brix and follow the prompts of the program. You can exchange components as many times as you want at the end of the program.

DEOILING SINGLE-STRENGTH JUICE

During certain times of the harvesting season, freshly extracted juice reaches a B/A ratio that is suitable for commercial juice as is without blending. This juice can be pasteurized to stabilize the enzyme activity or packaged as is, depending on the standard of identity employed. Freshly extracted juice, however, contains high levels of citrus oils that exceed the USDA grade standard of 0.035%. These oils generally require deoiling using a heated vacuum technique. This deoiling also serves as pasteurization and enzyme stabilization. New extraction techniques are emerging that minimize the oil levels of freshly extracted juice that may eliminate the need for deoiling in the future. During deoiling operations, the oil levels should be constantly monitored. If the oil levels are too high, more steam or slower flow rates should be used, and vice versa. Slight concentration of the juice during deoiling is commonly observed. The aqueous distillate containing aroma components is commonly returned to the deoiled juice in order to restore some of the flavor lost during deoiling as well as to counter the slight concentration effect.

OIL AND AROMA PRODUCTION

Citrus oils and aromas are generally the major by-products of citrus processing and are used in a variety of products. Eight types of oils and aromas are commonly manufactured, and their general characteristics, standards, uses, and method of processing will be discussed hereafter. Generally speaking, the term "aromas" refers to the aqueous phase separated from the oil phase in evapo-

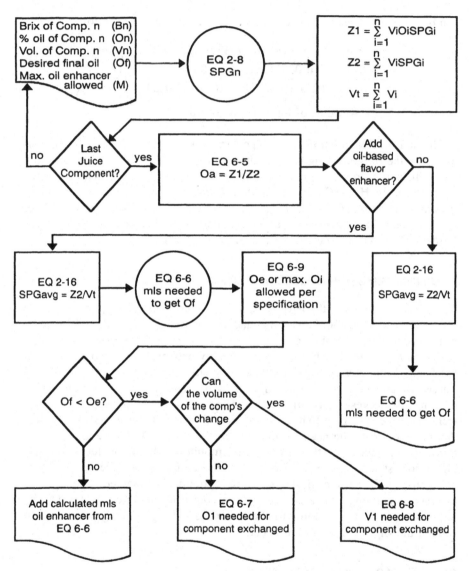

Fig. 6-3. Flow chart that can be used to program computers to calculate the milliliters of oil needed in a blend or the maximum allowable milliliters of oil-based flavor enhancer that can be added per specification.

rator and pasteurized condensates, which contains important volatile flavor and aroma components.

Cold Pressed Aroma and Oil

Cold pressed oils are recovered from the flavedo of citrus peel by rasping or rupturing the oil sacs close to the surface of the fruit just prior to or during juice extraction. Because heat easily removes the valuable volatile oil components, no heat is applied to the oil. As the oil is extracted, water is used to wash it from the peel to form a slurry. The use of too much water during this step causes excess peel material to get into the slurry, which can absorb important oil constituents, such as aldehydes, that are removed with the peel material. Excessive pressure in oil sac rasping or rupturing also will cause excess peel material to get into the slurry. The peel then is washed with water, and the washings are conveyed to a finishing press that separates the oil slurry from pieces of peel. Excessive finishing pressure will cause excessive insolubles to get into the oil emulsion, making it more difficult to break during polishing.

The oil slurry is centrifuged in a desludging centrifuge (8,000–10,000 rpm) that separates the slurry into an oil-rich emulsion, an aqueous discharge, and a semisolid sludge material. The sludge usually is discarded, but the aqueous discharge may be filtered and returned to wash the peel to form additional oil slurries. In this way, any small amounts of oil remaining in the aqueous discharge may be recovered through a recycle of the water. The oil-rich emulsion may be treated with enzymes for several hours, depending on the enzyme, or it may be placed in a freezer for about 30 days in order to help it break up for increased oil recovery. The emulsion then is centrifuged in a two-way separator or oil polisher (16,000–18,000 rpm) that centrifuges it into a discarded heavy phase and a light clear oil phase. Some plants polish the emulsion immediately after desludging. The product then can be treated enzymatically or stored in a freezer for about 30 days to facilitate the precipitation of waxes dissolved in the oil. After such "winterizing," the oil can be decanted and filtered and the waxy precipitates discarded. The oil should be stored in closed-head glass, tin-dipped, aluminum, wall-galvanized, or suitably lined drums under nitrogen or carbon dioxide. Citrus oils absorb readily into most plastics, so plastic containers should be avoided. Refrigeration and air exclusion in the headspace of the oil drums minimize oxidation. However, the precipitation of waxes may continue, as precipitation slows with time but never really stops at low temperatures. Storage temperatures of 60 to 75°F generally are recommended to keep the waxes in solution. Also, protection from light is advised because photochemical reactions with some oil components may detract from the quality of the oil.

Oil yields generally increase with maturity of the fruit but decrease right after periods of rainfall. Soft fruit generally gives poorer yields than firm fruit because of less efficient rasping or rupturing of the oil sacs. Grapefruit oils are best extracted from February through April and need to be cured after dewaxing for about 6 months (60–70°F). The oil starts out orange-like in nature and then undergoes a chemical change. Waxes take longer to precipitate in grapefruit oils than with orange oils even though 95% of the waxes are removed in the first two months of storage. However, waxes may continue to precipitate for up to 2 years. The lower the temperature is, the faster the precipitation.

Cold Pressed Oil Quality Control

To obtain the highest yield possible, it is important to perform quality control checks at different points of the oil extraction process. This can be done by taking samples of the oil slurry, the aqueous discharge from the desludger, the oil emulsion from the desludger, and the heavy discharge coming from the polisher. If yields seem unusually high or low, whole fruit can be analyzed for the total available oil in order to determine if yields are affected by biological or processing parameters. Analysis of the whole fruit is difficult, however, especially on a routine basis. Analyzing the samples described above generally will reveal processing problems, and comparison of overall oil production versus tons of fruit run generally will show problems in the fruit and the first stages of the oil process.

The efficiency of the desludger can be measured by taking 1.00 ml of the inbound oil slurry, adding about 25 ml of isopropyl alcohol, and titrating, using the Scott method described earlier in the chapter (O_s). Next one takes 1.00 ml of the aqueous discharge and performs the same oil determination (O_a). Then the following equation is used to calculate the desludger efficiency:

$$\% \text{ oil recovery} = (O_s - O_a)100\%/O_s \qquad (6\text{-}10)$$

The % recovery should be above 90%. The heavy aqueous discharge may be filtered and reused to wash more oil from the peel into the oil slurry in order to reclaim a portion of the oil remaining in the aqueous material.

The oil coming from the oil-rich emulsion from the desludger also is a parameter of the oil recovery efficiency. It can be measured by weighing 1-2 grams of the emulsion into a 100 ml volumetric flask and filling the flask to the mark with isopropyl alcohol. The weight of the sample must be recorded exactly. Then 1 ml of this dilution is added to 25 ml of isopropyl alcohol and titrated by the Scott method. The % oil in the emulsion can be calculated by using:

$$O_e = \frac{(\text{ml titrated})(0.001 \text{ ml oil}/\text{ml titrated})(100 \text{ ml flsk})(100\%)d}{(\text{wt of emulsion})}$$

(6-11)

where O_e is the % oil in the emulsion, and d is the density of the oil. For orange oil the density is 0.842 g/ml, and that for lemon oil is 0.853 g/ml. The oil level in the emulsion should be between 70 and 80% and can be changed by varying the flow rate through the desludging centrifuge, the length of time between bowl flushings, and the length of time the bowl remains open during the flush cycle. The oil emulsion then should be treated enzymatically and stored in cold storage, or it may feed into the polisher immediately, depending on the oil recoveries desired. Immediate treatment of the oil emulsion may result in some oil loss during polishing because the emulsion will be more difficult to break. The rate of flow into the polisher varies with equipment but should be about 1 to 1.5 gallons per minute. The heavy discharge from the polisher can be analyzed for oil by the Scott method in a manner similar to that for the oil emulsion; and it should not contain more than 5 to 7% oil. In larger operations more than one polisher may be required to minimize oil losses. A 100-mesh screen sometimes is used to remove extraneous debris from dewaxed or polished oils. Aroma oil is made from the distillation of the aqueous discharge from polishers.

Steam-Distilled Oils and Aromas

Steam-distilled oils generally are produced by steam injection or steam distillation of expressed sugar solutions (press liquor) obtained by pressing limed peel. Steam is injected into the press liquor and carries off d-limonene even though it has a higher boiling point than water (352°F compared to 212°F). The condensate separates into an oil layer on the top and an aqueous layer on the bottom. The oil layer, which is decanted, contains about 25% more aldehydes than cold press oil because the aldehydes are less volatile than d-limonene, which is lost during the heat treatment. The oil slurry feeding into the desludging centrifuge in cold pressed oil operations also may undergo steam distillation. The amount of steam used, as well as the amount of cooling in the condensers, may be adjusted to give a maximum yield with a minimum of oil remaining in the outgoing stripped press liquor or peel. These oils generally are used in the manufacture of paints, rubber, and textiles.

Folded Oils

Some cold pressed oils are further distilled in order to remove d-limonene while concentrating the less volatile oxygenated flavor compounds. These concentrated oils, referred to as folded oils, are valuable materials that may be added back to citrus juices to enhance their flavor without increasing oil levels as much as untreated cold pressed oils do. Some oils are stripped of all the terpenes and are referred to as terpeneless oil—a product that is very rich in flavor. Aromas from the aqueous layer in these steam distillations also may be recovered.

Essence Oils and Aromas

The natural fresh flavors and aromas characteristics of freshly extracted juice are largely removed during commercial evaporation. Essence oils are recovered from the condensate from these evaporators and differ from other citrus oils in the content of valencene, a sesquiterpene (0.5–2.0%), which is not found appreciably in other oils. Excess heat treatment during the recovery of essence oil will result in chemical conversion from valencene to nootkatone, a characteristic component of grapefruit oil, as well as breakdown of the esters, under the acidic conditions of the juice, to alcohols and acids, to give an off aroma commonly called a ''wet dog'' aroma.

There are basically two types of essence oil recovery systems. One uses a fractionation column to concentrate the volatile materials from the first effect or first vaporization in commercial juice evaporators, and the second fractionates the volatiles from both the first effect and the second effect; and then the final products from the two systems are mixed together. The aqueous phase, or aroma, primarily consists of alcohols and aldehydes (13% ethanol) (Johnson and Vora 1983). Yields are generally in the neighborhood of 0.20% of the inbound juice (11.8°Brix) feeding into the evaporator for aroma and 0.014% for essence oil.

Juice Oils

The term ''juice oils'' really has two meanings. Citrus oils not only are found in oil sacs in the flavedo of the fruit peels, but they occur in the juice cells themselves. About 0.005% of the oil in fresh single-strength juice comes from the juice cell. The juice cell oil is not exactly the same as that found in the fruit peel, as the juice oil is higher in esters and lower in aldehydes than cold pressed oil. The oil recovered from single-strength juice in deoiling operations also is called juice oil even though most of this oil comes from the peel. The latter definition is the one more commonly used in the industry. This juice oil is generally of better quality than that of essence oils, as in the corresponding aroma from the respective aqueous phases. As mentioned previously, the aqueous phase from deoiling operations generally is added back to the deoiled juice to restore flavor and to offset the slight concentration of the juice during deoiling.

Quality Control of Citrus Oils

The quality of citrus oils is centered primarily around their flavor attributes, which arises from a multitude of oil components in varying concentrations. These delicate flavor balances are little understood; so quality control or quality monitoring other than subjective organoleptic methods has been very difficult. However, other methods that are easily used in routine quality control have served as a means of estimating citrus oil quality, especially in regard to adulteration and oxidative tendency. Standards have been set by the United States Pharmacopeial Convention, Inc. (USP), which first met on January 1, 1820 and met every ten years thereafter until 1970, when the decision was made to update its information every five years. The information given below is based on the convention held in July, 1975. The *Food Chemicals Codex* (FCC), published by the National Research Council and officially recognized by the Food and Drug Administration, contains standards for citrus oil quality. The USP and FCC standards are summarized in Table 6-2. For more detailed descriptions of these standards the reader is referred to the original texts (*The United States Pharmacopeia 19th Revision* 1975; *Food Chemicals Codex* 1981).

Optical Rotation

Many chemical compounds occur in pairs that differ from one another only in the fact that they are mirror images and cannot be superimposed upon each other. These pairs are referred to as enantiomers, and their difference sometimes affects their chemical reactivity. If such a pair contains a chiral carbon (a carbon bonded to four different entities), they will rotate plane polarized light as shown in Fig. 6-4. If equal amounts of both enantiomers exist together, no net rotation of polarized light will be observed, as one enantiomer rotates the light to the right (dextrorotatory) and the other rotates the light to the left (levorotatory). The degree that the light is rotated as it passes through a solution containing the enantiomers is directly proportional to the concentration difference of one enantiomer compared to the other, according to:

$$R = alc \qquad (6\text{-}12)$$

Table 6-2. Summary of citrus oil specifications according to the United States Pharmacopeia (USP) (1975) and the *Food Chemicals Codex* (FCC) (1981).

	USP	FCC
Lemon Cold Pressed Oil		
aldehydes (as % citral)	2.2–3.8 (Cal)	2.2–3.8 (Cal)
	3.0–5.5 (Ital)	3.0–5.5 (Ital)
Optical rotation (+ degrees)	57.0–65.6	57.0–65.6
Refractive index (20°C)	1.473–1.476	1.473–1.476
Specific gravity	0.849–0.855	0.849–0.855
UV absorption (315 nm)	≥ 0.20 (Cal)	≥ 0.20 (Cal)
	≥ 0.49 (Ital)	≥ 0.49 (Ital)
Lemon Distilled Oil		
Aldehydes (as % citral)	none	1.0–3.5
Optical rotation (+ degrees)	none	55–75
Refractive index (20°C)	none	1.470–1.475
Specific gravity	none	0.842–0.856
UV absorption (315 nm)	none	≤ 0.01
Lemon Cold Pressed Oil Desert Type		
Aldehydes (as % citral)	none	≥ 1.7
Optical rotation (+degrees)	none	67–78
Refractive index	none	1.473–1.476
Specific gravity	none	0.846–0.851
UV absorption (315 nm)	none	≥ 0.20
Lime Cold Pressed Oil		
Aldehydes (as % citral)	none	4.5–8.5 (Mex)
		3.2–7.5 (Tahit)
Optical rotation (+ degrees)	none	35–41 (Mex)
		38–53 (Tahit)
Refractive index	none	1.482–1.486 (Mex)
		1.476–1.486 (Tahit)
Specific gravity	none	0.872–0.881 (Mex)
		0.858–0.876 (Tahit)
UV absoorption (315 nm)	none	≥ 0.45 (Mex)
		≥ 0.24 (Tahit)
Evaporative residue (%)	none	10.0–14.5 (Mex)
		5.0–12.0 (Tahit)
Lime Distilled Oil		
Aldehydes (as % citral)	none	0.5–2.5
Optical rotation (+ degrees)	none	34–47
Refractive index (20°C)	none	1.474–1.477
Specific gravity	none	0.855–0.863

Table 6-2. (*Continued*)

	USP	FCC
Orange Cold Pressed Oil		
Aldehydes (as % decanal)	1.2–2.5	1.2–2.5
Optical rotation (+ degrees)	94–99	94–99
Refractive index (20°C)	1.472–1.474	1.472–1.474
Specific gravity	0.842–0.846	0.842–0.846
UV absorption (315 nm)	≥ 0.130 (Cal)	≥ 0.130 (Cal)
	≥ 0.240 (Flor)	≥ 0.240 (Flor)
Evaporative residue	≥ 43 mg/3 ml	none
Orange Bitter Cold Pressed Oil		
Aldehydes (as % decanal)	none	0.5–1.0
Optical rotation (+ degrees)	none	88–98
Refractive index (20°C)	none	1.472–1.476
Specific gravity	none	0.845–0.851
Evaporative residue (%)	none	2–5
Orange Distilled Oil		
Aldehydes (as % decanal)	none	1.0–2.5
Optical rotation (+ degrees)	none	94–99
Refractive index (20°C)	none	1.471–1.474
Specific gravity	none	0.840–0.844
UV absorption (315 nm)	none	≤ 0.01
Grapefruit Cold Pressed Oil		
Optical rotation (+ degrees)	none	91–96
Refractive index (20°C)	none	1.475–1.478
Specific gravity	none	0.848–0.856
Evaporative residue (%)	none	5–10
Tangerine Cold Pressed Oil* (Dancy and closely related varieties)		
Aldehydes (as % decanal)	none	0.8–1.9
Optical rotation (+ degrees)	none	88–96
Refractive index (20°C)	none	1.473–1.476
Specific gravity	none	0.844–0.854
Evaporative Residue (%)	none	2.3–5.8
Mandarin Cold Pressed Oil* (*Citrus reticulata* Blanco var. Mandarin)		
Aldehydes (as % decanal)	none	0.4–1.8
Optical rotation (+ degrees)	none	68–78
Refractive index (20°C)	none	1.473–1.477
Specific gravity	none	0.847–0.853
Evaporative residue (%)	none	2–5

*Tangerine and mandarin oils refer to essentially the same fruit because Dancy tangerines are of the same species, *C. reticulata*, and there is no variety known as "mandarin" in the *C. reticulata* species (see Chapter 13).

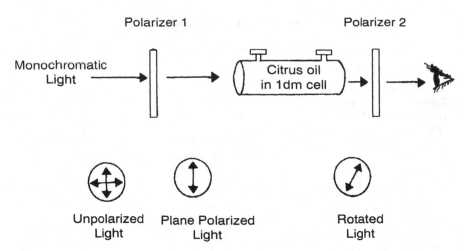

Fig. 6-4. Measurement of the optical rotation of citrus oils. The first polarizer polarizes the light, and the oil rotates the light. The second polarizer can be rotated to detect the degree of rotation.

where R is the observed optical rotation in degrees, a is the specific optical rotation, l is the length of the sample cell in decimeters, and c is the concentration difference between the enantiomers in g/ml.

Limonene, the primary component of citrus oils, occurs in enantiomeric pairs, but only the dextrorotatory enantiomer (d-limonene) occurs in citrus oils. The optical activity of d-limonene allows its measurement in citrus oils. Pure d-limonene has an optical rotation of +125.6° (20°C) using a sodium D line as the monochromatic light source. In citrus oils, the optical rotation that is observed ranges from +75° to +100° and constitutes a fair measure of the d-limonene content in the oil. It should be remembered the citrus oils contain components other than d-limonene that are optically active and may affect the optical rotation. However, this effect is believed to be small and is generally ignored by the industry. Also, most other optically active components of citrus oils generally occur as racemic mixtures, or mixtures containing both enantiomers, which result in no significant net rotation of light.

The purpose of measuring the d-limonene content in citrus oils is to estimate adulteration of the oils by using d-limonene from noncitrus sources. Also, folded oils can be characterized by the d-limonene content by polarimetry. Even though the Scott method also can be used to measure the d-limonene content, polarimetry is a much quicker and easier method. The optical rotation, which generally will decrease with fruit maturity, can be measured by the following procedure.

Determination of Optical Rotation

Equipment and Supplies

- Polarimeter accurate to $\pm 0.1°$ to $\pm 0.5°$. (Homemade polarimeters that meet these requirements can be made at much less cost than commercial instruments.)
- 1 dm sample cell if not included with polarimeter.
- Temperature bath and thermometer.
- Sodium lamp or soft yellow lamp.

Procedure

1. Most commercial polarimeters are preset to zero rotation when no sample is present. Polarimeters can be calibrated by filling the sample cell with alcohol or water and using the deviation from zero to correct the final rotation reading of the sample.
2. The oil sample is warmed to 25°C with a temperature bath and thermometer.
3. The sample cell is rinsed a few times with the oil that is to be analyzed, and then the cell is filled with the oil. One should make sure the cell is clean, especially the optical surfaces.
4. The cell is put into the polarimeter with the operator viewing the light source through both polarizers, with the sample cell aligned between them. One should rotate the second polarizer until the light source appears the darkest. This should be done with great care. The polarizer should always be rotated to the right, with the angular rotation read to the nearest degree.
5. Generally temperature corrections are not made with optical rotation measurements of citrus oils. For orange and grapefruit oils, 0.22° can be added or subtracted per degree centigrade above or below 25°C, and for lemon oils 0.14° can be similarly added or subtracted from the reading (Kesterson, Hendrickson, and Braddock 1971).
6. The final rotational reading usually is rounded off to the nearest degree if temperature corrections are applied.

Refractive Index

The principles behind measurement of the refractive index were discussed in Chapter 2. The refractive index scale and the Brix scale measure essentially the same physical parameter. It is redundant to perform a refractive index check as well as an optical rotation measurement because they measure essentially the same thing, the estimated *d*-limonene content of the oil. However, it would be very difficult to adulterate citrus oils with a component that had both the same optical rotation properties and the same optical density properties as *d*-limonene. Both tests are relatively quick and easy, so both are worthwhile if adulteration is a concern. Pure *d*-limonene has a refractive index of 1.4727 (20°C), and the refractive index of citrus oils ranges from 1.4720 to 1.4740 (20°C). Most processing plants already possess a refractometer for Brix measurements, making the refractive index of citrus oils an even easier parameter to define. In dry years the refractive index will increase with maturity. The following procedure can be used.

Determination of Refractive Index

Equipment and Supplies

- Refractometer that can read in units of refractive index.
- Dropper or plastic stirring rod to apply oil to the refractometer.
- Light source (soft yellow) if not included with the refractometer.
- Temperature bath or circulator with thermometer if not included with the refractometer.

Procedure

1. Carefully clean the prism of the refractometer with water or alcohol, and dry it with tissue paper.
2. Using the dropper or plastic stirring rod, apply a few milliliters of the sample to the prism so that it is completely covered. Care should be taken that no hard or abrasive object comes into contact with the prism.
3. For indirect scales, adjust the shadow to the center of the cross hairs, and read the refractive index on the corresponding scale. For direct scales, read the scale where it intersects the shadow.
4. It is recommended that all samples be measured at 20°C using a temperature bath and thermometer. However, if corrections are necessary, adjust the refractive index for orange oil by adding or subtracting 0.00045/°C, and for lemon oil by adding or subtracting 0.00046/°C if the temperature is above or below 20°C (Kesterson, Hendrickson, and Braddock 1971).

Specific Gravity

The specific gravity is the density of the material divided by the density of water at the same temperature. This test also is redundant if used with the refractive index and the optical rotation in estimating the *d*-limonene content of citrus oils and the general composition. Again, however, the greater the variety of tests performed, the greater the chance is of detecting adulteration. The specific gravity of *d*-limonene is 0.8419 (21°C), compared to that of citrus oils which ranges from 0.842 to 0.881. In determining the specific gravity, a pycnometer commonly is used to measure a certain volume of the oil that, along with the oil net weight, can be used to determine the specific gravity of the oil. The specific gravity generally will increase with fruit maturity.

Determination of Specific Gravity

Equipment and Supplies

- Pycnometer or beaker and volumetric pipette (10 ml or 25 ml).
- Laboratory balance.
- Temperature bath and thermometer.

Procedure

1. Rinse the pycnometer with a few portions of the oil that is to be analyzed, and fill it with the oil without putting the cap on the pycnometer; *or* place about 50 ml of the oil into a beaker.
2. Place the pycnometer or beaker in a temperature bath, and warm it to 20°C.
3. Place the cap on the pycnometer and wipe the excess oil from the exterior until it is clean and dry; *or* pipette 10 or 25 ml of the oil into a clean dry beaker. Weigh the pycnometer or beaker immediately, using the lab balance to ±1 mg.
4. Discard the sample, and clean and dry the pycnometer or beaker and reweigh it. Subtract the tare weight to get the sample net weight.
5. Repeat the procedure using distilled water. Divide the density of the sample (g/ml) by the density of distilled water (g/ml) to get the specific gravity. The density of the water need not be determined each time. A standard value of 0.998203 (20°C) can be used instead for the density of water, based on a water density of exactly 1 at 3.98°C.
6. Temperature corrections, if necessary, can be made by adding or subtracting 0.00078 (0.00077 for lemon oil) for each centigrade degree above or below 20°C (Kesterson, Hendrickson, and Braddock, 1971).

Aldehydes

Unlike the optical rotation, refractive index, and specific gravity, measurement of the aldehyde level in citrus oils is a direct measure of the flavor quality of those oils. For this reason, aldehyde measurements probably are the major quality parameter used in the marketing of these oils. Even though other oil constituents contribute significantly to the organoleptic characteristics, the aldehydes are considered the major contributor. In lemon oils the most abundant aldehyde is citral, which is actually the *cis–trans* mixture of neral and geranial and occurs at levels of about 2 to 4% in the oil. The most abundant aldehyde in other citrus oils is decanal, with levels ranging about 0.8 to 2.0% in the oil. Aldehyde analyses thus are based upon the citral or decanal equivalent of aldehydes.

There are several ways to measure the aldehyde content of citrus oils, including the Kebler method (Kebler 1921), the Kirsten modification of the Kebler method (Kirsten 1955), and the N-hydroxy benzenesulfonamide (or HBS) colorimetric method (Petrus, Dougherty, and Wolford 1970), as well as several forms of the hydroxylamine hydrochloride method (*J. of the AOAC* 1953; Carter 1981). The fastest, safest, and easiest method is that of Petrus, Dougherty, and Wolford (1970). Hydroxylamine hydrochloride is added to the oil and reacts with the aldehyde groups as follows:

$$\underset{\substack{\text{hydroxylamine}\\\text{hydrochloride}}}{(NH_3OH)^+Cl^-} + \underset{\text{an aldehyde}}{R\overset{\overset{\displaystyle O}{\|}}{C}H} \rightarrow \underset{\text{an oxime}}{R\overset{\overset{\displaystyle N-OH}{\|}}{C}H} + \underset{\substack{\text{hydrochloric}\\\text{acid}}}{HCl} \qquad (6\text{-}13)$$

The resulting acid can be determined by titration with a base.

Aldehyde levels generally increase as the season progresses. Also, wet weather will have a tendency to increase the level of aldehydes in citrus oils. Fruit stored for a long period of time generally will have lower aldehyde levels. The following procedure can be used for aldehyde determination.

Determination of Oil Aldehydes

Equipment and Supplies

- 0.5N hydroxylamine hydrochloride (34.75 g of hydroxylamine hydrochloride dissolved in 40 ml warm water and filled to the mark of a 1-liter volumetric flask with isopropyl alcohol; adjust the pH of the solution to 3.5, and store it in the refrigerator).
- Setup shown in Fig. 3-8.
- 30-minute timer.
- Laboratory balance.
- 50 ml graduated cylinder.

Procedure

1. Weigh about 5 g of the oil to be analyzed into the beaker shown in Fig. 3-8. Record the exact weight.
2. Add 35 ml of the 0.5N hydroxylamine hydrochloride solution to the flask using the graduated cylinder. Stir the mixture for exactly 30 minutes.
3. Using the setup in Fig. 3-8, titrate with 0.1562N NaOH solution to a pH endpoint of 3.5.
4. Calculate the % aldehyde using:

$$\% \text{ ald} = \frac{M(\text{ml titrated})(0.1562N)(1 \text{ liter}/1000 \text{ ml})100\%}{(\text{g of sample})} \quad (6\text{-}14)$$

where M is the molecular weight of the citral for lemon oils (152.23 g/mole) or for decanal for other citrus oils (156.27 g/mole).

Evaporative Residue

The residue left after evaporation of the volatile components of oil products has a tendency to stabilize the oil toward oxidation, one of the main causes of oil quality deterioration. For this reason, the amount of evaporative residue can serve as an estimate of the oxidative stability of the oil, which may be of occasional interest. The evaporative residue will generally increase with fruit maturity as well as with storage time of the fruit before processing. The following procedure can be used.

Determination of Evaporative Residue

Equipment and Supplies

- 100 ml Pyrex evaporating dish or equivalent.
- Steam bath or equivalent.
- Desiccator.
- Laboratory balance to ±0.001 g.
- Bunsen burner (optional).
- Oven (optional).

Procedure

1. Using the bunsen burner or steam bath, heat the evaporating dish; then dry and cool it in a desiccator for about 30 minutes.

2. Immediately weigh the dried evaporating dish as a tare weight.
3. Place 5 g (record the exact weight) of the oil into the dish, and heat it on a steam bath for 5 hours (6 hours for lime oils). The USP procedure requires orange oils to be heated in an oven at 105°C for an additional 2 hours (*The United States Pharmacopeia* 1975).
4. Cool the dish with the sample at room temperature in a dessicator, and reweigh it.
5. The % evaporative residue can be calculated from:

$$\%ER = (g\ residue)(100\%)/(g\ sample) \qquad (6\text{-}15)$$

Ultraviolet Absorption

Organic compounds that possess double bonds characteristically have electronic energy levels that correspond to the energy of ultraviolet light, so that these compounds absorb light in the UV region. Absorption peaks generally are very broad, making it difficult to use UV absorption as a definitive means of differentiating chemical structures. However, like many of the previous tests for citrus oils, UV absorption can be useful in conjunction with other tests in estimating general oil quality, detecting adulteration, and helping to differentiate between oil varieties (Kesterson, Hendrickson, and Braddock 1971). Lemon and lime oils have UV absorption peaks around 315 nm. Grapefruit oils have two peaks, one around 318 nm and one around 268 nm. Orange oils have a peak around 330 nm, and tangerine oils have two peaks, at about 330 nm and 318 nm. The intensities of these peaks have been used as a general guide to the quality of the oils, and the USP and FCC standards both include UV absorption parameters. The intensity of the main absorption peak is determined by means of the following procedure.

Determination of UV Absorption

Equipment and Supplies

- UV spectrophotometer with a 1 cm cell and graph paper or printout.
- Laboratory balance.
- 100 ml volumetric flask.
- 100+ ml isopropyl alcohol.

Procedure

1. Add 0.250 g of oil to the 100 ml volumetric flask, fill it to the mark with isopropyl alcohol, and mix the solution.

2. Rinse and fill the 1 cm cell of the spectrophotometer, and measure the absorbance between 260 nm and 400 nm at intervals of at least 5 nm to within 12 nm of the peak. Then read the absorbance at intervals of at least 3 nm three times, and then at 1 nm intervals until 5 nm beyond the peak maximum. Thereafter absorbance readings should be made at least every 10 nm.

3. Draw a tangent to the areas of minimum absorbance. Drop a vertical line from the main peak maximum to the tangent line, and determine the difference in absorbance that this *CD* line represents (absorbance at *C* at the peak minus absorbance at *D* or the tangent line).

4. If a sample is used that is not exactly 0.250 g, a weighted ratio can be applied to the *CD* value. For example, if a 0.255 g sample is used, you should make the following calculation:

$$(CD \text{ value observed})(0.250 \text{ g})/(0.255 \text{ g}) = CD \text{ value} \qquad (6\text{-}16)$$

QUESTIONS

1. What imparts the majority of the flavor to citrus juices next to sugars and acids?
2. What is the major constituent of citrus oils and at about what percent?
3. What compounds characterize grapefruit oil and tangerine oil?
4. What terpene (besides *d*-limonene) is found in lemon and lime oils?
5. What contribution does *d*-limonene make to citrus flavor?
6. What four compounds have been used to duplicate up to 87% of fresh orange juice flavor?
7. What are the basic principles behind the Scott method of oil analysis?
8. Why is the methyl orange indicator effective in the Scott method?
9. Why is a bromide–bromate salt solution used in the Scott method rather than a straight bromine solution?
10. Can any condenser be used in the Scott method? Why?
11. How do you determine the end of the distillation in the Scott method, and why?
12. What is the purpose of the HCl in the Scott method?
13. What are the industrial ranges of oil in commercial orange juices?
14. What is the difference between citrus oils, aromas, and folded oils?
15. What is "winterizing," and why is it performed?
16. Is it possible to directly distill citrus oils out of aqueous solutions?
17. What declaration must appear on citrus juice labels if citrus oils or aromas are added?
18. What is the purpose of determining the optical rotation, refractive index, and specific gravity of citrus oils?

19. What are the purposes of measuring the evaporative residue, ultraviolet absorption, and aldehydes in citrus oils?
20. What is the best way to determine the quality of citrus oils and aromas?

PROBLEMS

1. If you titrate the distillate of several juice samples using the Scott method, what are the % oils if the following quantities of the bromide–bromate solution are used?

ml bromide–bromate solution

1.67
2.72
3.11
4.76
6.01

2. What is the average oil in a blend consisting of the following blend components?

Lot #	Gallons	Brix	Lb Sol/Gal	% Oil
1	250	60.9	6.576	0.029
2	1675	58.7	6.259	0.005
3	1497	64.2	7.021	0.011
4	902	65.6	7.221	0.018
5	3676	59.9	6.423	0.015

3. Suppose that you wanted to increase the oil level in the above blend to 0.020% oil. How much oil would you need to add?
4. Suppose that in the above blend you want to add 2.5 ml of oil flavor enhancer per gallon of concentrate. How much would you want to add?
5. In the above blend, suppose that you wanted to add the flavor enhancer determined in the previous problem but not to exceed 0.020% in the final blend according to specification. What average oil level must the blend have before the addition of the oil?
6. In comparing the answers in problems 2 and 5, we see that one of the blend components must be changed in order to meet the 0.020% oil specification and allow for the flavor enhancer. Suppose we could replace blend component 3 with the same amount of 64.9°Brix concentrate with an oil level of 0.003%. Would this solve the problem?
7. Suppose that in the above blend you eliminate lot 5 and increase the volume of lot 2 to 3000 gallons. Would this solve the problem stated in problem 6?
8. What would be the oil extraction efficiency in the manufacture of cold pressed oils if you titrated 4.40 ml for a heavy discharge from the desludger sample, using the Scott method and the procedure outlined in the chapter, and 25.52 ml for a sample from the inbound oil slurry? Would this efficiency be acceptable according to information in the chapter?

9. What % oil would there be in the oil emulsion discharge from the desludger in a cold pressed orange oil operation if 1.984 g of oil were analyzed with 18.46 ml titrated, using the Scott method according to the procedures explained in the chapter? According to information in the chapter, would this be an acceptable level of efficiency?

10. In the aldehyde determination for lemon oil, suppose you weighed 4.568 g of the oil and titrated it with 6.16 ml of 0.1562N NaOH according to the procedure in the chapter. What would the % aldehyde be, and would this be good or poor quality oil?

REFERENCES

Ahmed, E. M., Dennison, R. A., and Shaw, P. E. 1978. Effect of selected oil and essence volatile components on flavor quality of pumpout orange juice, *J. Agric. Food Chem., 26*, 368–372.

Carter, B. 1981. Private communication. Ventura Coastal Corp.

Food Chemicals Codex, 1981. National Academy Press, Washington, D.C., 140, 168, 169, 170, 172, 209, 210, 319.

Johnson, J. D. and Vora, J. D. 1983. Natural citrus essences, *Food Tech., 12*, 92–93.

1953. *J. of the AOAC*, 119.

Kebler, 1921. *J. of the A.O.A.C., 4*, 474.

Kesterson, J. W., Hendrickson, R., and Braddock, R. J. 1971. *Florida Citrus Oils*, Bulletin 749 (technical). University of Florida, Gainesville, Fla., 24–27, 114–127.

Kirsten, 1955. *J. of the A.O.A.C., 38*, 738.

Petrus, D. R., Dougherty, M. H., and Wolford, R. W. 1970. A quantitative total aldehydes test useful in evaluating and blending citrus essences and concentrated citrus products, *J. Agric. Food Chem., 18*, 908–910.

Shaw, P. E. 1977. Essential oils. In *Citrus Science and Technology Vol. 1*, Steven Nagy, Philip Shaw, and Matthew Veldhuis, eds. The AVI Publishing Company, Inc., Westport, Conn., 430–435.

Scott, W. C. and Veldhuis, M. K. 1966. Rapid estimation of recoverable oil in citrus juices by bromate titration. *J. of the A.O.A.C., 49*, 628–633.

The United States Pharmacopeia 19th Revision, 1975. United States Pharmacopeial Convention, Inc., Rockville, Md., 560, 563.

Chapter 7

Citrus Juice Pulp

Most juices, such as apple, grape, and berry juices, are preferred by consumers in a filtered and clarified form. However, citrus juices are preferred in a pulpy and opaque form. Most of the opaque nature of citrus juices is attributed to the colloidal cloud material, which will be described in Chapter 8. The solid particles of the fruit that will eventually settle out, which are primarily juice sac and membrane material, impart a turbidity or mouth-feel that characterizes citrus juices as well as giving a natural appearance. This citrus pulp can be divided into two main groups—sinking, or spindown, pulp and floating pulp.

SINKING PULP

Sinking or spindown pulp generally is made up of the smaller or finer pulp particles that form a sediment on standing because of juice saturation and/or because they have slightly higher densities than that of the juice itself. The major contributions of sinking pulp to citrus juice quality include contributing to an opaque appearance and increasing the juice turbidity or mouth-feel. This sediment is readily visible in glasspack or clear pack containers of single-strength juices. Even though this pulp imparts a desirable mouth-feel to the juice, it may impart an undesirable sludge appearance in clear containers. This problem has been overcome in part by using opaque containers or using wraparound labels at the level of the sediment on clear containers. Clear containers have the advantage of portraying the brilliant natural color of citrus juices even though a proper package design on an opaque container can enhance product appeal as well.

Although the precise amount of sinking pulp is elusive, the following industrially accepted method of measuring the % pulp can be used for routine quality control. This procedure is somewhat empirical in nature. The diameter and the speed of the centrifuge along with the duration of centrifuging can dramatically affect the results. For this reason, standards have been set regarding speed versus centrifuge diameter and time of centrifuging. Also, the temperature may affect the results. In order to see how these parameters affect a particular juice, it is recommended that tests be done that vary the speed, duration, and temperature used in the procedure.

Determination of Spindown Pulp

Equipment and Supplies

- Lab centrifuge.
- Tachometer.
- Two 50 ml conical centrifuge tubes. (Clear plastic tubes are the safest and easiest to use.)
- Temperature bath and thermometer.

Procedure

1. Using the temperature bath and thermometer, bring the 11.8°Brix juice sample to 78°F (26°C).
2. Pour the temperature-adjusted juice into a clean and dry 50 ml centrifuge tube with mixing to the 50 ml mark.
3. Put the tube into the centrifuge, along with another tube filled with another sample or water to 50 ml on the opposite side to balance the load. More than two samples can be analyzed at once, depending on the capacity of the centrifuge. However, samples should be properly balanced in the centrifuge. Place the tubes so that the graduations are facing the direction of rotation. This will allow an average reading of the pulp directly on the graduations if the surface of the pulp is uneven.
4. Using the tachometer, bring the rpm to 1500 for a centrifuge measuring $11\frac{1}{2}$ inches from the bottom of one centrifuge tube to the bottom of the other on the opposite side when both tubes are extended in the horizontal position. For centrifuges with diameters that are different from $11\frac{1}{2}$ inches, the following equation can be used to determine the rpm needed to exert the same centrifugal force on the sample:

$$\text{rpm} = 1500 \sqrt{11.5/d} \qquad (7\text{-}1)$$

For example, a centrifuge measuring 12.4 inches in diameter as described above would require an rpm value of:

$$1500 \sqrt{11.5/12.4} = 1445 \text{ rpm}$$

5. Centrifuge at the proper rpm for 10 minutes.

6. After the centrifuge comes to a complete stop, read the pulp level on the graduation on the centrifuge tube halfway between the highest and lowest levels of the separated pulp sediment. If the graduations are hard to see, use a felt-tip pen to color the raised portions of the graduations.
7. The % pulp is found from:

$$\% \text{ pulp} = (\text{ml pulp})(100\%)/50 \text{ ml sample} \qquad (7\text{-}2)$$

The USDA grade standards can be found in Table 7-1. The industrial range for orange juices is generally 8 to 12% pulp. Grapefruit juice usually has about 2% less pulp than orange juices have.

Pulp levels can be controlled to some extent through control of the finishing pressure used just after extraction. Higher finisher pressures will generally result in higher pulp levels. Finishing pressures can be monitored by % pulp measurements. If the % pulp is consistently running high, the pressure can be lowered, and vice versa. The chapter on statistics (Chapter 23) can be used to define how many consecutive high or low pulp levels determine a nonrandom effect. High finisher pressures often mean higher juice yields, but juice quality may suffer, especially in the early part of the season. Excess loss of juice yields can be estimated by examining the moisture content of the pulp expelled by the finisher. If it is too wet, too much juice is being lost with the pulp, and the finisher pressure should be increased, and vice versa. A more exact way to determine the juice content in expelled pulp is to perform a quick fiber test on it. This test is, again, somewhat empirical but less so than the spindown pulp procedure.

Quick Fiber Test

Equipment and Supplies

- Mechanical shaking screen (20 mesh) with drain pan.
- Large serving spoon.
- 1-liter beaker.
- Magnetic stirrer.
- Lab balance or triple beam balance.
- 250 ml graduated cylinder.

Procedure

1. Weigh 200 g of well-mixed pulp into a 1-liter beaker using a large serving spoon. Add 200 g of water and stir for 1 minute, let the mixture sit for 3 minutes, and then stir it for another 1 minute.

Table 7-1. USDA grade standards for maximum free and suspended pulp levels in citrus juices.

	Grade A	Grade B
Grapefruit Juice (48FR 9/12/83 21, 52.1226–52.1227)		
Grapefruit juice	10	15
Grapefruit juice from concentrate	10	15
Frozen concentrated grapefruit		
juice (sweetened)	10	15
Grapefruit and Orange Juice (6FR 11/1/72, 52.1289)	12	18
Tangerine Juice (2FR 7/1/69, 52.2931–52.2941)		
Canned tangerine juice	7	10

2. Pour the contents of the beaker into the shaker screen and shake it for 3 minutes.
3. Weigh the liquid from the drain pan. The grams of liquid is a measure of the juice content of the pulp. Optimum values depend on the commercial machinery used and the manufacturing objective. High plant juice yields mean high pulp levels and perhaps some loss in flavor quality, and vice versa.

Industrial ranges are classified into three categories: tight (< 150 g liquid), moderate (150–180 g liquid), and loose (180–200 g liquid). In California it is difficult to get quick fiber pulp measurements of less than 180, with 160 being the standard target value. In Florida the quick fiber usually ranges from 90 to 140 g liquid.

The physical condition of the fruit and the variety will also have an effect on the pulp level. Fruit has been known not only to break down and go soft after peak maturity, but to do so during certain times of the early and mid seasons as well. This often causes sudden shifts in pulp levels. If more than one variety is being run, the varieties will most likely be at different stages of maturity, in addition to differing in physical composition. Therefore, it is necessary to closely monitor and adjust finisher pressures. For all these reasons, quality control personnel need to monitor pulp levels, juice yield, and juice quality in maintaining the ideal finisher pressure. It should be noted that if pulp washing is being done and added back on-line in accordance with federal standards of identity for 100% juice products, juice yields may become less important because juice lost to the pulp expelled by finishers will be recovered in the pulp wash operation.

Processing Effects on Juice Pulp

Another consideration that is important to the pulp quality of citrus juices concerns the changes that take place in the pulp during processing. The spindown pulp of the juice leaving the finisher needs to be about 12 to 20% for it to be 8 to 12% in the final 60°Brix concentrate. The reason for this is that the heat applied in pasteurization and evaporation operations, as well as the chopping action of pumps and the juice itself, induces a degree of breakdown in the pulp particles that results in lower pulp levels after processing. Such breakdown can be illustrated by considering the change in the pulp level in freshly extracted juice before and after evaporation to 60 to 65°Brix. A rule of thumb is that for each effect in the evaporator (E), the final pulp level in the concentrate (P_f) can be found from the pulp level of the freshly extracted juice (P), by using the following equation:

$$P_f = P (0.6973 - 0.0265E) \qquad (7\text{-}3)$$

If the juice were reconstituted and reevaporated, E would be twice the number of effects used in the evaporation, and so on. Citrus juices are rarely reconstituted and reevaporated more than once, if at all. This step is sometimes taken to remove contaminates such as potassium citrate crystals or burnt pulp particles, to produce low-pulp products, or to debitter citrus concentrates.

It should be kept in mind that the above equation gives only approximate results. The nature of the pulp and the empirical nature of the procedure for measuring the pulp make it difficult to predict precise results. For example, if the pulp level of the juice coming out of the finisher is 17%, and it is concentrated to 60°Brix using a three-effect evaporator, the expected approximate pulp level in the final concentrate is:

$$(17\%)(0.6973 - 0.0265(3)) = 10.5\% \text{ pulp}$$

If this concentrate were for some reason reconstituted to 12°Brix and reevaporated to the same concentration in the same evaporator, the pulp level in the second concentrate would be:

$$(17\%)(0.6973 - 0.0265(6)) = 9.2\% \text{ pulp}$$

In normal juice processing, the finisher can be adjusted by monitoring the pulp levels in the final concentrates. However, in the manufacturing of low-pulp products, changes in pulp levels with evaporation become more important.

Low-Pulp Juices

The pulp in citrus juices sometimes presents a problem for citrus processors and their customers. The pulpy nature of the juices requires greater space between plate heat exchangers in pasteurizers, chillers, and evaporators, not to mention nozzles in fillers, filters, and other machinery having restricted spaces that is commonly used to process juice. The use of natural juices in various drinks is increasing in popularity, with efforts to increase the health appeal of many beverages. Also, low-pulp juices can be evaporated to higher Brix levels because of lower viscosities that can save on storing and shipping costs. Moreover, many dairies and other nonjuice processors are processing juices with equipment designed for milk and other nonpulp products. The plugging of such equipment has prompted many users to seek citrus juices with lower pulp levels. Once a piece of machinery becomes a site for pulp collection, microbial spoilage can easily set in, as well as pulp burning in pasteurizers, not to mention the restriction or stoppage of product flow. The increased use of ion exchange and adsorption resins for acid reduction and debittering also has increased the demand for depulping technology.

Where pulp specifications exist, predictions of final pulp levels are necessary to control the operation of the industrial centrifuge. By using the pulp levels desired in the final concentrate, the pulp levels required before evaporation can be estimated, and from these values the required pulp levels before centrifugation can be determined. If freshly extracted juice were to be centrifuged and concentrated to make low-pulp products, Equation 7-3 could be used to determine the needed P value from the P_f specification. However, if reconstituted juice were to be used, Equation 7-3 would pose problems because centrifugation takes place between the first and the second evaporations. To estimate the centrifuge pulp levels needed before evaporation based on final pulp specifications, the following alternate form of Equation 7-3 can be used:

$$P = kP_f \qquad (7\text{-}4)$$

For freshly extracted juice a factor of 1.619 for k can be used, and for reconstituted juice a factor of 1.189 has been used for a three-effect evaporator. It should be noted that the number of effects of the evaporator is not used in Equation 7-4 for lack of data. Because each processing system is different, good record keeping should help you to determine the exact factors for your system. For example, if a final pulp specification in 60°Brix concentrate were 2.0 to 3.0% pulp, we would want to shoot for 2.5% pulp. Using Equation 7-4, we would want to shoot for 4.0% pulp coming out of the centrifuge for freshly extracted juice and 3.0% pulp for reconstituted juice. Again, flow rates can be adjusted to target these levels. Because centrifugation is a fairly rapid process, pulp tests are best done continuously during depulping. Part of the operation of most centrifuges is the periodic opening of the inner bowl in order to discharge the collected heavy layer, or pulp in this case. The length of time that the bowl is open and the length of time between bowl openings affect not only the pulp level of the outbound juice but the amount of juice lost with the expulsion of the pulp. Optimum flush cycles should be determined in order to have no more than 10% loss of juice soluble solids, and thereafter the pulp levels are best controlled by adjusting the flow rates through the centrifuge. The expelled pulp can be added back downstream, collected, or added to other products in order to minimize the loss of juice soluble solids.

In light of the fact that juice pulp is heterogeneous and amorphous in nature, difficulty in predicting the exact nature and level of pulp in citrus juices should be expected. The control of pulp in juices has been described as more an art than a science, and periodic deviations in expected results should be considered normal. Depulped juice is characteristically less viscous and can be concentrated to higher levels than conventional juices.

Blending of Pulp Levels

The blending of juices to meet pulp specifications may be of interest at times, especially in the manufacture of low-pulp concentrates. Lots that do not meet specification can be blended to specification by using the following equations, which are similar to Equations 6-5, 6-7, and 6-8. The average pulp level for a blend can be found as follows:

$$P = \frac{V_1 P_1 S_1 + V_2 P_2 S_2 + V_3 P_3 S_3 + \cdots + V_n P_n S_n}{V_1 S_1 + V_2 S_2 + V_3 S_3 + \cdots + V_n S_n} \qquad (7\text{-}5)$$

where P is the final average pulp level and V_n, P_n, and S_n are the volumes, pulp levels, and the pounds of solids per gallon of the various components of the blend. As before, if the Brix levels of all the components are similar, the S values can be eliminated from the equation. If the final pulp level is too high or too low, a lot with a different pulp level can be substituted, or the amount of a particular lot can be varied to get the desired pulp level. If the first lot is substituted, the required pulp level of the new component needed to achieve a certain final pulp level can be calculated by using the following:

$$P_1 = P_f + \frac{S_2 V_2 (P_f - P_2) + S_3 V_3 (P_f - P_3) + \cdots + S_n V_n (P_f - P_n)}{S_1 V_1}$$

$$(7\text{-}6)$$

If the volume of lot 1 is to be varied, the required volume to achieve a certain final pulp level can be found as follows:

$$V_1 = \frac{S_2 V_2 (P_f - P_2) + S_3 V_3 (P_f - P_3) + \cdots + S_n V_n (P_f - P_n)}{P_1 S_1 (1 - P_f/P_1)} \qquad (7\text{-}7)$$

In blending low-pulp juices, care should be taken not to use uncentrifuged juices to bring up the pulp levels because the higher-pulp juices contain larger pulp particles that may cause problems for low-pulp users even though the spindown pulp levels are within specifications.

As an example of pulp blending, suppose that we wanted to blend the following components:

Lot 3	Gallons	Brix	Sol/Gal	% Pulp
1	567	60.5	6.505	1.6
2	432	59.6	6.382	2.7
3	167	65.2	7.164	3.7
4	546	62.0	6.712	2.0
5	321	61.3	6.615	4.2

By using Equation 7-5, the final average pulp of this blend can be calculated as follows:

$$\frac{\begin{matrix}567(6.505)1.6 + 432(6.382)2.7 + 167(7.164)3.7 \\ + 546(6.712)2.0 + 321(6.615)4.2\end{matrix}}{\begin{matrix}567(6.505) + 432(6.382) + 167(7.164) + 546(6.712) \\ + 321(6.615)\end{matrix}} = 2.5\% \text{ pulp}$$

This would fit the 2.0 to 3.0% pulp specification mentioned in the last section.

Suppose, however, that 2.8% pulp was the desired pulp level. In order to meet this specification, lot 1 could be substituted for a lot with a pulp level determined as follows by using Equation 7-6:

$$2.8 + \frac{\begin{matrix}432(6.382)(2.8 - 2.7) + 167(7.164)(2.8 - 3.7) \\ + 546(6.712)(2.8 - 2.0) + 321(6.615)(2.8 - 4.2)\end{matrix}}{567(6.505)}$$

$$= 2.6\% \text{ pulp in lot 1}$$

If, instead of exchanging lot 1 for another lot, you wanted to change the amount of lot 1 to achieve the 2.8% final pulp level in the blend, Equation 7-7 could be used as follows:

$$\frac{\begin{matrix}432(6.382)(2.8 - 2.7) + 167(7.164)(2.8 - 3.7) + 546(6.712) \\ (2.8 - 2.0) + 321(6.615)(2.8 - 4.2)\end{matrix}}{1.6(6.505)(1 - 2.8/1.6)} = 108 \text{ gal}$$

Again, such equations are easily programmed into computers or programmable calculators. You may notice that the equations for adjusting the pulp levels in juice blends are identical to many of the equations used to adjust the oil in citrus blends. Much of the same computer logic and programming can be used for both. A flow chart that can be used to adjust pulp levels in citrus juice blends is shown in Fig. 7-1. An example of the use of this flow chart can be found in the GWBASIC program found in Appendix B. A similar HP-41C program can be found in Appendix C.

JUICE SACS OR FLOATING PULP

Juice sacs refer to the coarse membrane material that is screened from the juice during juice finishing operations. Citrus juices generally go through two stages of finishing prior to evaporation. The reason for this is to prevent coarse pulp from sticking to heated surfaces in the evaporator with resultant burning or the formation of brown or black flakes in the juice. Coarse pulp particles also may clog or become entrapped in the evaporator, thus restricting flow and inducing the same type of burning. The pulp that is removed can be washed with water in order to leach out additional juice solids to make pulp wash juice. Instead of being washed, the juice sacs can be frozen or dried. The latter products usually are referred to as commercial juice sacs and can be added back to juices after evaporation to provide floating pulp or a fresh juice appearance and mouth-feel. Juice sac material also can be used as an extender or texturizer in other food products.

Frozen Juice Sacs

Juice sacs generally require pasteurization in order to reduce microbial activity and, more important, to deactivate pectinase enzymes that can induce gelation and cloud loss, which are of concern when the juice sacs are to be used in citrus juices or concentrates. Because juice sac material, as it is expelled from juice finishers, is viscous in nature, shell-in-tube type heat exchangers generally are required for pasteurization. The juice sacs can be frozen at $-10°F$ ($-23°C$) in polyethylene bags inside 5-gallon cans during storage, or they can be frozen with dry ice sprays and packaged in polyethylene bags in 5-gallon cardboard boxes. Larger containers, such as 55-gallon drums, require longer freezing and thawing times, which may result in the development of off colors, off flavors, and/or spoilage as the frozen mass develops or thaws unevenly. The use of choppers to break up frozen material can reduce the thawing times.

Dried Juice Sacs

Dried juice sacs can be made from washed pulp by using pulp expelled in pulp wash operations as well as from pulp expelled from juice finishers. Dried juice sacs are not usually added back to citrus juices because such actions generally will violate standards of identity for 100% juice products. Juice sac material is believed to be responsible for much of the air pollution in feedmill operations.

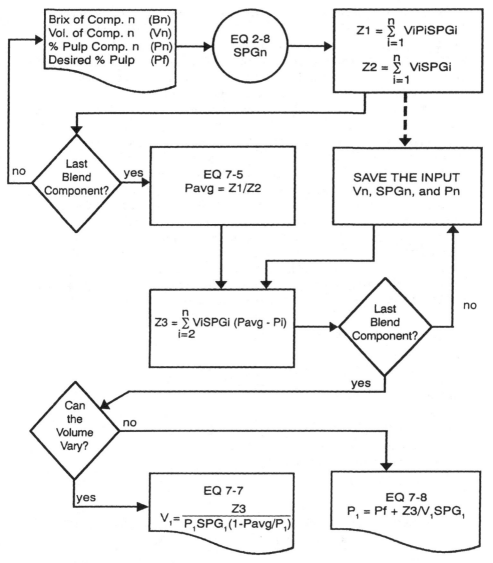

Fig. 7-1. A flow chart that can be used to program computers to calculate the average pulp and pulp level adjustments in citrus juice blending.

If the juice sac material were dried separately, this pollution would be reduced considerably. Unwashed juice sacs can be dried in foam mats or freeze-dried under a vacuum. The latter method is qualitatively best but is the costliest to perform. The sugar content of unwashed juice sacs produces a darker caramelized color than that of washed juice sacs upon drying. Washed juice sacs can be dried in drum dryers or direct fire dryers. The final moisture content of the dried sacs must be below 10%. (See Chapter 21 for moisture tests on dried citrus peels.) A third method of drying juice sacs involves the leaching of the juice sacs with acetone. The resulting juice sacs are lighter in color and fluffier, looking more like individual juice cells when reconstituted with water, compared to juice sacs produced by the other drying methods.

Dried juice sacs can be milled to any consistency down to fine flour. The common types are grits, flakes, and flour. Grits are milled to a mesh of −16 to +30. Flakes are milled to pass through a −4-mesh sieve, and flour is milled to pass through a 100-mesh screen.

The color of dried juice sacs may fade but is not an important quality parameter. Juice sac composition is illustrated in Table 7-2. Antioxidants may have to be added to prevent off flavor development in the dried sacs. Ambient temperature storage for longer than one year may result in weevil infestation.

The measurement of floating pulp in juices is a little less precise than the measurement of spindown pulp. Screening of the pulp followed by weighing of the larger screened pulp particles can provide for an estimate of the floating pulp level. This method measures the floating pulp only and is somewhat comparable to the quick fiber test mentioned previously, but requires less costly equipment, is more rapid, and thus is better suited to routine quality control. The procedure is outlined below.

Determination of Floating Pulp or Juice Sacs

Equipment and Supplies

- 20-mesh sieve.
- Laboratory balance or triple beam balance.
- 1-liter volumetric or Erlenmeyer flask.
- Stirring rod or spatula.

Table 7-2. Some characteristics of dried washed juice sacs (Kesterson and Braddock 1973).

Parameter	Orange	Grapefruit
% Crude fiber	18.9	—
% Protein	9.0	—
% Ash	20.6	—
% Moisture	8–10	8–10
% Fat	1.2–2.1	0.7–0.8
Bulk density (lb/ft^3)	9.3	9.3
Water holding capacity	10/1 to 13/1	14/1 to 15/1
Fat holding capacity	4/1 to 5/1	4/1 to 5/1

(Source: *Food Technology* 1973. 27(2): 50, 52, 54. Copyright © by the Institute of Food Technologists.)

Procedure

1. Weigh 1 liter of well-mixed 11.8°Brix juice into a 1-liter flask while swirling it so that the juice contains a representative amount of pulp.
2. While swirling, pour the juice through the 20-mesh sieve over a sink and shake the screen until the pulp retained by the screen "balls up" and is free of excess juice.
3. Carefully turn the ball with the spatula or stirring rod into one mass and then onto the balance pan and weigh it. Disregard any slight amount of pulp that may cling to the screen.
4. Divide the weight of the recovered pulp by the weight of the juice, and multiply by 100% to get the % floating pulp. In Florida the floating pulp is reported by use of the following:

$$(\% \text{ floating pulp})(7.49) = g/6 \text{ fl oz of FCOJ}$$

or:

$$(\% \text{ floating pulp})(19.77) = g/64 \text{ fl oz carton juice}$$

Pulp levels of 0.75 to 0.80% are ideal in most situations. This is equivalent to roughly 4 gallons of juice sac material per 1000 gallons of single-strength juice.

PULP WASH

The pulp expelled from primary and secondary finishers contains about 80% juice. These juice solids can be recovered by washing and refinishing the pulp several times. The resulting juice is commonly referred to as pulp wash or water-extracted soluble orange or grapefruit solids (WESOS or WESGS). In the first stage, or the first time that the juice pulp is washed and refinished, about 50% of the available juice from the expelled pulp can be recovered. If two stages are used, 63% of the available juice can be recovered, and if three stages are used, 75% of the available juice in the pulp can be recovered. A four-stage system can recover up to 80% of the available juice from the pulp. More stages than this generally do not recover enough juice to be economically feasible. The efficiency of a pulp wash system depends on a balance of parameters including the water-to-pulp ratio and the number of stages. Figure 7-2 shows the results of a study performed by FMC using its equipment. As can be seen in the figure, the ideal water-to-pulp ratio is between 1.5 and 2.0, with the number of stages exceeding 3 to 4 contributing little to the % recovery of juice from the pulp. If 100 lb of pulp can be produced per ton of fruit processed, a pulp wash system can increase the juice yield by 5 to 8 lb of soluble solids/ton of fruit. Usually 80 to 120 lb of soluble solids/ton can be obtained in normal juice processing without pulp washing, depending on the time of year and the variety

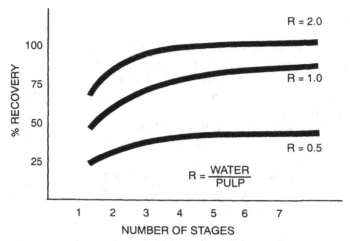

Fig. 7-2. The efficiency of an FMC pulp wash system as a function of the water-to-pulp ratio and the number of stages (Ballentine and Ferguson 1989).

of fruit. This increase in yield due to pulp washing can represent a significant increase in juice yield, especially when juice is in high demand.

Pulp wash juice generally is considered inferior to extracted juice in both color and flavor. The quality control of pulp wash juice generally consists of the measurement of limonin (see Chapter 10). The limonin content in pulp wash juices is responsible for the major effect of pulp wash on citrus juice quality, as pulp wash can contain as much as twice the limonin found in conventional juices. Concerns over juice quality and excessive juice supplies have led to restrictions on the use of pulp wash in 100% juice products. The federal standards of identity permit the adding of pulp wash on-line to freshly extracted juice, but Florida law prohibits any adding of pulp wash to 100% juice products. In order to enforce this law, Florida requires the addition of a sodium benzoate tracer to all pulp wash products made in the state so that detection of their illegal use is easier. Florida also requires that drums of pulp wash be clearly labeled and encircled by a 5-inch yellow band.

Pulp wash juice contains a greater amount of high molecular weight compounds, such as pectin, which increase the viscosity of resulting concentrates. The added viscosity can pose problems in conveyance during evaporation and chilling, so enzymes sometimes are added during the first stage of pulp washing to help break down the high molecular weight compounds. Thus these pectinase or polygalactonase enzymes have a maximum amount of time to work to decrease the viscosity. Without enzyme treatment, concentrate levels of over 40°Brix may present difficulty in the concentration of pure pulp wash juices. An alternative enzyme treatment consists of enzyme addition to the final pulp wash and holding the juice for about an hour at 115°F. The pulp wash may be pasteurized and/or centrifuged prior to evaporation if necessary. Centrifuged and enzyme-treated pulp wash juice can be concentrated to 65°Brix. Again, the state of Florida requires the addition of sodium benzoate as a tracer to detect unauthorized use of pulp wash in 100% citrus juice products, which is usually done just prior to evaporation.

The most efficient multistage pulp wash system is a countercurrent one where fresh water is used in the last stage and then is used in succeeding prior stages until the first stage. This reduces the time needed for the juice in the pulp and in the washing solution to equilibrate.

Pulp wash also is used as a carbohydrate source in drink bases and as a clouding agent. Quality control tests sometimes used to measure the quality in such products include % light transmission (ability to act as a clouding agent), % spindown pulp (mouth-feel), diacetyl or plating (microbiological growth), and viscosity (ability to reconstitute).

CORE WASH

Core wash is the same as pulp wash except that it is obtained from the core material expelled from the orifice tubes in the FMC extractor. This core material consists of much of the rag or membrane material and seeds formerly contained inside the fruit. The rag and seed material is very high in limonin, and successive stages of washing and finishing of this material can result in a juice extremely high in limonin (up to three to five times more than in conventional juices, depending on the variety and the time of the season). Core wash juice is very opaque and is used as a clouding agent in drink bases. Core washing methods generally employ gentle washing using fewer stages than in pulp washing and gentle treatment to avoid breaking the seeds and releasing large amounts of limonin. Commercial debittering systems have successfully debittered such juices, as explained in Chapter 10, with other characteristics remaining similar to those of pulp wash concentrates.

QUESTIONS

1. What are the advantages and disadvantages of sinking or spindown pulp in relation to juice quality?
2. If the pulp level is too low, should the finisher pressure be increased or decreased?
3. What would be the effect on juice yield if the pulp level were too low?
4. What are some of the reasons why pulp levels change during processing?
5. What benefits would there be in manufacturing low-pulp juices? What are the disadvantages?
6. How can the depulping efficiency be regulated in depulping operations?

7. What are the advantages and disadvantages of a pulp wash system?
8. What are some of the differences between sinking and floating pulp?
9. Besides using pulp wash in 100% juice products according to federal standards, what are some other uses that may be possible for pulp wash juice?
10. How could a quick fiber test be of use in pulp wash processing?

PROBLEMS

1. What speed, in rpm, should a lab centrifuge use if the distance from bottom to bottom between two centrifuge tubes in the horizontal position in a lab centrifuge is $13\frac{1}{2}$ inches?

2. What is the % pulp in each of the following spindown pulp tests according to the procedure in the text?

Ml Pulp in Tube

1.6
2.5
4.8
3.6
5.1

3. If the pulp level of freshly extracted juice is 19.3%, what would be the expected spindown pulp level after evaporation using a seven-effect evaporator? What would be the expected pulp level if this same concentrate were reconstituted and reevaporated a second time in the same evaporator?

4. If a customer wanted juice with a pulp level of 2.8 to 3.0%, what pulp level would you want to shoot for coming out of the centrifuge using freshly extracted juice?

5. In blending pulp levels, what would be the final average pulp level using the following blend components?

Lot #	Drums	Brix	Sol/Gal	% Pulp
1	5	60.2	6.464	3.6
2	19	60.1	6.450	1.0
3	6	59.8	6.409	0.4
4	8	60.4	6.491	2.8
5	15	59.6	6.382	2.3

6. In the above blend, suppose you want to exchange lot 5 for a lot with a pulp level that would give a final pulp level of 2.0%. What is the required pulp level for this new blend component, assuming the same Brix and volume for lot 5?

7. If the number of drums in lot 5 in problem 5 are to be changed to meet the 2.0 specification in the previous problem rather than substituting a new lot, what is the required number of drums of lot 5?

8. What would be the % floating pulp in 1045 g of single-strength juice that yielded 8.6 g of screened pulp?

Chapter 8

Juice Cloud

In citrus juices, unlike other juices, an opaque nature is considered a desirable characteristic. Citrus juice itself comes primarily from the juice cell food storage vacuoles, where it exists in a clear cloudless form. As the juice cell is ruptured during juice extraction, high molecular weight compounds from the organelles and cytoplasm of the juice cell become suspended in the juice, along with membrane and pectin material. This colloidal suspension, which gives citrus juices its cloud, is comprised of about 30% proteins, 20% hesperidin, 15% cellulose and hemicellulose, and 5% pectin (Bennett 1987). The content of the other 30% of the juice cloud remains a mystery.

Proteins come primarily from the organelles and cytoplasm in the juice cell, such as the mitochondria. Semisoluble membrane material is believed to be responsible for the cellulose and hemicellulose material in the cloud, in addition to being a common component of citrus peel. Hesperidin remains in a soluble form within the intact juice cell vacuole; however, juice extraction causes a change to take place rendering it insoluble and inducing precipitation that contributes to the juice cloud.

The suspended cloud and spindown pulp are closely related. It is believed that the pulp breakdown discussed in the previous chapter contributes to the suspended cloud material. It also has been shown that as pulp levels increase, so does the amount of cloud material (Rouse, Atkins, and Huggart 1954).

PECTIN

Even though pectin comprises a small portion of the cloud material, and although juice cloud stability can be achieved in its absence, the effect of pectin on juice cloud stability is dramatic. The term "pectin" refers to a class of high molecular weight compounds, with molecular weights of 100,000 to 200,000, consisting of 150 to 1500 galacturonic acid units linked together via $\alpha(1 \rightarrow 4)$ glycoside bonds with side chains of rhamnose, arabinans, galactans, xylose,

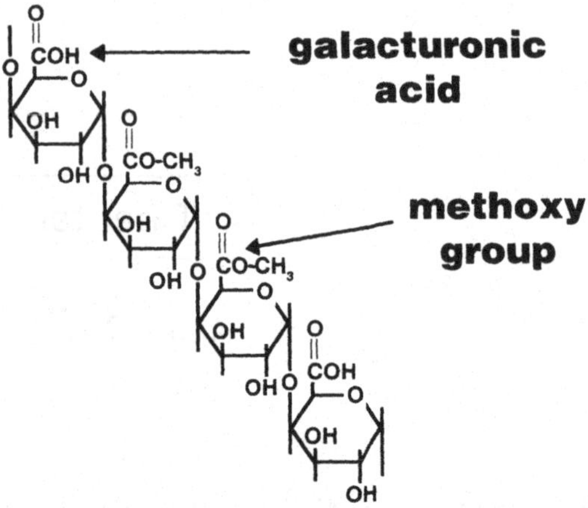

Fig. 8-1. Segment of a pectin molecule showing the characteristic galacturonic acid and esterified methoxy groups.

and fucose (Stevens 1941). Many of the carboxyl groups are esterified with methanol to form methoxy groups, as shown in Fig. 8-1. These methoxy groups block many reactions, including polymerization or gelation; so the degree of esterification (D.E.) is a measure of the gelling ability of the pectin or the grade of pectin, and ranges from 1 to 100%. The grade also may be determined according to the percentage of methoxy groups, from 0 to a maximum of 16.32%. See Chapter 20 for a more detailed description of commercial pectin recovery and characteristics.

GELATION AND CLOUD LOSS

There are two basic types of reactions that result in gelation in citrus concentrates or cloud loss in single-strength citrus juices. One reaction is the reversible acid- or base-catalyzed esterification of the carboxyl groups of the galacturonic acid components of pectins, according to:

$$CH_3 + R-\overset{\overset{\displaystyle O}{\|}}{C}OH \underset{H^+ \text{ or } OH^-}{\overset{H^+ \text{ or } OH^-}{\rightleftharpoons}} R-\overset{\overset{\displaystyle O}{\|}}{C}OCH_3 + H_2O + \text{Heat} \qquad (8\text{-}1)$$

The equilibrium of this reaction depends on many factors, including pH, water content, heat, and secondary reactions. In the presence of sugars, under the proper conditions, polymerization can occur in citrus concentrates, resulting in gelation. Even though the methoxy groups themselves prevent such polymerization, the reverse reaction in equation 8-1 allows it to occur. Such polymerization is not common in citrus concentrates.

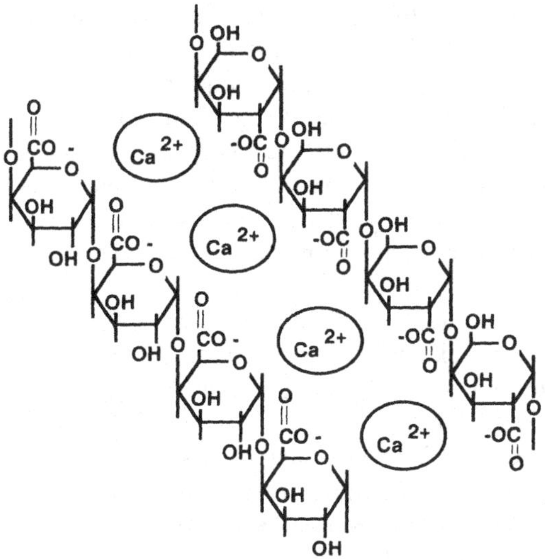

Fig. 8-2. Calcium pectate structure responsible for most gelation and cloud loss in citrus juices.

A more common form of gelation occurs when there are sufficient galacturonic acid groups to form pectates, which link together via divalent cations such as calcium, as shown in Fig. 8-2. This can occur only if there are enough deesterified carboxyl groups available. In freshly extracted citrus juices, the D.E. values are usually over 50%. Levels below about 20% may be sufficient to cause the formation of calcium pectates, which in turn results in gelation in concentrates and cloud loss in single-strength juices. This precipitation in single-strength juices seems to entrap or affect other cloud components as well, resulting in total cloud loss. Also, the length of the pectic chain or the molecular weight of the pectin affects the cloud stability. Juices with low molecular weight pectins generally will not gel. If too much heat is used during lemon and lime juice processing, the acid hydrolysis described in the reverse reaction of Equation 8-1 will occur (because these juices contain more acid than other citrus juices), which will result in polymerization and/or calcium pectate formation or gelation.

PECTINASE ENZYMES

Another type of deesterification that can lead to cloud loss and/or gelation through calcium pectate formation results from the pectinase enzymes that naturally occur in citrus juices. These enzymes, also referred to as pectin esterases or pectin methylesterases, are the primary cause of gelation and cloud loss in commercial juices. In order to prevent the removal of the methoxy groups by this enzyme, heat treatment is used to deactivate or destroy the enzyme. In most juices a temperature of 150°F (66°C) is generally sufficient for pasteurization; but 190°F (88°C) or more for about 10 to 15 seconds is required to deactivate most of the pectinase enzymes, with lemon and lime juices requiring only 165–190°F (74–88°C) to deactivate the enzyme because of their higher acid content. Unpasteurized "fresh squeezed" juice products are easily recognized by severe cloud separation, which may be partially overcome by opaque containers or labels, or, as done in the food service industry, shaking the juice to disperse the cloud before it is served to a customer. Heat treatment is the only commercial way to prevent this cloud loss.

In monitoring the stability of juice cloud, several methods have been used. Before 1952 the Stevens cloud test was extensively used (Rouse, Atkins, and Huggart 1954), which involved the addition of excess pectin followed by a period of incubation, generally a couple of days, and measurement of the volume of the clarified layer which resulted from enzymatic activity. Usually a volume of 10 ml of clarified juice per 100 ml of single-strength juice represented an unstable cloud. More rapid techniques have since been developed, which are more conducive to routine quality control operations. These rapid techniques center around the measurement of the rate of acid formation of the reverse reaction in Equation 8-1 or:

$$
\underset{\text{}}{R-\overset{\overset{\textstyle O}{\|}}{C}OCH_3} + H_2O \xrightarrow[\text{enzyme}]{\text{pectinase}} \underset{\text{acid}}{R-\overset{\overset{\textstyle O}{\|}}{C}OH} + CH_3OH \tag{8-2}
$$

The rate of this reaction can be expressed in the form of a first-order rate law:

$$
R = k(\text{acid}) = \frac{(\text{volume titrated})(\text{NaOH normality})}{(\text{reaction time})(\text{sample volume})} \tag{8-3}
$$

where the rate R is expressed in terms of the acid concentration and the rate constant k. The concentrations of other species, assuming a negligible effect on the reaction rate, are incorporated in the rate constant. This equation suggests that a measurement of the change of acid concentration or pH can be used as an estimate of the reaction rate or the enzymatic activity. The absolute value of R usually is expressed in standard scientific units. However, the enzymatic activity, a relative value, is of interest here; so convenient units can be used to calculate a final reaction rate, which can be expressed in arbitrary units established by standard industrial methodology.

The change in pH of a citrus juice sample is difficult to detect with the small amount of pectin naturally present in the juice. In order to speed up the reaction or to render the pH change large enough to measure, excess pectin is added to the juice sample. The pH of the juice is adjusted to near neutrality with a base; and then either a standard amount of base is added and the time needed for the enzymatically produced acid to restore the sample to the original pH is measured, or a standard pH is maintained and the amount of base needed to accomplish this is measured for a given period of time. It is important that a proper balance of enzyme and pectin exist. Samples of 10 to 30 ml of single-strength juice (or the equivalent weight of soluble solids) are commonly used with 40 to 100 ml of standard 1% pectin solution. The final PEU (pectin esterase units), or arbitrary reaction rate units, can be expressed in various units, including ml NaOH at a certain normality per milliliter, gram, or pound soluble solids per second or minute. Another novel method is to add methyl red pH indicator to the pectin–juice mixture. A negative test occurs if the solution maintains color for 30 minutes. In routine quality control, the least tedious method that produced an adequate positive/negative test follows. This PEU test is a pass/fail test having little value in further refinement of the results.

PEU Test

Equipment and Supplies

- pH meter
- 10 ml volumetric pipette
- 100 ml beaker
- 250 ml beaker
- 50 ml graduated cylinder
- Temperature bath
- Two disposable Pasteur pipettes
- Graduated 1 ml pipette
- Stopwatch or timer
- Magnetic stirrer
- 1% pectin–salt solution made by mixing 10 g of pectin and 15.3 g NaCl in a one-liter container and filling it with distilled water. (Heat may be needed to get the pectin into solution.)
- 2.0N NaOH (20 g NaOH in 250 ml distilled water) protected from atmospheric carbon dioxide and moisture. (See section on acid titrations in Chapter 3.)
- 0.05N NaOH (32.0 ml of 0.1562N NaOH mixed with 100 ml distilled water) protected in the same way as the above basic solution.

Procedure

1. Pipette 10 ml of single-strength juice (11.8°Brix) into a 100 ml beaker.
2. Add 40 ml of 1% pectin–salt solution to the sample using a 50 ml graduated cylinder, and add a magnetic stirring bar.
3. Place the 100 ml beaker containing the sample inside a 250 ml beaker containing enough water to insulate the inner 100 ml beaker and sample.
4. Warm the beakers in an 86°F (30°C) temperature bath until the sample in the 100 ml beaker is at the temperature of the bath as measured by the thermometer.
5. Place the beakers on a magnetic stirrer, along with a thermometer, and insert a pH electrode as shown in Fig. 8-3.
6. While stirring and maintaining constant temperature, add 2.0N NaOH solution until the pH is stable at 7.0, using a disposable Pasteur pipette.
7. Add 0.05N NaOH using a second Pasteur pipette until the pH is between 7.6 and 7.8, and record the exact pH.
8. Using a graduated 1 ml pipette, add exactly 0.10 ml of the 0.05N NaOH solution and start and stopwatch or timer.
9. Record the time it takes to regain the pH recorded in step 7 above.
10. The rate of acid formation can be calculated in pectin esterase units as follows:

$$PEU = \frac{(0.05N \text{ NaOH}) (0.10 \text{ ml NaOH})}{(10 \text{ ml sample}) (\text{minutes})} \qquad (8\text{-}4)$$

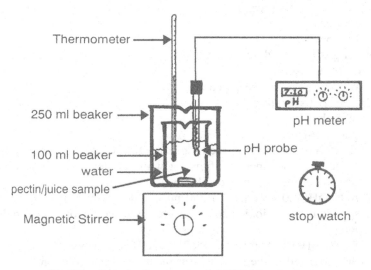

Fig. 8-3. Setup for the recommended PEU determination.

where the minutes refers to the time it took to regain the pH recorded in step 7. For example:

$$PEU = \frac{(0.05)(0.10)}{(10)(6.0)} = 8.33 \times 10^{-5}$$

Most juices processed in modern evaporators and pasteurizers have PEU values of 1×10^{-6} to 1×10^{-4}. Those with levels much higher than this are susceptible to gelation and/or cloud loss, depending on pH, temperature, and other related conditions. In routine operations, rather than having a maximum allowable PEU level, it may be more convenient to have a minimum time interval for the pH recorded in step 7 to be regained, such as 2.5 minutes. If the recorded pH is regained in 2.5 minutes or less, the cloud should be considered unstable, and further heat treatment should be applied. In the above process it should be noted that with stirring of the sample during the timing of the pH change, the incorporation of atmospheric CO_2 into the sample will result in some drop in pH. However, this should not prevent the determination of a stable or an unstable cloud test. It should be kept in mind, however, that even with no enzyme activity the pH will drop because of this CO_2 effect.

It is possible to have a gelled or clarified citrus juice with low enzyme activity because deesterification can take place prior to heat treatment. Hence, one should be careful not to allow excess time between extraction and heat treatment. Sometimes when a juice concentrate gels, measurement of the enzyme activity is of little value because deesterification could have taken place prior to heat treatment. In such a situation, direct measurement of gelation is in order. For frozen concentrated orange juice (FCOJ) in 6 oz cans, the following guide can be used in grading the degree of gel when it is poured out onto a tray at room temperature:

#1 gel—Questionable whether gel is present or not.
#2 gel—Definite gel lumps.
#3 gel—Definite gel that holds to the shape of the can but breaks up partially upon pouring.
#4 gel—Definite gel that retains the shape of the can upon pouring.

A #1 or #2 gel can usually be salvaged by reconstitution and centrifuging, but a #3 or #4 gel usually cannot be salvaged and must be discarded.

VISCOSITY

Cloud and pulp material contribute strongly to the viscosity of citrus juices, especially at high Brix levels. The cold temperatures required for storage and shipping of these concentrates also add greatly to the viscosity. Excessively viscous concentrates put a heavy burden on processing pumps, and viscosity is usually the limiting factor in the degree of concentration of citrus juices. If the viscosity can be reduced, higher concentration levels can be achieved with greater savings in storage and shipping costs. However, the use of enzymes in primary citrus products is not allowed in Florida.

The viscosity can be reduced by centrifugation and controlled enzyme treatment, used to break down juice pulp and cloud material. Polygalacturonase has been used for the latter purpose (Baker and Bruemer 1971); it cleaves primary pectin chains between adjacent nonesterified carboxyl groups. This reaction, which is carried out at 86°F (30°C), is terminated by heat treatment to deactivate the enzyme. Enzyme addition is commonly used during the first stage of pulp wash systems for the same purpose, and it also increases the total soluble solids and thus the juice yield. Ultrasonic and irradiation treatments have also been used to break down juice cloud and pulp in order to reduce the viscosity of the concentrate (Berk 1964). Complete enzyme clarification sometimes is desired for lemon and lime juices.

QUESTIONS

1. What is the source of most of the cloud material found in citrus cloud?
2. What is the relationship between cloud material and pulp?
3. What is the main cause of cloud loss or gelation in citrus products?
4. What role do pectinase enzymes play in cloud stability, and how can gelation be prevented commercially?
5. Why is excess pectin added to juice samples in the PEU test?

PROBLEMS

1. Suppose that you open a drum of concentrate obtained from another processing plant and notice a high viscosity. You take a sample and discover that the Brix is not excessively high, but gel lumps are found. What is the gel number? Does this mean that the concentrate must have a high PEU value? What can be done to correct the situation?
2. What would be the PEU value, according to the procedure in this chapter, for a sample that took 2 minutes and 23 seconds to return to the original pH? Would this be considered cloud-stable juice?
3. What would be the PEU value of a juice sample that took 6 minutes and 10 seconds to return to the original pH? Is this juice cloud stable?
4. Suppose you used 1.0 ml of 0.01562N NaOH in the procedure for 10 ml of sample, and it took 28 minutes and 53 seconds to return to the original pH. What would be the PEU value, and would this be considered stable juice cloud?

5. What would be the PEU value per gram of soluble solids per second using 0.5 ml of 0.0312N NaOH for 15 ml of sample (12.0°Brix) that took 25 minutes and 16 seconds to return to the original pH? Would this be considered stable juice cloud?

REFERENCES

Baker, R. A. and Bruemer, J. H. 1971. *Proc. Florida St. Hort. Soc.*, *84*, 197.

Bennett, R. D. 1987. From presentation to the Citrus Products Technical Committee at the USDA Fruit and Vegetable Laboratory, Pasadena, Calif., March 13, 1987.

Berk, Z. 1964. *Food Tech.*, *18*, 1811.

Rouse, A. H., Atkins, C. D., and Huggart, R. L. 1954. Effect of pulp quantity on chemical and physical properties of citrus juices and concentrates, *Food Tech.*, *8*, 431–435.

Stevens, J. W. 1941. U.S. Patent 2,267,050, December 23, 1941.

Chapter 9

Color of Citrus Juices

The natural bright color of citrus juices has long been regarded as one of their major qualitative advantages over other food products. Their carotenoidal pigmentation has been associated with the color of the sun, producing a bright and cheerful effect that complements the sweet and tart flavors and pleasant aromas of the juices. The yellow-to-orange-to-pink-to-red colors have been used extensively in marketing through transparent packaging and the use of sliced fruit in advertising. The USDA holds color equal to flavor in its quality scoring system, an indication of the importance of color in the quality of commercial citrus products.

The main carotenoids responsible for the orange color of orange and tangerine juice are α-carotene, β-carotene, zeta-antheraxanthin (yellowish), violaxanthin (yellowish), β-citraurin (reddish orange), and β-cryptoxanthin (orange) (Stewart 1980), with the red or pink color of the pigmented grapefruit juice varieties due to the presence of lycopene (Khan and MacKinney 1953). The red color in blood oranges and related varieties is due to the presence of anthocyanins (Chandler 1958).

In dry cooler Mediterranean climates, as in California, fruit pigmentation is well developed. In hotter and/or more humid areas, such as Florida, the coloration appears more dilute. For this reason the measurement of juice color plays a greater role in juice processing in Florida than it does in California.

Color enhancement can occur only through blending with juices of higher coloration. The federal code permits the blending of up to 10% of tangerine juice (*Citrus reticulata*) and its hybrids with orange juice while still allowing the product to be called 100% orange juice, in an effort to enhance juices with weak color.

Off colors due to Maillard oxidation (as a result of scorching or excessive heat applied to concentrates over too long a time) detract from the color quality of the juice and can result in lower color scores. Such browning reactions dull the natural citrus color. However, off flavors from these reactions develop before the visible effect occurs and further render the juice inferior. Also, white hesperidin flakes or brown or black flakes from evaporator burn are readily visible in citrus juices and mar their appearance. See Chapter 18.

USDA COLOR SCORING

The USDA has established a standard for the grading of the colors of orange juice using six plastic color tubes, OJ1 (lightest) to OJ6 (darkest). One-inch-diameter plastic tubes matching these color standards can be obtained from a licensed manufacturer (such as Fiper/Magnuson, P.O. Box 11427, Reno, Nev. 89510, 1-800-648-4737) or by contacting:

Processed Products Branch
Fruit and Vegetable Quality Division, AMS
U.S. Department of Agriculture
Washington, D.C. 20250

These tubes can be used for direct color comparisons or to calibrate approved colorimeters. Like most colors, these tubes are sensitive to light and may fade on prolonged exposure; so they should be stored in a dark place. The following procedure should be observed in using the tubes for direct comparison to citrus juices.

Color Tube Test

Equipment and Supplies

- USDA color tubes (OJ1–OJ6)
- 45° angle tube rack with dull gray background and space between each standard tube for sample tube. (See Fig. 9-1.)
- One-inch-diameter screw-cap culture tube for juice samples.
- Standard light source of approximately 150 candles with a color temperature of 7500 ± 200°K or a Macbeth Color Box model EBA-220, which approximately simulates daylight under a moderately overcast day. Subdued northern daylight may be used if care is exercised to minimize the effects of reflected light from colored objects. Ordinary incandescent or fluorescent light is not satisfactory.
- 80°F temperature bath.
- Deaeration setup shown in Fig. 2-3.

Procedure

1. Deaerate the juice as explained in Chapter 2.
2. Place the reconstituted juice (11.8°Brix) in a one-inch-diameter test tube, and warm it to 80 ± 2°F (27 ± 1°C) in the temperature bath.

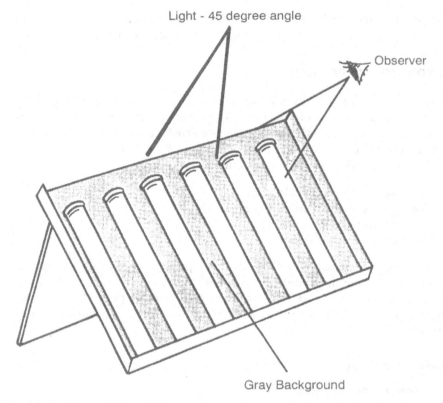

Fig. 9-1. Proper rack arrangement for direct comparison of orange juice color with USDA standard color tubes.

3. Place the sample tube between the standard color tubes that are just lighter than and darker than the sample tube.
4. Use Table 9-1 to determine the USDA color score. Table 9-2 gives the corresponding USDA grade standards for citrus juices.

TRISTIMULUS TEST

The standard method of color scoring in the food industry recommended by the International Committee of Illumination is based on the ''standard observer'' or simulated standard eye and consists of three primary colors or filters, referred to as X (amber), Y (green), and Z (blue). Table 9-3 correlates these parameters with the wavelength of visible light usually associated with color.

In 1952 Hunter came up with a scheme that measured redness (a), yellow-

Table 9-1. USDA color scores from comparison with plastic USDA color tube standards (C = juice color).

	Frozen conc. orange juice, reduced acid, and canned conc. orange juice	Pasteurized orange juice and orange juice from concentrate	Canned orange juice and conc. orange juice for manufacturing	Dehydrated orange juice
$C > OJ1$	40	40	40	40
$OJ2 < C < OJ1$	40	40	40	40
$C = OJ2$	40	40	40	40
$OJ3 < C < OJ2$	39	39	39	39
$C = OJ3$	39	39	39	39
$OJ4 < C < OJ3$	38	38	38	38
$C =$ or slightly $> OJ4$	37	37	38	37
$OJ5 < C < OJ4$	36	36	37	36
$C = OJ5$	36*	36	37	36
$OJ6 < C < OJ5$	35	36*	36	35
$C = OJ6$	34	35	36*	34*
$C < OJ6$	33 or less	34 or less	35 or less	33 or less

*Limits for Grade A.

ness (b), and lightness (L), which can be depicted as shown in Fig. 9-2 (Hunter 1958). These Hunter parameters can be related to the tristimulus parameters according to (Kramer and Twigg 1970):

$$L = 100 \sqrt{Y} \tag{9-1}$$

$$a = 175(1.20X - Y)/\sqrt{Y} \tag{9-2}$$

$$b = 70(Y - 0.847Z)/\sqrt{Y} \tag{9-3}$$

The Hunter citrus colorimeter further simplifies the color measurement by measuring the citrus redness (CR) and the citrus yellowness (CY), which are related to the tristimulus parameters as follows:

$$CR = \frac{(1.277X - 0.213Z - 1)}{Y} 200 \tag{9-4}$$

$$CY = 100 \left(1 - 0.847(Z/Y)\right) \tag{9-5}$$

The Florida Department of Citrus adopted the Hunter D45 colorimeter as the official method of color measurement, using the following procedure.

Table 9-2. USDA color grades for citrus juices. (No color standards exist for concentrated grapefruit juice for manufacturing, dehydrated grapefruit juice, tangerine juice or its hybrids, or any lemon or lime juices.)

	Grade A	Grade B
Orange Juice (47FR 12/10/82, 52.1557)		
Frozen concentrated orange juice	36–40	32–35
Canned concentrated orange juice	36–40	32–35
Reduced acid orange juice	36–40	32–35
Pasteurized orange juice	36–40*	32–35
Orange juice from concentrate	36–40*	32–35
Concentrated orange juice for manufacturing	36–40	32–35
Canned orange juice	36–40	32–35
Dehydrated orange juice	34–40	33 or less
Grapefruit Juice (48FR 9/12/83, 52.1224, 52.1226)		
Grapefruit juice	18–20**	16–17***
Grapefruit juice from concentrate	18–20**	16–17***
Frozen concentrated grapefruit juice	18–20**	16–17***
Grapefruit and Orange Juice Blend (6FR 11/1/72, 52.1284, 52.1288)		
Grapefruit and orange juice blend	18–20**	16–17***

* > OJ6.
**Means color representative of juice from mature, well-ripened grapefruit; the juice may show fading and lack luster.
***Means juice may be slightly, but not materially, affected by scorching, oxidation, or caramelization.

Hunter Colorimeter Test

Equipment and Supplies

- Hunter D45 or D45D2 colorimeter or other approved colorimeter.
- Temperature bath (80°F or 27°C).
- Deaeration setup shown in Fig. 2-3.
- One-inch-diameter screwcap culture tube.
- USDA standard OJ4 color tube.

Table 9-3. Reflectance values for *X, Y, Z* of the standard observer at various wavelengths from average daylight illumination (Kramer and Twigg 1970).

Wavelength nm	X	Y	Z
380	4		20
90	19		89
400	85	2	404
10	329	9	1570
20	1238	37	5949
30	2997	122	14,628
40	3975	262	19,938
450	3915	443	20,638
60	3362	694	19,299
70	2272	1058	14,972
80	1112	1618	9461
90	363	2358	5274
500	52	3401	2864
10	89	4833	1520
20	576	6462	712
30	1523	7934	388
40	2785	9149	195
550	4282	9832	86
60	5880	9841	39
70	7322	9147	20
80	8417	7992	16
90	8984	6627	10
600	8949	5316	7
10	8325	4176	2
20	7070	3153	2
30	5309	2190	
40	3693	1443	
650	2349	886	
60	1361	504	
70	708	259	
80	369	134	
90	171	62	
700	82	29	
10	39	14	
20	19	6	
30	8	3	
40	4	2	
750	2	1	
60	1	1	
70	1		

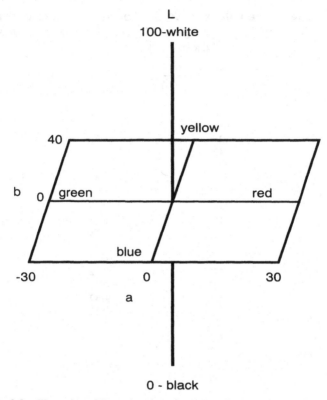

Fig. 9-2. Illustration of Hunter's color scheme based on the parameters a, b, and L.

Procedure

1. Reconstitute the juice to the following Brix values:
   45°Brix FCOJ to 12.8°Brix
   42°Brix FCOJ to 11.8°Brix
   COJFM to 12.3°Brix
   or to the standard 11.8°Brix as in other quality tests.
2. Deaerate the juice sample for at least 3 minutes using the procedure found in Chapter 2.
3. Warm the sample in the temperature bath to 80 ± 2°F (27 ± 1°C) in the one-inch-diameter culture tube.
4. The colorimeter should be kept on standby with the power connected at all times. If the instrument is off, turn it on, and let it warm up for about 10 minutes.
5. Turn the sensitivity knob as far to the right as possible.

6. Standardize *CR* and *CY* with an OJ4 USDA color tube as indexed by the instrument.
7. Place the culture tube containing the sample at the indexed position, and read the *CR* and *CY* values. Orange juices yield *CR* values of 26 to 45 and *CY* values of 74 to 89.
8. Apply the *CR* and *CY* values to the nomograph in Fig. 9-3, or use the following equation to calculate the equivalent color number (*CN*):

$$CN = 0.165CR + 0.111CY + 22.510 \qquad (9\text{-}6)$$

The Hunter D45D2 automatically calculates the color number.
9. The calculated color number then can be converted into the USDA color score by using the information in Table 9-4 for the various types of orange juices.

Other colorimeters have since been approved by the Agricultural Research Service of the USDA in cooperation with the Florida Department of Citrus, including the following:

- *Minolta Chroma Meter II/Reflectance/CR100*—provided that the instrument is equipped with an adapter for one-inch-diameter test tubes and is calibrated by using an OJ4 color standard tube according to plans available from the ARS. The color number then can be determined by using:

$$CN = 43.85 + 1.07X - 0.61Y - 2.74Z \qquad (9\text{-}7)$$

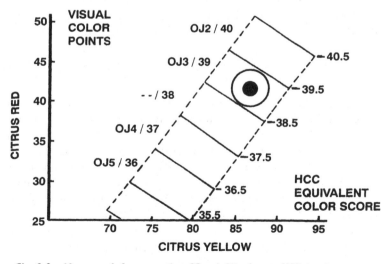

Fig. 9-3. Nomograph for converting *CR* and *CY* values to USDA color scores.

Table 9-4. Conversions from color number to USDA color score (Florida Dept. of Citrus 1975).

Color number	Pasteurized OJ and OJ from concentrate	Frozen conc. OJ and conc. canned OJ	Conc. OJ and conc. OJ for manufacturing
32.5–33.4	33	33	35
33.5–34.4	34	34	36
34.5–35.4	—	35	36
34.5–34.9	35	—	—
35.0–36.4	36	—	—
35.5–36.4	—	36	37
36.5–37.4	37	37	38
37.5–38.4	38	38	38
38.5–39.4	39	39	39
39.5–40.4	40	40	40

- *Macbeth Color-Eye Model 1500*—provided that the instrument is equipped with an adapter for one-inch-diameter test tubes, according to plans available from the ARS, and is calibrated by using a white tile. The color number then can be calculated by using:

$$CN = 39.35 + 1.74X - 1.28Y - 0.94Z \qquad (9\text{-}8)$$

- *HunterLab LabScan Model LS-5100*—provided that the instrument is equipped with an adapter for one-inch-diameter test tubes, according to plans available from the ARS, and is calibrated by using an OJ4 color standard tube. The color number then can be calculated by using:

$$CN = 40.10 + 1.90X - 1.55Y - 1.45Z \qquad (9\text{-}9)$$

Table 9-4 can be used to convert these color numbers into USDA color scores.

There are no color grade standards for grapefruit juices or other citrus juices. The Texas Citrus Exchange (U.S. Expressway 83 at Mayberry, Mission, Tex. 78572) has established a grapefruit color scoring system that may be of value to other grapefruit juice processors. Pigmented citrus juices containing lycopene (pink grapefruit) and anthocyanins (blood oranges) have a tendency to undergo a color change during processing to a brownish or dull color. This color change generally renders the juices inferior as processed products even through pigmented citrus fruits are said to be the most delicious. Decolorization of these pigmented juices is the subject of ongoing research.

QUESTIONS

1. What main carotenoids are responsible for the colors observed in citrus juices?
2. What is the difference between California and Florida orange juice color, and why?
3. How much tangerine juice can be added to orange juice while it still may be called 100% orange juice, and why is this done?
4. What processing problems can occur to mar or degrade the color appearance of citrus juices?
5. What colorimeters are approved for measurement of the color of citrus juices?

PROBLEMS

1. What would be the L, a, and b Hunter color parameters for orange juice reflecting average daylight at a wavelength of 620 nm, which represents pure orange color?
2. What would be the CR and CY values for the light in problem 1, and how does this compare to CR and CY values found from orange juice measurements?
3. What would be the USDA color score for FCOJ with a CR value of 34 and a CY value of 78? What would be the USDA color score for COJFM with the same CR and CY values?
4. What USDA score and grade would FCOJ have if it were slightly lighter in color than an OJ5 USDA standard color tube?
5. What would be the color number and corresponding USDA color score and grade for concentrated orange juice for manufacturing using the Minolta colorimeter if the tristimulus values were $X = 555$, $Y = 321$, and $Z = 148$?

REFERENCES

Chandler, B. V. 1958. Anthocyanins of blood oranges, *Nature*, *182*, 993.

Hunter, R. S. 1958. Photoelectric color difference meter, *J. Opt. Soc. Am.*, *48*(12), 985–995.

Khan, M. U. D. and MacKinney. 1953. Carotenoids of grapefruit, *Plant Physiol.*, *28*, 550–552.

Kramer, A. and Twigg, B. A. 1970. *Quality Control for the Food Industry Vol. I*. The AVI Publishing Company, Inc., Westport, Conn., 28, 32.

Stewart, I. 1980. Color as related to quality in citrus. In *Citrus Nutrition and Quality*, S. Nagy and J. A. Attaway, eds. The American Chemical Society, Washington, D.C., 129–150.

Chapter 10

Bitterness in Citrus Juices

As mentioned in Chapter 4, taste is a four-dimensional phenomenon consisting of sweet, sour, salty, and bitter sensations. Even though a single taste receptor on the tongue can respond to three or four of these modalities, the taste sites on the tongue can be divided up into: the filiform papillae on the front tip and sides of the tongue, which are particularly sensitive to salty and sweet tastes; the fungiform papillae on the middle and sides of the tongue, which are particularly sensitive to sour tastes; and the circumvallate papillae on the back of the tongue, which are particularly sensitive to bitter tastes. A strong bitter taste is a result of electrical stimuli of monopolar-monohydrophobic or bihydrophobic contacts with the tongue receptor (Belitz et al. 1979). The bitter taste in foods is sometimes a delayed reaction that "comes back" or forms an aftertaste that can linger for long periods of time. Thus, when performing organoleptic analysis of citrus juices for bitterness, one should pay attention to poignant sensations on the back of the tongue and delayed responses. When one is tasting more than one sample, crackers or some other tongue scrubber should be used to remove the lingering bitterness of the previous sample.

In citrus juices, two compounds impart essentially all of the bitter taste: limonin and naringin. In some grapefruit juices, a degree of bitterness is sometimes desirable in order to contribute to a "wide awake" flavor along with the sourness of the acids. However, most juices, especially orange juices, are known for their sweetness. Bitterness thus is considered detrimental to quality in these products.

NARINGIN

One class of compounds associated with citrus juices is the flavonoids, which have the following characteristic molecular structure:

These flavonoids are differentiated from each other by the carbohydrate linkup, R. In sweet oranges (*Citrus sinensis* (L.) Osbeck), as well as in most other citrus species, the predominant flavonoid is the tasteless and odorless hesperidin. In grapefruit varieties (*Citrus paradisi* Macf.) the primary flavonoid is naringin, a bitter principle. As with other flavonoids, naringin is primarily found in the membranes and albedo of the fruit and contributes to the bitterness of fresh fruit as well as to the juice. The taste threshold of naringin is about 50 ppm even though levels of about 500 ppm are considered ideal, and levels of over 900 ppm can occur.

The naringin content can be measured in several ways (Ting and Rouseff 1986). However, the spectrophotometric Davis test is the method most commonly used. The Davis test involves the reaction of naringin, or any flavonoid, with diethylene glycol in dilute alkali to the corresponding chalcone, which is yellow in color. The reaction is as follows:

Davis Test for Naringin (Davis 1947)

Equipment and Supplies

- Spectrophotometer with necessary accessories.
- Six test tubes.
- Six 100 ml volumetric flasks.
- 0.1 ml pipette.
- 0.5 ml pipette.
- Centrifuge and accessories.
- 90% diethylene glycol (2,2'-dihydroxyethyl ether) (reagent grade) in water.

- 4N sodium hydroxide.
- Graph paper and/or calculator.
- 200 mg naringin for standards in distilled water.
- 50 ml graduated cylinder.

Procedure

1. Centrifuge the juice sample, using the procedure for determining the % pulp.
2. Add 10 ml of 90% diethylene glycol to a test tube along with 0.1 ml of the centrifuged juice serum. Add 0.1 ml of the NaOH solution, mix the solution, and let it stand for 10 minutes.
3. Measure the sample's absorbance on the spectrophotometer at 420 nm, comparing it to a distilled water blank. It may be easier to measure the % transmittance and convert it to absorbance by using:

$$\text{Absorbance} = \log(100/\%T) \qquad (10\text{-}1)$$

4. The naringin content can be found by comparing the absorbance to a graph of absorbance versus naringin concentration values, C, or by calculation using Beer's law:

$$\text{Absorbance} = kC \qquad (10\text{-}2)$$

5. The standard graph or the factor k can be found from the following calibration or standardization.

Calibration

1. Prepare a stock solution by dissolving 200 mg of naringin in 100 ml of distilled water.
2. Make five standards by pipetting, or by using a graduated cylinder to transfer, 5.0, 12.5, 25.0, 37.5, and 50.0 ml of the stock solution into each of five 100 ml volumetric flasks, filling them to the mark to obtain 10, 25, 50, 75, and 100 mg/ml standard solutions.
3. To each of five test tubes add 10 ml of the 90% diethylene glycol, and then add 0.1 ml of each standard to separate glycol-containing test tubes.
4. Add 0.1 ml of the NaOH solution to each standard-containing test tube, and after 15 minutes measure the absorbance at 420 nm using the spectrophotometer, or measure the % transmittance and convert that value to absorbance using Equation 10-1.
5. Either plot the absorbance versus mg naringin/100 ml (10 mg/100 ml of standard being equal to 100 ppm naringin in the juice sample, so that

to get ppm one may multiply the corresponding mg naringin/100 ml by 10), or calculate the average value of k using Beer's law (Equation 10-2).

6. Another suggested approach is to use five times the concentrations of the above chemicals and sample (Ting and Rouseff 1986).

Another method commonly used is high performance liquid chromatography (HPLC), which can be used effectively to measure the levels of other flavonoids separately. The benefit of HPLC analysis is that naringin can be differentiated from other flavonoids. Usually this is not necessary because naringin predominates in grapefruit juices.

HPLC Naringin Test (Fisher and Wheaton 1976)

Equipment and Supplies

- Isocratic or single pump HPLC system with UV detector and chart recorder. Gradient or multipump systems are useful when one is analyzing for more than one flavonoid at one time. Integrators or computers linked to the HPLC are helpful but not necessary.
- 20:80 acetonitrile–water mobile phase.
- Lab centrifuge.
- Syringe filter (2 μm).
- 50 μl HPLC syringe.
- Same naringin standards used in the Davis test.

Procedure

1. Centrifuge the reconstituted or single-strength juice (11.8°Brix) by using the procedure for the measurement of the % pulp.
2. Filter the supernatant with the syringe filter.
3. Inject 25 μl of the filtered sample or standard into the HPLC system set at 1.5 ml/min; the UV detector should be set at 280 nm and 0.1 absorbance unit, full-scale range.
4. Elution times are about 12 to 16 minutes for naringin and 16 to 18 minutes for hesperidin, depending on the C18 column used.
5. Using 25 μl of the standards described in the Davis test, the mg of naringin per unit peak height (usually in cm) can be determined by averaging as follows:

$$\left(\sum_{i=1}^{i=n} (N_i/P_i) \right)/n = F \qquad (10\text{-}3)$$

where N is the mg naringin in the five standards, n is the total number of standards (five here), and F is the factor for calculating naringin levels. P is the peak height.

6. A plot of N versus P can be used to compare with unknown juice samples, or the factor F can be used to calculate the naringin concentrations (mg naringin/100 ml) by multiplying F times the peak height. The ppm of naringin can be found from the mg of naringin/100 ml by multiplying by 10.

There are no USDA standards for naringin content in citrus juices. Naringin bitterness is accounted for in flavor scores. However, Florida law (Official Rules Affecting the Florida Citrus Industry 1975) requires grade A grapefruit juice to have less than 600 ppm naringin and grade B juice to have less than 750 ppm naringin.

LIMONIN

Limonin, which is the principal bitter component in most citrus juices, including grapefruit juice, is a product of a series of reactions that originate in the trunk of citrus trees (Hasegawa et al. 1986). These reactions generate a class of compounds known as limonoids, which are similar to each other in chemical structure and characteristics. The limonoid nomilin is formed in the trunk of the tree and translocated to the fruit, where it is converted into several limonoids including the tasteless limoninoate A-ring lactone, the open ring structure of limonin. In most citrus varieties, this limonin precursor concentrates in the seeds within mature citrus fruit. However, in some orange varieties, such as navel oranges (*Citrus sinensis* (L.) Osbeck), the A-ring lactone remains in the neutral environment of the juice cell cytoplasm or membrane. Upon the rupture of this membrane during juice processing, the A-ring lactone encounters the acid environment of the juice, which gradually catalyzes the closing of the ring to form limonin according to the following reversible reaction:

LIMONIN A-RING LACTONE
(non-bitter)

LIMONIN
(bitter)

The rate of this reaction is primarily heat-dependent, with some effect due to the juice pH. At cold temperatures viscosity can also affect this reaction in high-Brix concentrates. The rate of this reaction is slow enough that it does not affect the quality of fresh fruit, which generally is consumed soon after it is cut. However, navel juice that is stored overnight in a refrigerator will turn bitter if it is from early or midseason fruit. Even though navel juice is considered substandard because of this bitterness problem, the navel orange is excellent as a fresh fruit. It has an early growing season, outstanding fresh flavor, and a low juice content that makes it easy to eat as a fresh fruit. Because more money can be made in the fresh markets than with processed products (up to 10 times as much), both fruit variety and production are dominated by the fresh markets in areas such as California where navel oranges can be grown of satisfactory quality. Such fresh market domination has left juice processors in such areas with the dilemma of either selling the resulting bitter juice at a lower price or turning to commercial debittering techniques.

Another problem associated with the delayed development of limonin has to do with premature testing for limonin before development is complete. The rate of limonin formation is affected primarily by acidity and heat, as mentioned previously. Also, it has been shown that limonin solubilities range from 1 to 18 ppm in aqueous solutions containing 0 to 10% sucrose (Chandler 1971). Pectin and other complex components of citrus juices increase this solubility, but it is clear that high limonin levels (up to 30 ppm) can easily be affected by the solubility equilibria of limonin in citrus juices. It also is safe to assume that upon juice extraction not all the limonin in the juice cell membranes is leached out into the juice at one time. The leaching has its own rate, which can affect detectable limonin amounts during storage. Also, it is possible that continued biochemical reactions may interfere with limonin levels, including formation from precursors, limonin metabolism, and equilibria with aqueous-phase limonoid glucosides.

In the absence of a more thorough knowledge of these effects, a good rule of thumb is that you should boil freshly extracted juice four minutes for every expected ppm; reconstituted juices require about one minute of boiling for every expected ppm. Overheating of juice has from time to time resulted in sudden drops of limonin as well as Maillard abuse of the sample, and should be avoided. Figure 10-1 shows some general behavior of the limonin levels in freshly extracted California Washington navel juice. From the figure it can be seen that a significant amount of heat is required to completely develop limonin in freshly extracted juices. Many observers have assumed that the heat applied during normal juice processing is sufficient to fully develop the limonin in citrus juices. However, it is common in California for early and midseason navel concentrates to develop significant amounts of limonin even after complete heat treatment in evaporation and relatively long storage periods.

As mentioned previously, carbohydrates increase limonin solubility. This ef-

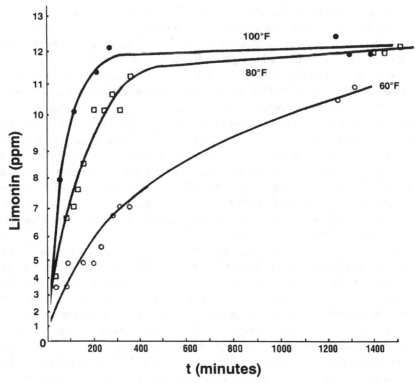

Fig. 10-1. The development of limonin in freshly extracted Washington navel orange juice after time *t* at the temperature given.

fect is a result of hydrogen bonding. Such complexing reactions also have been shown to affect the ability of the tongue to detect limonin. Guadagni et al. showed that increasing the sweetness of the juice decreases one's ability to taste the limonin bitterness (Guadagni, Maier, and Turnbaugh 1974). Organic acids also interfere with the detection of limonin by taste, primarily through pH effects. A pH of 3.8 has been shown to be the ideal pH for suppressing limonin detection (Tatum and Berry 1973) (Maier et al. 1977).

Several methods have been developed to measure limonin levels in citrus juices, including spectroscopy (Wilson and Crutchfield 1968), thin layer chromatography (TLC) (Tatum and Berry 1973), gas chromatography (GC) (Kruger and Colter 1972), radio immunoassay (RIA) (Weiler and Mansell 1980), enzyme-linked immunoassay (EIA) (Jourdon et al. 1984), and high performance liquid chromatography (HPLC) (Rouseff and Fisher 1980). The two most widely used in commercial plants are EIA and HPLC; the EIA method has the advantage of being able to analyze several samples simultaneously and thus is faster, whereas the HPLC has the advantage of being more accurate and reliable.

Enzyme-Linked Immunoassay

EIA is based upon several reactions involving anti-limonin antibodies produced in rabbits and limonin–alkaline phosphate tracer. The tracer is synthesized by the manufacturer of EIA limonin test kits (Idetek, Inc., San Bruno, Calif.) by refluxing limonin with aminooxy acetic acid to give limonin-7-O-carboxy-methyl oxime, which then is coupled with the amino groups of the alkaline phosphatase enzyme via carbodiimide to give a limonin–enzyme linked compound. This tracer is mixed with a limonin standard or unknown juice sample in a micro test tube, and the mixture then is pipetted into an antibody-coated well and incubated. The free limonin and the limonin-linked enzyme tracer compete for the fixed number of antibodies attached to the well's surface. The more free limonin that exists from the juice or limonin standard, the less enzyme-linked limonin tracer there will be to react and become fixed to the immobilized antibodies on the wall of the wells.

Of the other limonoids that are similar in chemical structure and characteristics to limonin, only deoxylimonin reacts similarly to the antibodies and thus can give artificially high limonin results. However, this limonoid occurs at levels of less than 0.5% of the limonin levels in citrus juices and generally should not affect citrus limonin tests. Aqueous-phase limonin glucosides, on the other hand, occur at levels as high as 500 ppm in citrus juices and have been found to interfere with EIA results, giving artificially high limonin determinations. This interference, failure to properly calibrate EIA test systems, and irreproducibility of results in juice samples analyzed at different Brix levels are believed to be the primary reasons for inconsistency between HPLC and EIA determinations.

After incubation, the wells are washed free of excess nonbonded material. Then the wells are filled with p-nitrophenyl phosphate, which acts as a substrate for the limonin–linked enzyme conjugate. The intensity of the colored product is directly proportional to the amount of enzyme–linked limonin bound to the antibodies on the wall of the reaction well and inversely proportional to the amount of dormant limonin displacing the active enzyme–linked limonin. In other words, the lighter the color, or the less the absorbance or optical density, the more limonin there is present.

EIA test kits require the use of standards for every test. Nonjuice limonin standards are difficult to prepare because the complex components of the juice give limonin its solubility. Juice limonin standards also often vary, as already mentioned, making accurate calibration difficult. Idetek's test kit claims validity in the 1 to 10 ppm limonin range, as higher limonin levels yield a curve instead of a line in their Beer's Law calibration plot. This constraint poses a problem for navel orange juice, which may contain as much as 30 ppm limonin. Idetek suggests diluting the sample and calculating the equivalent limonin level for 11.8°Brix juice. This dilution places the sample under different limonin solubility conditions, which can result in significant errors. The lower the Brix is, the less soluble the limonin; and once the limonin crystallizes, it cannot be detected by the Idetek method. In order to avoid these solubility problems, all limonin analyses should be done with samples of at least 11.8°Brix. Higher-Brix samples also upset the solubility equilibria and should be avoided as well. When using standard solutions of limonin for calibration, one should check the effect of the organic solvent used in the standard on the procedure. This can be done by adding similar amounts of the pure solvent without limonin to a juice sample and observing the calibration curve followed by using the standard with the limonin. Generally, the more solvent that is used, the more distorted the calibration will become. It is also recommended that when one is calibrating the Idetek system, a standard should be added to a juice sample before analysis rather than used directly. Instructions for the use of the Idetek kit are included with it.

High Performance Liquid Chromatography

HPLC has undergone dramatic development in the past several decades, having become a major analytical tool in the food industry. It is possible to vary columns, detectors, and mobile phase, and to make slight modification of the procedures, so that few compounds exist that cannot be detected by this method.

HPLC analysis can be divided into four main sections: sample preparation, compound separation, detection, and calculation or display. Because orange juice contains many insoluble and colloidal macromolecules and much particulate matter, sample preparation usually is mandatory to prevent column plugging. This preparation can range from simple filtering to elaborate extractions and preliminary separations.

Limonin is readily soluble in organic solvents, so the use of such solvents is common in preliminary separations. The elimination of aqueous material simplifies the HPLC procedure and provides for a cleaner chromatograph, void of extraneous interfering peaks. Three organic solvents have a high solubilizing power for limonin—acetone, acetonitrile, and chloroform; and chloroform is the solvent of choice for several reasons. First, its immiscibility with aqueous solutions makes it a strong extracter, whereas acetone and acetonitrile dissolve in juice samples and are much weaker in pulling the limonin from the aqueous solution. Of the three, chloroform is the most volatile, a property that facilitates vacuum evaporation when chloroform extracts are concentrated to raise the limonin level so that it is high enough to detect. Also, acetonitrile and acetone sometimes extract other compounds that interfere with limonin peaks. There are three different ways in which the limonin can be extracted from the juice: one way is to use a simple chloroform extraction; another is to use special diatomaceous earth in a 60 ml coarse-porosity fritted disc funnel to extract the limonin from a juice–diatamaceous earth cake using vacuum filtration; a third method consists of the use of a C_{18} "Sep-Pak" cartridge attached to a 10 ml syringe. Whichever method is used, the final solution should be filtered prior to injection into the HPLC system.

As with any other analytical technique, proper calibration is crucial to accurate results. It is not too difficult to analyze each of the extractions above or multiple extractions to ensure that all the limonin is being extracted. Using diatomaceous earth (see procedure below), we found that the first extraction removes 75% of the limonin, the second about 25%, the third about 5%, and the fourth a trace, and the fifth gives no limonin peak. Analyzing juice samples at various times of preheating will show how much boiling of the sample is needed to fully develop the limonin. Also, analyzing the same sample at different Brix levels will clarify limonin solubility effects. Repeat analyses will illustrate precision. Analysis of the same sample under different storage conditions will reveal storage effects. All these influences may dramatically affect limonin results and should be checked regardless of the procedure used.

For routine limonin determinations, a simple isocratic or single-pump system usually suffices. Multiple pumps or gradient systems are useful when one wants to look at several components at once that require different or changing mobile phases. Cyano (CN) columns usually are used for normal phase limonin separations because their intermediate polarity most closely matches that of limonin. Reverse phase methods employing C_{18} columns are used with the "Sep-Pak" sample preparations mentioned above. Short precolumns can be used in addition to or instead of regular columns to protect the longer more expensive analytical columns or by themselves to shorten retention times. The solvent system used in the determination must contain miscible components. The normal and reverse phase mobile solutions commonly used contain solvents with a wide range of polarity. The ratio of these components can be varied to resolve the limonin peak. Degassing and filtering of the solvents helps to protect against column plugging and/or spurious gas peaks that may interfere with limonin peaks.

The method of choice generally is ultraviolet detection. Limonin has an absorption maximum at 207 nm. However, many other compounds extracted with limonin have significant absorption at this wavelength, causing a low signal-to-noise ratio. For this reason, limonin HPLC determinations usually are performed at higher wavelengths such as 215 or 220 nm. Because UV lamps have a limited lifetime and can undergo degradation, monthly calibration of the HPLC procedure is recommended. A good lamp can last indefinitely, but the usual lifetime is about 200 hours of use. It is wise to keep a spare on hand so that loss of a lamp will not interfere with quality control needs.

The means of displaying the detected limonin is very important. It can range from a simple chart recorder to a sophisticated computer display. HPLC seldom is used in routine quality control, so excessive expenditures may not be warranted. For this reason, the author considers a simple chart recorder the most cost-effective device. Comparing sample peak heights to the peak heights of standards gives acceptable results. However, chart recorder integration and computer analysis are much more convenient to use if they are economically feasible.

Even though HPLC is considered slower than the EIA method, many determinations usually are not necessary in most quality control systems. If many limonin determinations were required, we would recommend that both methods be used, the HPLC to check accuracy and the EIA system to produce the bulk of the determinations. The industry often refers to two limonin levels—the HPLC value and the EIA value. In reality there should be only one, and proper calibration should yield only a single value, which should lean definitively toward the HPLC results. The following is a guideline for the development of an HPLC procedure.

HPLC Limonin Determination

Equipment and Supplies

- Isocratic HPLC system, UV detector (207–230 nm), chart recorder, 25 cm × 4.6 mm Zorbax CN column with a 5 cm × 4.6 mm Zorbax CN precolumn for the normal phase or 22 mm × 2.1 mm C_{18} column for the reverse phase.
- HPLC solvent system consisting of ethylene glycol monomethyl ether, 2-propanol, and n-heptane (10:15:25 v/v) or heptane, 2-propanol, and methanol (11:12:2 v/v) for the normal phase or acetonitrile, tetrahydrofuran, and water (17.5:17.5:65 v/v) for the reverse phase (Shaw and Wilson 1984).
- 20 μl HPLC syringe.
- For normal phase separation, the setup shown in Fig. 10-2, including a 60 ml coarse-porosity fritted disc funnel connected to a rotary evaporator with glass tubing, a heating source, about 30 ml diatomaceous earth (amorphous diatomaceous silica and crystalline silica obtained from swimming pool supply venders), and about 300 ml chloroform.
- For reverse phase separation, a C_{18} "Sep-Pak" (Waters Associates, Milford, Mass.) attached to a 10 ml syringe containing a small piece of silanized glass wool conditioned with about 2 ml acetonitrile followed by 5 ml water.
- 10 ml graduated cylinder.
- 10 ml pipette.
- Stirring rod.
- About 60 ml hexane (optional).

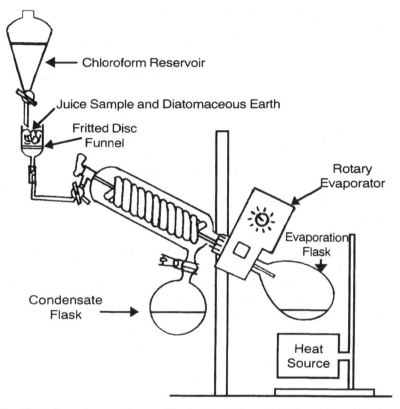

Fig. 10-2. Setup for sample preparation for normal phase HPLC limonin determination.

- 15 ml screw-cap test tube and 2 μm filter syringe (optional for the normal phase) or a 0.45 μm nylon 66 Millipore filter for the reverse phase.
- Limonin chloroform standard (25 mg/100 ml).
- 1 ml graduated pipette or 0.1 ml autopipette.
- 25 ml pipette and extra 100 ml volumetric flask for reverse phase separation.

Procedure (Normal Phase)

1. Add about 30 ml of diatomaceous earth to the fritted disc funnel, pipette 10 ml of 11.8°Brix juice into the funnel, and mix the contents well with a stirring rod. See Fig. 10-2.
2. Without removing the stirring rod, fill the funnel with hexane, stir it, and vacuum-extract the hexane into the evaporation flask of the rotary evaporator. Discard the hexane extract. This step is done to remove extraneous

material that may interfere with the limonin peak. If there is no interference, this step can be ignored.

3. Without removing the stirring rod, fill the funnel with chloroform, stir it, and vacuum-extract the chloroform into the rotary evaporator, which is already in motion with a heat source applied. A three-way stopcock, as shown in Fig. 10-2 should be open during the extraction and closed when one is adding new solvent and stirring.

4. Repeat step 3 four more times while concentrating the chloroform in the rotary evaporator. Continue concentrating it until less than 2 ml of chloroform remains in the evaporation flask. The rate of evaporation can be enhanced by stronger vacuum, greater heat, and/or the use of refrigerated coolant in the condenser. It is recommended that one use moderate heat, compressed air, aspirated vacuum, and tap water in the condenser. Too much heat results in hot glassware that is difficult to handle.

5. Remove the evaporation flask, rinse it with four or five 2 ml portions of chloroform into a 10 ml graduated cylinder, and record the volume (ml GC). Be careful not to lose any of the limonin-rich chloroform rinses.

6. Immediately filter the rinses with a syringe filter, place the chloroform solution in a screw-cap test tube, and seal it.

7. Rinse the 20 μl syringe several times with the chloroform extract, and inject 15 μl of the solution into the HPLC injector using the normal phase mobile phase described above. Start the chart recorder (10 mV 5 mm/min), and inject the sample at the same time. The pump speed should be 1.5 ml/min.

8. The peak height (P) can be combined with the ml GC, the density of the sample, d (g/ml), and the pounds soluble solids/gal, S, in the following equation:

$$\text{limonin (ppm)} = F(\text{ml GC})P/dS \qquad (10\text{-}4)$$

where F is a calibration factor. The S and d factors are used to calculate the limonin content in juice samples on an 11.8°Brix juice basis. The d and S values for 11.8°Brix juice are 1.0446 g/ml and 1.029 lb sol/gal. If the juice sample has a slightly different concentration, Equations 2-7 and 2-8 should be used to calculate the d and S values. Figure 10-3 gives an example of what the chromatograph should look like.

Procedure (Reverse Phase)

1. Pass 2.5 ml of 11.8°Brix juice through a conditioned "Sep-Pak" cartridge followed by 2.5 ml of water.

2. Extract the limonin from the cartridge using 2.5 ml of acetonitrile, filter-

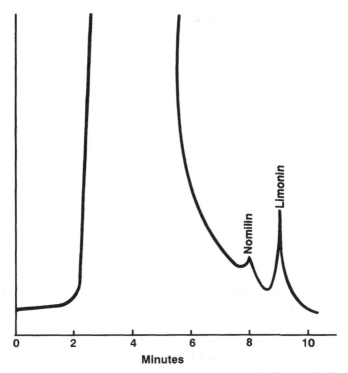

Fig. 10-3. HPLC chromatograph of limonin and nomilin peaks using the normal phase method.

ing through a 0.45 μm nylon 66 Millipore filter, place it in a screw-cap test tube, and seal it.
3. Follow the procedure given above for the normal phase using 2.5 ml for the ml GC and a 22 mm × 2.1 mm C_{18} column and the reverse phase mobile phase described above.

Calibration

1. Add 0.1, 0.3, 0.5, 0.8, and 1.0 ml of the standard to 10 ml each of 5 identical 11.8°Brix juice samples for normal or reverse phase calibration, and proceed with the analysis as explained above.
2. The μg limonin per unit of peak height h can be found as follows:

$$\mu g/h = \frac{(0.025 \text{ g lim})(10^6 \text{ }\mu g)(? \text{ ml std.})(1 \text{ ml})(15 \text{ }\mu l)}{(100 \text{ ml chlor})(1 \text{ g})(? \text{ ml GC})(10^3 \text{ }\mu l)(\delta P)} \quad (10\text{-}5)$$

where δP is the difference in the peak height of the unspiked juice sample compared to the spiked juice sample.

3. The average $\mu g/h$ from the five standards can be used to calculate the factor F used in Equation 10-5:

$$F = \frac{(\mu g/h)(1.029 \text{ lb sol/gal } @11.8°\text{Brix})}{(15 \ \mu l \text{ inject})(1 \text{ ml}/1000 \ \mu l)(? \text{ ml juice sample})} \quad (10\text{-}6)$$

5. In order to check the calibration, the factor calculated above can be used in Equation 10-5 to determine the expected limonin levels of each of the standard tests. This is done by first calculating the limonin level of the unspiked juice sample (L), using the average F value from Equations 10-6 and 10-5. Then the expected limonin levels of the spiked samples can be calculated:

$$E = (? \text{ ml std})C_s(1/? \text{ ml juice})(1/d)(10^6 \ \mu g/1 \text{ g}) + L \quad (10\text{-}7)$$

where C_s is the concentration of the standard (0.025 g limonin/100 ml, the ml juice is 10 ml for the normal phase and 2.5 ml for the reverse phase, and d is the density of the juice (1.0446 g/ml for 11.8°Brix juice).

6. The expected limonin levels from Equation 10-7 and the measure limonin levels from Equation 10-4 using the average F value are compared in the sample calibration illustrated in Table 10-1. The validity of the calibration can readily be seen by comparing the two sets of values.

A limonin determination can be performed in 30 to 40 minutes. The USDA has no grade standards for limonin levels, even though bitterness is incorporated into the flavor score. Florida law sets a maximum of 5.0 ppm limonin for Grade A canned, chilled, and frozen concentrated grapefruit juices and a maximum of

Table 10-1.   Calibration of the normal phase HPLC limonin determination based on 11.8°Brix, assuming 10.0 ml GC for each of the five spiked analyses and the unspiked analysis.

Sample	Peak ht. in chart units	Change in peak ht.	μg/unit	ppm lim	Expected ppm lim
Navel juice	1.65	—*	—*	13.8	—*
+0.1 ml std.	1.93	0.28	0.13	16.2	16.2
+0.3 ml std.	2.55	0.90	0.13	21.4	21.0
+0.5 ml std.	3.10	1.45	0.13	26.0	25.8
+0.8 ml std.	4.03	2.38	0.13	33.7	32.9
+1.0 ml std.	4.50	2.85	0.13	37.7	37.7
			Average = 0.13		
			$F = 0.9$		

*Does not apply to unspiked sample.

7.0 ppm limonin for Grade B (Florida Dept. of Citrus). These grade standards apply to concentrate (not finished products) processed between December 2 and July 31. They also apply to finished products processed from concentrate or bulk single-strength juice between December 2 and July 31, provided that any juice or concentrate blended with it that is processed between August 1 and December 1 meets the permissible limits of limonin for the desired grade.

Limonin TLC Determinations

The principle of thin layer chromatography (TLC) determinations is similar to that of HPLC. However, instead of using a packed column to separate the components of a sample, the separation takes place on a flat silica-coated plate. Instead of pumping an eluting solvent through a column, the eluting solvent travels through the silica by capillary action. Instead of detecting the limonin by means of a UV detector, the limonin is developed chemically to enable visual comparison with standards. Most of the materials needed for TLC limonin analysis are available in most citrus laboratories with no need for major capital investments; thus the major advantage of the method is its low cost. The main disadvantages include the method's less objective visual comparison with standards and the somewhat tedious nature of the determination.

Several procedures exist for limonin TLC analysis. Juice samples can be extracted with chloroform in a manner similar to that described for the HPLC procedure or its equivalent, and the extractant can be spotted on the TLC plate and eluted. Direct spotting of the juice onto the TLC plate has been shown to be successful, but it requires more drying time on the plate compared to organic extractants. Several elution solvents have been used, including: chloroform and acetic acid in a 98 : 1 ratio (Chandler and Kefford 1966); a top layer of toluene, ethanol, distilled water, and glacial acetic acid in a 200 : 40 : 15 : 1 ratio after overnight standing (Hasegawa, Patel, and Snyder 1982), ethylacetate and cyclohexane in a 3 : 2 ratio (Hasegawa 1980); and acetone elution followed by elution with benzene, hexane, acetone, and acetic acid in a 65 : 25 : 10 : 3 ratio (Tatum and Berry 1973). (The last reference also contains a study of 16 other solvent systems used for TLC analysis of limonin in citrus juices.) After elution, two main methods of development have been used. One consists of spraying the dried plate with a 10% sulfuric acid solution in ethanol and heating it to develop a dark colored spot. The other involves the use of Ehrlich's solution (0.395 g p-dimethylamino-benzaldehyde and 0.789 g sulfuric acid in 50 ml ethanol) and exposing the sprayed dried plate to chlorine gas, which can be generated in a well-ventilated area by adding sulfuric acid to ammonium chloride. The result is a brownish limonin spot. Considering safety, speed, and simplicity, the author recommends the following procedure.

TLC Limonin Determination (Tatum and Berry 1973)

Equipment and Supplies

- Silica-coated TLC plate approximately 12–14 cm × 8.5 cm.
- 50 μl syringe with flat unbeveled tip.
- 10 μl syringe with flat unbeveled tip.
- Warm air stream.
- Limonin standard solution (0.010 g limonin/100 ml in ethanol or acetonitrile).
- TLC solvent tank with paper lining (second tank optional).
- Acetone for solvent tank to a depth of about $\frac{1}{8}$ inch.
- Eluting solution for solvent tank to a depth of about $\frac{1}{8}$ inch, consisting of benzene, hexane, acetone, and acetic acid in a ratio of 65:22:10:3 by volume.
- 10% sulfuric acid solution.
- Chromist TLC sprayer.
- Oven to develop plate (125°C).
- UV light source (optional).
- Syringe filter (10 μl).

Procedure

1. Filter the juice, using the syringe filter, and spot 25 μl of filtered single-strength or reconstituted juice (11.8°Brix), using the 50 μl syringe, about 1 cm from the bottom of the TLC plate. Multiple samples can be analyzed at one time on wider plates. Spots should be about 1 cm from the edge of the plate and/or 2 cm apart.
2. Dry the spot with a warm air stream, respot with another 25 μl sample of the juice over the first spot, and redry the plate.
3. Using the 10 μl syringe, spot 1, 2, 3, 4, and 5 μl of the limonin standard solution about 1 cm from the bottom and 2 cm apart on the TLC plate. For estimated limonin levels over 10 ppm, spot 6, 8, 10, 12, and 14 μl of the standard instead.
4. Dry the TLC plate again with warm air for at least 1 minute. Incomplete drying will result in streaked spots.
5. Place the plate in the solvent tank containing acetone with the liquid level about halfway between the bottom of the plate and the spot. Elute until the liquid level travels about 3 cm, remove the plate, and dry it. Uneven liquid levels on the plate indicate insufficient drying of the sample spot.
6. Place the plate in a paper-lined solvent tank containing the benzene solution, and elute to the top of the plate.

7. Remove the plate, dry it, and repeat the elution two more times.
8. Spray the plate with the sulfuric acid solution, using the chromist sprayer, until it is lightly wet.
9. Incubate the plate in the oven (125°C) for 6 minutes, and then remove and cool it.
10. The limonin spot will appear dark in color and rectangular in shape if it was dried properly during sample preparation.
11. The unknown sample can be compared to the standards visually by viewing them against a light source from the back or glass side of the TLC plate. They also can be viewed by using an ultraviolet light source under the plate, in which case the spots appear reddish brown.
12. The limonin level can be determined by comparison with the closest standard and multiplying the microliters of the standard by 2.

PROCESSING CONSIDERATIONS

Even though juice bitterness is induced naturally, processing can have an important effect on it. For example, it has been shown that ethylene gas, commonly used to degreen early-season fruit for the fresh markets, can trigger a mechanism that reduces limonin levels (Maier, Brewster, and Hsu 1973). It has been suggested that leaving fruit so treated in storage bins for as long as possible will help to decrease the bitterness. Even so, early-season navel juices are usually too bitter for this precaution to have a significant effect on final juice quality.

Because both naringin and limonin enter the juice primarily from membrane material in the fruit, extraction and finishing pressures and the length of time that the juice is in contact with the membrane material can affect the degree of bitterness of the resulting juice. Such variations are small, becoming important only when the limonin and naringin levels are reduced through natural maturation to levels close to the taste thresholds. Most juice extractors can be adjusted to give a soft or a hard squeeze, and finishers usually can be adjusted to give a soft or a hard finish. In either case, hard extraction or finishing usually results in greater bitterness and greater juice yield. Therefore, in the early season, extraction and finishing pressures should be set for hard squeezing, as it gives greater yields, and the bitterness is too high for a soft squeeze to yield good-quality juice anyway. Later in the season, when bitter taste thresholds are approached, the squeeze should be changed to a softer squeeze in an effort to produce a lesser or a nonbitter juice. Lesser or nonbitter juice products usually bring in more than enough money to offset losses of yield due to softer squeezes. Also, immediate centrifuging of the juice to remove naringin-laden pulp has been shown to reduce naringin levels in grapefruit juices by as much as 10 to 20% (Berry and Tatum 1986).

The common units of bitter component levels, as mentioned previously, are parts per million (ppm), a weight/weight parameter similar to the Brix. For this reason, in blending limonin (or naringin) levels, equations similar to Equations 2-16 and 2-18 can be used. For example, suppose that we wanted to blend the following lots:

Lot 3	Drums	Brix	lb/gal	ppm Limonin
1	5	65.1	10.983	8.1
2	10	60.2	10.737	16.2
3	4	59.8	10.717	10.5
4	3	61.7	10.811	3.6
5	8	62.4	10.846	4.7

The lb/gal can be found from the Brix by using Equation 2-8 or 2-10 or sucrose density tables. We can find the final weighted average limonin level of the blend by using the following:

$$L_{total} = \frac{L_1 D_1 V_1 + L_2 D_2 V_2 + L_3 D_3 V_3 + \cdots + L_n D_n V_n}{D_1 V_1 + D_2 V_2 + D_3 V_3 + \cdots + D_n V_n} \quad (10\text{-}8)$$

where L_n is the limonin level of component n in ppm, D_n is the density of component n in lb/gal found from Equations 2-9 or 2-11 according to Brix, and V_n is the volume of component n, in units of drums in the above example. Substituting the data in the example into Equation 10-8 gives:

$$\frac{8.1(10.983)5 + 16.2(10.737)10 + 10.5(10.717)4 + 3.6(10.811)3 + 4.7(10.846)8}{(10.983)5 + (10.737)10 + (10.717)4 + (10.811)3 + (10.846)8}$$

$$= 9.7 \text{ ppm}$$

This value is well over the 7 ppm taste threshold. Suppose you wanted to reduce the number of drums of lot 2 so that the final blend would be at or below the threshold. The number of drums of lot 2 needed for this could be calculated as follows:

$$V_2 = \frac{D_1 V_1 (L_1 - L_f) + D_3 V_3 (L_3 - L_f) + \cdots + D_n V_n (L_n - L_f)}{D_2 (L_f - L_2)} \quad (10\text{-}9)$$

where L_f is the desired final limonin level in the blend.

Using the above example gives:

$$\frac{5(10.983)(8.1 - 7.0) + 4(10.717)(10.5 - 7.0) + 3(10.811)(3.6 - 7.0) + 8(10.846)(4.7 - 7.0)}{(10.737)(7.0 - 16.2)}$$

$$= 1 \text{ drum}$$

Therefore, if you use 1 drum of lot 2 instead of 10 drums, you will have a final weighted average limonin level in the blend of exactly 7.0 ppm.

Rather than change the number of drums of lot 2, you may wish to exchange lot 2 for the same number of drums with a lower limonin level in order to achieve 7.0 ppm limonin in the blend. The limonin level needed in such a lot can be calculated as follows:

$$L_2 = L_f + \frac{D_1V_1(L_f - L_1) + D_3V_3(L_f - L_3) + \cdots + D_nV_n(L_f - L_n)}{D_2V_2}$$

$$(10\text{-}10)$$

which in the example gives:

$$7.0 + \frac{5(10.983)(7.0 - 8.1) + 4(10.717)(7.0 - 10.5) + 3(10.811)(7.0 - 3.6) + 8(10.846)(7.0 - 4.7)}{(10.737)10}$$

$$= 7.9 \text{ ppm limonin}$$

Thus if the 10 drums of lot 2 at 16.2 ppm were exchanged for 10 drums of a lot with 7.9 ppm and the same Brix, the weighted average limonin level in the final blend would be the desired 7.0 ppm.

It must be remembered that the juices from all citrus varieties contain limonin to some degree, and it cannot be assumed that they contain no limonin even if they come from what is considered a nonbitter juice such as Valencia. It is interesting to note that Valencia orange juice, the most popular orange juice in the world, is considered a nonbitter juice even though early-season California Valencia juice contains limonin levels as high as 15 ppm. However, the acid content of this early season juice is so high that by the time the acid levels are blended to commercial levels, the limonin levels are blended to below taste threshold values. Hence, limonin levels, even in traditionally nonbitter juices, should not be treated lightly in blending.

As with other citrus juice parameters, the adjustment of limonin levels lends itself to computer application. Limonin levels usually are not determined as often as other juice parameters, so blending or adjusting citrus juices to a specified limonin level does not have the same import as adjusting the Brix, ratio, oil, or pulp levels. However, where debittering is not done, the use of the above equations may be warranted. Figure 10-4 shows a flow chart that can be used to program computers to make the above calculations. Also, a sample GWBASIC program can be found in Appendix B.

DEBITTERING OF CITRUS JUICES

Methods to debitter citrus juices have been under investigation for many years, including: ethylene gas treatment (Maier, Brewster, and Hsu 1973); the use of

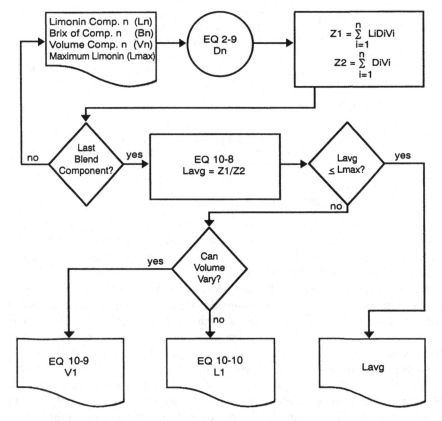

Fig. 10-4. Flow chart that can be used to program computers to adjust the limonin level in citrus blends.

the nonbitter flavone neodiosmin, which competes with limonin for receptor sites on the tongue (Guadagni, Maier, and Turnbaugh 1976); the use of immobilized bacteria and enzymes (Hasegawa and Maier 1983); the use of supercritical carbon dioxide (Kimball 1987); the use of auxin sprays on trees, which interfere with the biosynthesis of limonin in the fruit (Hasegawa et al. 1986); and the use of adsorption and ion exchange resins (Shaw and Buslig 1986; Maeda et al. 1984; Johnson and Chandler 1985). The last has proved to be the only commercially feasible method (Kimball and Norman 1990a,b; Norman and Kimball 1990). Ion exchange resins not only have the capacity to remove limonin and naringin, but they can remove substantial quantities of organic acids as well. Early-season bitter juice may not need to be debittered to below the taste threshold because of the high acid content of the juice; as with the Valencia juice mentioned above, acid blending will dilute out some of the limonin content. If deacidification and debittering take place simultaneously, however, limonin levels probably will need to be reduced to below taste threshold values.

Commercial Absorption Debittering

Commercial debittering of citrus juices consists of juice preparation, resin treatment, and column regeneration. In preparing the juice for resin treatment, several factors must be considered. First, freshly extracted juice must be treated differently from reconstituted juice, which is generally low in oil and pulp content, whereas the freshly extracted juice requires temperature elevation to about 120°F and pulp removal down to about 1%. Pulp removal can be accomplished by using commercial centrifuges or ultrafiltration membranes. Freshly extracted juice requires heat treatment for resin efficiency, which serves also to stabilize pectinase enzymes and pasteurize the juice. Freshly extracted juice also must be deoiled because the citrus essential oils are absorbed by the resin and can decrease its efficiency; oil levels of 0.015% or less are recommended. The heating of the juice also increases centrifuging efficiency. Freshly extracted juices contain almost twice the spindown pulp as pasteurized or evaporated juices. Because efficient resins are small in size and weight, down-flow resin beds generally are required. The use of up-flow systems results in costly resin loss.

After resin treatment, the juice has only a trace pulp content and is lower in oil content. Insignificant amounts of organic acids, including vitamin C, are lost during resin treatment. Some of the vitamin C loss is due to the increased heat applied to the juice. Acid treatment of the resin during regeneration prevents acid reduction during the resin treatment. Navel juice processors prefer high acid levels, as there is usually an overabundance of low-acid late-season navel juice that needs the early-season high-acid juice for blending.

When one is using absorption or ion exchange resins, saturation of the resin eventually requires its regeneration. In commercial debittering the entrapment of pulp particles results in pressure buildup in the column, which necessitates regeneration before the resin becomes saturated with limonin or naringin. Pulp and cloud clarification by means of ultrafiltration reduces the concern for column clogging, so that regeneration can be less frequent. Regeneration generally consists of a series of backflushes of basic solutions, clear water rinses, air scrubs, and, if acid reduction is to be minimized, a final acid treatment. Because of the difficulty of disposing of sodium hydroxide, potassium hydroxide or ammonium hydroxide generally is preferred. Also, sulfuric acid is considered to be the acid of choice as far as waste disposal is concerned. If regeneration wastes are to be disposed of through irrigation, then sodium bases, nitric acid, and hydrochloric acid should be avoided, as they are considered detrimental to agricultural soils.

QUESTIONS

1. When tasting juice for bitterness, what should one look for?
2. What two compounds cause citrus juice bitterness, and in what varieties do they occur?
3. What is the taste threshold of naringin in grapefruit juice, and what is the ideal level?
4. What are the USDA and Florida grape standards for naringin and limonin?
5. Why is it that limonin bitterness does not affect the fresh fruit industry but naringin bitterness does?
6. What affects the rate of limonin development?
7. How can you be sure that the limonin in a juice sample is completely developed?
8. What are the advantages and disadvantages of the limonin EIA, HPLC, and TLC methods of analysis?
9. What is the commercially accepted means of debittering citrus juices, and why has this method been successful?
10. What are some ways of reducing bitterness in citrus juices during processing?

PROBLEMS

1. In the HPLC normal phase limonin determination, finish the calibration table (p. 159), assuming that the sample is 11.8°Brix. What is the factor F?

Sample	P	ml GC	δP	μg/P	ppm LIM	EXPT ppm LIM
Lot 1	0.85	9.4				
+ 0.1 ml std	1.05	9.4				
+ 0.3 ml std	1.35	9.4				
+ 0.5 ml std	1.80	9.4				
+ 0.8 ml std	2.30	9.4				
+ 1.0 ml std	2.70	9.5	———			
			AVG =			

2. Using the factor calculated above, what is the limonin level in a 12.1°Brix juice sample with a peak height of 1.56 and a GC value of 9.6 ml?
3. What would be the expected limonin level after blending the following components? Would this be considered bitter juice?

Lot	Gallons	Brix	Limonin ppm
1	467	65.1	16.1
2	1102	12.6	8.2
3	136	59.6	3.7
4	48	61.9	2.1
5	2676	58.2	5.5

4. Supposed you wanted to make the same amount of juice in the blend in problem 5. What would be the maximum limonin level of a 62.1°Brix lot be if you exchanged it with lot 1 to get a final limonin level of 7.0 ppm?
5. Suppose you had no other lot to exchange with the components in problems 3 and 4, but you wanted to reduce the limonin level to 7.0 ppm. How many gallons of lot 1 would you need to pull from the blend?
6. What would be the expected naringin level after blending the following components? What would be the Florida grade?

Lot	Drums	Brix	Naringin ppm
1	16	58.2	546
2	19	57.0	772
3	5	46.4	534
4	8	62.1	817
5	10	13.6	907

7. Suppose you wanted to make the same amount of juice, exchanging lot 2 above for another lot of the same Brix. What would be the maximum naringin content allowed in order to get grade A juice by Florida standards?
8. Suppose you wanted to keep the number of drums of lots 1 and 3 and use one of the other lots to adjust the naringin content to the maximum level for grade A in the state of Florida. How many drums of lots 2, 4, or 5 would you need?

REFERENCES

Belitz, H. D., Chen, H., Jugel, H., Treleano, R., Wieser, H., Gasteiger, J., and Marsili, M. 1979. Sweet and bitter compounds: Structure and taste relationship. In *Food Taste Chemistry*, ACS Symposium Series 115, J. C. Boudreau, ed. The American Chemical Society, Washington, D.C., 125.

Berry, R. E. and Tatum, J. H. 1986. Bitterness and immature flavor in grapefruit: analyses and improvement in quality, *J. Food Sci.*, *51*, 1368–1369.

Chandler, B. V. 1971. Some solubility relationships of limonin. Their importance in orange juice bitterness, *CSIRO Food Res. Q.*, *31*, 36–40.

Chandler, B. V. and Kefford, J. F. 1966. The chemical assay of limonin, the bitter principle of oranges, *J. Sci. Food Agric.*, *17*, 193.

Davis, W. B. 1947. Determination of flavanones in citrus fruits, *Anal. Chem.*, *19*, 476–478.

Fisher, J. F. and Wheaton, T. A. 1976. A high pressure liquid chromatographic method for the resolution and quantitation of naringin and naringenin rutinoside in grapefruit juice, *J. Agric. Food Chem.*, *24*, 898–899.

Florida Dept. of Citrus, *Official Rules Affecting the Citrus Industry*, Lakeland, Fla., 20-64.03(5), 20-64.09(7).

Guadagni, D. G., Maier, V. P., and Turnbaugh, J. G. 1976. Effect of neodiosmin on threshold and bitterness of limonin in water and orange juice. *J. Food Sci.*, *41*, 681–684.

Guadagni, D. G., Maier, V. P., and Turnbaugh, J. G. 1974. Some factors affecting sensory thresholds and relative bitterness of limonin and naringin, *J. Sci. Food Agric.*, *25*, 1199–1205.

Hasegawa, S. 1980. Private communication. Biosynthesis of limonoids in *Citrus*: sites and translocation.

Hasegawa, S. and Maier, V. P. 1983. Solutions to the limonin bitterness problem of citrus juices, *Food. Tech.*, *37*(6), 73–77.

Hasegawa, S., Herman, Z., Orme, E. D., and Ou. P. 1986. *Phytochemistry*, *25*, 542, 1323, 2783.

Hasegawa, S., Patel, M. N., and Snyder, R. C. 1982. Reduction of limonin bitterness in navel orange juice serum with bacterial cells immobilized in acrylamide gel, *J. Agric. Food. Chem.*, *30*, 509–511.

Johnson, R. L. and Chandler, B. V. 1985. Debittering and deacidification of fruit juices, *Food Tech. in Australia*, *38*, 294–297.

Jourdon, P. S., Mansell, R. L., Oliver, D. G., and Weiler, E. W. 1984. Competitive solid phase enzyme-linked immunoassay for the quantification of limonin in citrus, *Anal. Biochem.*, *138*, 19–24.

Kimball, D. A. 1987. Debittering of citrus juices using supercritical carbon dioxide, *J. Food Sci.*, *52*, 481–482.

Kimball, D. A. and Norman, S. I. 1990a. Changes in California navel orange juice during commercial debittering, *J. Food Sci.*, *55*, 273.

Kimball, D. A. and Norman, S. I. 1990b. Processing effects during commercial debittering of California navel orange juice, *J. Agric. Food Chem.*, *38*, 1396–1400.

Kruger, A. J. and Colter, C. E. 1972. Gas chromatographic identification of limonin in citrus juice, *Proc. Fla. State Hort. Soc.*, *85*, 206–210.

Maeda, H., Takahashi, Y., Miyake, M., and Ifuku, Y. 1984. Studies on the quality improvement of citrus juices and utilization of peels with ion exchange resins. 1. Removal of bitterness and reduction of acidity in Hassaku (*Citrus hassaku* hort. ex. Tanaka) juice with ion exchange resins and adsorbents, *Nippon Shokuin Kogyo Gakkaishi*, *31*, 413–420.

Maier, V. P., Bennett, R. D., and Hasegawa, S. 1977. Limonin and other limonoids. In *Citrus Science and Technology Vol. I*, S. Nagy, P. E. Shaw, and M. K. Veldhuis eds. The AVI Publishing Company, Inc., Westport, Conn., 379, 382.

Maier, V. P., Brewster, L. C., and Hsu, A. C. 1973. Debittering citrus juices, *Citrograph, 58*, 403-404.

Norman, S. I. and Kimball, D. A. 1990. A commercial citrus debittering system, *Proc. of the 36th Citrus Eng. Conf.*, Florida section of the ASME, Lakeland, Fla., March 29, 1990, 1-31.

Official Rules Affecting the Florida Citrus Industry, January 1, 1975, State of Florida, Dept. of Citrus, 20-65.01 to 20-65.05.

Rouseff, R. L. and Fisher, J. F. 1980. Determination of limonin and related limonoids in citrus juices by high performance liquid chromatography, *Anal. Chem., 52*, 1228-1233.

Shaw, P. E. and Buslig, B. S. 1986. Selective removal of bitter compounds from grapefruit juice and from aqueous solution with cyclodextrin polymers and with Amberlite XAD-4, *J. Agric. Food Chem., 34*, 837-840.

Shaw, P. E. and Wilson, C. W. 1984. A rapid method for determination of limonin in citrus juices by high performance liquid chromatography, *J. Food Sci., 49*, 1216-1218.

Tatum, J. H. and Berry, R. E. 1973. Method for estimating limonin content of citrus juices, *J. Food Sci., 38*, 1244-1246.

Ting, S. V. and Rouseff, R. L. 1986. *Citrus Fruits and Their Products*. Marcel Dekker, Inc., New York, 108-113.

Weiler, E. W. and Mansell, R. L. 1980. Radioimmunoassay of limonin using a titrated tracer, *J. Agric. Food Chem., 28*, 543.

Wilson, K. W. and Crutchfield, C. A. 1968. Spectrophotometric determination of limonin in orange juice, *J. Agric. Food Chem., 16*, 118-124.

Chapter 11

Nutritional Content of Citrus Juices

Natural foods, especially citrus products, have always been highly regarded as excellent sources of human nutrition. Nutrition is the ability to engender growth, and many of the components of citrus products contribute to the growth and well-being of the human body. In order to grow, the human body requires certain dietary components, and this need has given rise to the regulation of nutritional labeling by many governments. Such labeling, which provides nutritional information to consumers, is of great value, especially for persons who must monitor their nutritional intake for medical or dietary purposes. Nutritional labeling also gives an aura of professionalism and quality to a product, especially one that contains significant amounts of nutrients.

There are two basic types of nutrients. The *macronutrients* are those that occur as major components of the food product, such as sugars or carbohydrates, salts, acids, protein, and fats. These nutrients provide energy for the body in addition to regulating the absorption rates of various body chemicals and water retention in the body. Proteins act as catalysts for certain life-giving chemical reactions and as building blocks for many parts of the body. The *micronutrients* are those that occur in trace amounts. They either regulate specific chemical reactions, like the proteins, or they form important organometallic complexes such as the iron complexes that make up the blood.

To establish a uniform way of reporting nutritional information, laws regulating labeling have been passed in various countries. In the United States, the Food and Drug Administration (FDA) first published nutritional labeling regulations in the *Federal Register* in March of 1973 (U.S. Food and Drug Administration 1973). Since that time, changes in the regulations have occurred, and they continue to occur because of constant research and new understanding about human nutrition. The current FDA regulations can be found in the *Code of Federal Regulations* (CFR), Title 21, 101.9–101.13. For natural products where no nutrients have been added, nutritional labeling is optional. About two-thirds of the 100% citrus products on the market do not include any nutritional labeling. If nutritional labeling is used, the rules and format outlined in the federal code must be followed.

With respect to labeling, there are two classes of nutrients. Group I nutrients are those added to the product, which must be present at least at the levels declared on the label. Group II nutrients are those that occur naturally, which must be present in at least 80% of the levels declared on the label. Excess amounts of nutrients are allowed within good manufacturing practice except that the calorie, carbohydrate, sodium, and fat levels cannot exceed 20% of the value declared on the label. Official compliance with declared label values is determined by taking 12 samples, selected from 12 separate shipping cases at random, and analyzing them by AOAC or other suitable methods.

Even though FDA is reviewing the commerical nutritional labeling format that will most likely contain changes; the format that the nutritional information must presently follow is illustrated in Table 11-1. All nutritional levels are expressed on a per-serving basis. A serving of citrus juice usually is designated as 6 fluid single-strength ounces (177 ml) with 113 g designated for citrus sections. The servings per container is the number of reconstituted or consumable servings rounded off to the nearest whole number.

Table 11-1. Sample format for nutritional labeling of a 6-ounce can of frozen concentrated orange juice. The words "per serving" in the first line are optional, as are the sodium and potassium labeling.

Nutritional Information Per Serving

Serving Size	6 fl. oz.
Servings per Container	4
Calories	90
Protein	1 g
Carbohydrates	21 g
Fat	0 g
Sodium	0 mg (optional)
Potassium	260 mg (optional)

Percentage of U.S. Recommended Daily Allowances (U.S. RDA)

Vitamin C	120%
Thiamin	8%

Contains less than 2% of the U.S. RDA of Protein, Vitamin A, Riboflavin, Niacin, Calcium, and Iron.

REPORTING OF MACRONUTRIENTS

Calories

Calories, in reference to nutritional labeling, represent the amount of energy available for human metabolism. Calories can be determined by using accepted analytical methods such as the AOAC (1980) or the Atwater method (USDA 1955). However, a more common and simple way to determine the caloric content of citrus juices is to use a mathematical approach on the basis of 4, 4, and 9 calories per gram of protein, carbohydrates, and fat, respectively. For example, if we use the example in Table 11-1, the calories per serving could be calculated as follows:

1 g protein	× 4 =	4 calories
21 g carbohydrates	× 4 =	84 calories
0 g fat	× 9 =	0 calories
	Total =	88 calories

Calorie values then are rounded off to the nearest even number up to 20 calories, to the nearest increment of 5 from 20 to 50 calories, and to the nearest increment of 10 from there on up. Because, in the above sample, the number of calories is greater than 50, the calorie value is rounded off to the nearest 10, or to 90 calories, which appears on most nutrition labels of 100% orange juice.

Protein

Proteins are enzymes or biological catalysts that govern the biochemical reactions occurring in the body, and provide energy like the carbohydrates. The human body manufacturers many of the proteins it needs. However, it requires an outside source for some of them. Some parts of the body, such as hair, are themselves made up of proteins. Proteins are comprised of long amino acid polymeric chains that are cross-linked in various ways. For the purposes of nutritional labeling, the protein level is calculated by multiplying the weight percent of total nitrogen by 6.25. The nitrogen content in citrus juices has been reported to be 60 to 120 mg per 100 ml of single-strength juice (Ting 1967), with higher values reported in California orange juices (Tockland 1961). The nutritional quality of proteins should aslo be taken into consideration. This quality is dependent on the protein efficiency ratio (PER). A PER value of at least 20% of casein is required for a food to be considered a significant source of protein. Citrus products generally fall short of this and thus are considered to be insignificant or poor sources of protein.

The Kjeldahl procedure is generally used to measure the total nitrogen level in citrus juices, but it is a lengthy and involved procedure for routine quality control work and may account for only 11 to 20% of the total juice protein (Attaway et al. 1972). Also, at least 1 gram of protein is required for nutritional labeling, and citrus juices just meet this requirement. Given these circumstances, protein analysis usually is not performed on citrus products in quality control laboratories. When protein is reported on nutritional labels, the value should be rounded off to the nearest gram per serving, which is usually 1 gram. The USRDA for protein depends on the gender and age of the consumer, ranging from 23 grams for children 1 to 3 years old to 56 grams for adult males 23 to 50 years old.

Carbohydrates

Carbohydrates, or sugars, are abundant in citrus juices and products, comprising about 80 to 90% of the soluble solids found in citrus juices (except lemon and lime) and containing high levels of potential energy in their chemical bonds. Thus citrus juices contain a high level of calories, the only nutritionally undesirable characteristic of citrus products. About half of the carbohydrates are in the form of sucrose, an oligosaccharide or disaccharide, consisting of one fructose and one glucose molecule. The remaining carbohydrates are essentially equal portions of glucose and fructose, the natural products of acid or enzymatic hydrolysis of sucrose. The bond between the glucose and fructose molecules in sucrose contains energy; so when the bond has been broken through hydrolysis, some of the energy or calorie content of the sucrose is lost. For this reason, in citrus juices, where half of the sucrose has been hydrolized, less potential energy is available, and the calorie content is reduced. Complex carbohydrates or polysaccharides such as pectin, hemicellulose, and cellulose occur in insignificant amounts in citrus juices compared to the primary sugars mentioned above.

Carbohydrates generally are measured according to Brix levels, as described in Chapter 2. The Brix measurement includes organic acids, soluble salts, and other soluble nitrogenous and pectic substances. Corrections are commonly made to the Brix for acid levels as citric acid equivalents, which, for nutritional labeling purposes, are considered part of the carbohydrates. Thus the acid-corrected Brix measurement can be used to determine the nutritional carbohydrate level in citrus juices.

Nutritional labeling laws require that the carbohydrate level be reported in grams, rounded off to the nearest gram, per serving. In order to convert the % sugar (Brix measurement) into grams per 6-ounce serving, the following can be used.

$$g_{carbohydrates}/6 \text{ oz} = (\text{Brix})(177 \text{ ml per serving})(d)/100\% \quad (11\text{-}1)$$

where d is the density of the single-strength juice in g/ml. For example:

$$(11.8°\text{Brix}) (177 \text{ ml/serving}) (1.0446 \text{ g/ml}) /100\%$$
$$= 21.8 \text{ g carbohydrates}$$

Rounding this off to the nearest gram gives 22 grams, which would be reported on the nutritional label. If nutritional sweeteners are added to make sweetened juice, they must be included in the carbohydrate level in nutritional labeling.

Fat

Fats are lipids (natural compounds that are soluble in hydrocarbons but not in water), and are generally considered to be detrimental to health. In citrus juices there are essentially no fats, and that can generally be declared on nutritional labels. If fats were to exist in significant quantities, their level would be reported to the nearest gram per serving.

Sodium and Potassium

Sodium and potassium labeling are optional under the present federal code. When reported, the sodium or potassium level is given to the nearest milligram per serving. Sodium generally is considered to be detrimental to health, whereas potassium is considered to be advantageous. Citrus juices generally contain insignificant amounts of sodium and can be so labeled. However, high levels of potassium are common; levels of 200 mg to 2000 mg per serving can be found in commercial juices, depending on the geographical location (Harding, Winston, and Fisher 1940). Also, legally added water in reconstituted products may be rich in potassium and/or sodium and contribute to the presence of these nutrients. Again, as mentioned earlier, naturally occurring nutrients can legally exist in levels of at least 80% of the declared value, with no maximum limit for potassium. The natural sodium or potassium levels in water used to legally reconstitute citrus products fall in this category. Because sodium levels are low in citrus products, routine quality control of those levels is seldom performed. Also, because potassium is plentiful in many foods, its presence in citrus products does not constitute an anomaly and rarely justifies routine monitoring. On those infrequent occasions when it is required, professional laboratories, using ion chromatography or atomic absorption, can be used.

REQUIRED REPORTING OF MICRONUTRIENTS

When nutritional labeling is used, seven vitamins and minerals as well as protein must be mentioned on the label: vitamin A, ascorbic acid (vitamin C), thiamin (vitamin B_1), riboflavin (vitamin B_2), niacin (vitamin B_4), calcium, and iron. These values must be reported as a percentage of the U.S. recommended daily allowance (USRDA): in 2% increments up to and including 10%, in 5% increments from 10% to 50% of the USRDA, and in 10% increments thereafter. When a nutrient occurs at a level of less that 2% of the USRDA, the product can be labeled as containing less that 2% of the nutrient . If five or more of these eight nutrients occur below 2% of the USRDA, the manufacturer can state that collectively these nutrients occur in levels less than 2% of the USRDA, rather than list each of the eight nutrients separately. For citrus juices, generally only vitamin C and thiamin occur in levels above 2% of the USRDA. The federal code indicates that no food can be considered a significant source of a particular nutrient that does not occur at least at a level of 10% of the USRDA, and that no food can be considered a superior source of a particular nutrient unless it contains 10% more of the USRDA than another product does.

Vitamins and minerals, other than the ones mentioned above, do not occur in citrus juices to any significant degree and generally do not appear on citrus nutritional labels.

Vitamin C

Vitamin C, also known as ascorbic acid, hexuronic acid, cevitamic acid, and antiscorbutic acid, has long been associated with citrus nutrition. Limes once were used extensively to stave off scurvy attacks on board seagoing vessels, so much so that British sailors were labeled "Limeys." Even though other fruits and vegetables have higher levels of vitamin C than citrus juices, few are as attractive in color, taste, and thus popularity as citrus. A single 6-ounce serving (177 ml) provides more than 100% of the USRDA of vitamin C; so foods containing higher levels of the vitamin offer no real nutritional advantage over citrus. Also, vitamin C is very stable in citrus juices and degrades very little with storage, another nutritional advantage.

L-Ascorbic acid has the following chemical structure:

It is easily oxidized to L-dehydroascorbic acid, which has the following chemical structure:

$$
\begin{array}{c}
\text{OH} \\
| \\
\text{H}-\text{C}-\text{H} \\
\text{H}-\text{C}-\text{OH} \\
\text{C}-\text{H} \\
\text{C}=\text{O} \\
\text{C}=\text{O} \\
\text{C}=\text{O}
\end{array}
$$

L-Dehydroascorbic acid is less stable than L-ascorbic acid, but the two have nearly the same nutritional effect. Vitamin C contributes to: iron absorption; cold tolerance; maintenance of the adrenal cortex; antioxidizing activity; metabolism of tryptophan, phenylalanine, and tyrosine; body growth; wound healing; synthesis of polysaccharides and collagen; formation of cartilage, dentine, bones, and teeth; and capillary maintenance. Humans cannot synthesize vitamin C, and must depend on an external source for its supply.

The vitamin C level in citrus decreases with maturity. Florida oranges go from about 50 mg/100 ml of single-strength juice at the beginning of the season to about 30 mg/100 ml by the end of the season (Harding, Winston, and Fisher 1940). During an average marketing period, the loss of vitamin C amounts to less than 10%, which is indicative of the stability of vitamin C in citrus juices. Atmospheric oxygen is responsible for most vitamin C loss during long-term storage. Many polymeric containers readily admit oxygen, which degrades vitamin C as well as contributing to the development of off colors and flavors. Grapefruit juice generally contains slightly less vitamin C (45 mg/100 ml) than orange juice (50 mg/100 ml) and lemons (60 mg/100 ml). Tangerines and limes contain even less vitamin C (about 30 mg/100 ml). These vitamin C levels represent approximate averages and indicate general relative comparisons. Actual vitamin C levels may vary widely, by as much as 50%.

Even though vitamin C is a major nutrient in citrus juices and is considered to be an important quality parameter, its stability and generally consistent levels render routine analysis unimportant. Industrially, vitamin C levels are not considered an important parameter. However, vitamin C levels can be of interest on occasion, especially in establishing nutritional labeling; and the following two methods of analysis can be used.

Vitamin C Indophenol Titration

Equipment and Supplies

- 50 ml buret, magnetic stirrer, and 125 ml Erlenmeyer flask for titrating.
- 100 ml volumetric flask.
- 10 ml pipette.
- 50 ml beaker.
- 150 ml beaker.
- 500 ml volumetric flask.
- Dye solution: Add 0.250 g sodium 2,6-dichlorobenzenoneindophenol or sodium 2,6-dichlorophenolindophenol to 500 ml warm water (60°C) (or 0.050 g sodium 2,6-dichloroindophenol to 200 ml distilled water) and add approximately 100 mg sodium bicarbonate ($NaHCO_3$). Filter the solution, and store it in a sealed amber container in the refrigerator until it is used. When the dye fails to give a distinct endpoint, discard it.
- Acid solution: 10 g of metaphosphoric acid (HPO_3) in warm distilled water (60°C) in successive portions until the acid will not dissolve. Transfer the solution to a 500 ml volumetric flask, cool it to room temperature, and add 40 ml glacial acetic acid. Fill the flask to the mark with distilled water. The solution should be good for 7 to 10 days when stored in the refrigerator. HPO_3 slowly changes to H_3PO_4.
- Ascorbic acid standard solution: 0.100 g of pure ascorbic acid (melting point 190–192°C) in a 50 ml beaker. Add 10 ml acid solution, and carefully transfer the solution to a 100 ml volumetric flask. Fill the flask to the mark with distilled water. This solution contains 1 mg ascorbic acid/ml. Make a fresh solution for each titrating session. Do not store it.

Standardization Procedure

1. Bring the dye solution to room temperature. Transfer 10 ml of the fresh ascorbic acid standard solution to a 150 ml beaker, and add 10 ml of acid solution.
2. Titrate the ascorbic acid solution rapidly with the dye solution until a light but distinct rose pink color persists for at least 5 to 10 seconds.
3. Repeat at least once until the milliliters titrated are within 0.10 ml for each titration.
4. The dye concentration can be calculated by using:

$$\text{Dye conc.} = (10 \text{ mg ascorbic acid})/(\text{ml titrated}) \quad (11\text{-}2)$$

For example:

$$(10 \text{ mg ascorbic acid})/(21.42 \text{ ml}) = 0.4669 \text{ mg/ml}$$

Titration Procedure

1. Add 10 ml of reconstituted or single-strength juice (11.8°Brix) to 10 ml of the acid solution.
2. Titrate immediately with the standardization dye solution until a rose pink color persists for 5 to 10 seconds.
3. Calculate the ascorbic acid (AA) or vitamin C level by using:

$$\text{mg AA}/100 \text{ ml} = \frac{(\text{dye conc.})(\text{ml titrated})(100 \text{ ml juice})}{(10 \text{ ml sample})} \quad (11\text{-}3)$$

For example:

$$(0.4669)(9.46 \text{ ml titrated})(100 \text{ ml})/(10 \text{ ml}) = 44.2 \text{ mg AA}/100 \text{ ml}$$

The following HPLC procedure uses ion pairing and can be used to determine both the L-ascorbic acid level and the dehydroascorbic acid level. Dehydroascorbic acid is derivatized with 1,2-phenylenediamine in order to increase its ultraviolet absorbance.

HPLC Vitamin C Determination (Keating and Haddad 1982)

Equipment and Supplies

- Isocratic or single-pump HPLC system with UV detector (290 nm and 348 nm), 30 cm × 4 mm C_{18} column, 20 μl syringe, and chart recorder (integrator is optional).
- 0.45 μm filter.
- 100 ml volumetric flask.
- 200 ml methanol.
- Standard solution: Dissolve 25 mg ascorbic acid, 5 mg dehydroascorbic acid, and 180 mg 1,2-phenylenediamine dihydrochloride (PDA · 2HCl) in 50 ml of methanol in a 100 ml volumetric flask, and fill the flask to the mark with methanol. Let it stand in the dark for 80 minutes in order to complete the derivatization.
- Mobile phase: methanol in water (60:40 v/v) containing 2.5×11^{-4} M hexadecyltrimethylammonium bromide ion pairing agent.
- Lab centrifuge.

Procedure

1. Add 200 mg of PDA · 2HCl to 70 ml of reconstituted or single-strength juice (11.8°Brix) or standard solution to a 100 ml volumetic flask. Dissolve it, and fill the flask to the mark with methanol.
2. Centrifuge for 5 minutes, filter, and allow the solution to stand in the dark for 1 hour. Filter it with a 0.45 μm filter.
3. Inject 15 μl of this solution into the HPLC system with a pump flow rate of 1.0 ml/min after conditioning the column with the mobile phase for 3 hours.
4. Set the wavelength of the detector to 348 nm initially, and change it to 290 nm after the dehydroascorbic acid peak has eluted. Detector attenuation is 0.1 aufs throughout the run.
5. The ascorbic acid and dehydroascorbic acid level can be calculated by using:

$$C_j = P_j C_s / P_s \qquad (11\text{-}4)$$

where P_j and P_s are the peak heights (or areas under the peak if an integrator is used) of the juice and standard samples, respectively, and C_j and C_s are the concentrations of the vitamin C in the juice and standard samples, respectively. In the standard, the ascorbic acid concentration is 25 mg/100 ml and the dehydroascorbic acid concentration is 5 mg/100 ml.

Nearly 75% of the vitamin C in oranges and over 80% of the vitamin C in grapefruit can be found in the peel. The USRDA for vitamin C is 60 mg.

Thiamin

This vitamin is also known as vitamin B_1, aneurin, antineuritic factor, antiberiberi factor, and oryzamin. The metabolically acitve form of thiamin is thiamin pyrophosphate.

This molecule acts as a coenzyme which participates in decarboxylation of α-keto acids in the initial phases of the citric acid cycle. Thiamin deficiency is responsible for the development of beriberi.

Orange juice has the highest level of thiamin (170 μg/6 oz serving or 11% of the USRDA), followed by tangerine juice (100 μg/6 oz serving or 7% of the USRDA) and grapefruit juice (70 μg/6 oz serving or 5% of the USRDA). Unlike vitamin C, thiamin levels increase with fruit maturity, expanding from as low as 45 μg/100 g of single-strength juice to 90 μg/100 g of juice throughout a given season. The USRDA for thiamin is 1.5 mg.

Thiamin determinations are commonly performed by using a combination of enzymes, packed open columns, and fluorimeters (Ting and Rouseff 1986). This procedure is generally too tedious and complex for routine quality control analysis. Because thiamin determinations are performed infrequently, professional laboratories usually are consulted.

Calcium

Even though calcium occurs in very small amounts naturally in citrus juices, calcium-added products have increased in popularity. Any nutrient that is added must be declared on the nutritional label, and nutritional labeling becomes mandatory. As mentioned earlier, most minerals are determined by using ion exchange or atomic absorption. This is usually best done in professional laboratories because of the infrequent need for such determinations and the cost of the instrumentation. The USRDA for calcium is 1 gram. Calcium hydroxide is easily added to citrus juices and neutralizes the acidity, producing a sweeter juice at the same time. However, it should be remembered that such calcium additions violate the standards of identity for 100% juice products, rendering them ''drinks'' rather than ''juices.''

Pectin and Flavonoids

Even though there are no labeling requirements or formats in the federal code regarding pectins and flavonoids, it should be noted that these compounds, found abundantly in citrus products, are believed to be linked to the reduction of serum cholesterol, a rare and important characteristic in foods. Like many natural products, the nutritional label, as outlined in the federal code, falls far short of completely describing the nutritional advantages of citrus products. Many such advantages have yet to be discovered and understood. Even though nutritional analysis plays a minor role in overall citrus quality control, it should be remembered that the nutritional value of citrus products is one of their major selling points, and quality control personnel should properly represent it as such to the industry and to the consumer.

QUESTIONS

1. What is the difference between macronutrients and micronutrients?
2. What is the difference between class I and II nutrients?
3. If no nutritional labeling appears on a 6-ounce can of frozen concentrated orange juice, is the law being violated? Why or why not?
4. What are some of the advantages and disadvantages of using nutritional labeling on 100% citrus juice products?
5. What nutrient must appear on the label twice?
6. What is the most abundant nutrient in single-strength juices?
7. How do the acids in citrus juices fit into nutritional labeling?
8. What micronutrients must be reported when nutritional labeling is used?
9. Which fruit has the most vitamin C and thiamin, and does the vitamin C and/or thiamin increase or decrease as the season progresses?
10. How much of the USRDA of vitamin C and thiamin is commonly declared on most 100% orange juice products?

PROBLEMS

1. Suppose that super citrus scientist Dr. VanOhjay has discovered a new orange variety that produced juice of a higher nutritional quality. This new juice contains 11 grams of protein, 8 grams of carbohydrates, and 1 gram of fat per 6 ounces (177 ml). What calorie level would be declared on the label?
2. In the indophenol titration for vitamin C, suppose that you titrated 19.64 ml in the standardization and 11.46 ml in the sample titration. What would be the dye concentration, the vitamin C level (mg/100 ml juice), and the highest declarable % USRDA value on the label?
3. How many grams of carbohydrates would be declared on a label of product whose serving size is 100 ml of single-strength juice at 12.0°Brix?
4. What would be the maximum declarable amounts on a nutritional label (serving 6 ounces or 177 ml) for a juice with the following analysis results. Assume USRDA levels for an adult male.
   - 0.501 g protein
   - 20.4 g carbohydrates
   - 35.6 mg/100 ml vitamin C
   - 52.3 μg/100 g thiamin
   - 63.1 mg/serving calcium added to the juice
5. If the peak height of a standard ascorbic acid solution, using the HPLC method, were 1.46 and the dehydroascorbic acid peak height 0.32, what would be the resulting levels of the two if the juice sample peaks were 2.23 for ascorbic acid and 0.36 for dehydroascorbic acid? What would be declared on the label?

REFERENCES

Attaway, J. A. et al. 1972. Some new analytical indicators of processed orange juice quality, *Proc. Fla. State Hort. Soc.*, *85*, 192–203.

AOAC. 1980. *Official Methods of Analysis*, *13th Edition*, 10.031, Association of Official Analytical Chemists, Washington, D.C.

Harding, P. L., Winston, J. R., and Fisher, D. F. 1940. Seasonal changes in Florida oranges. *USDA Tech. Bulletin*, 753.

Keating, R. W. and Haddad, P. R. 1982. Simultaneous determination of ascorbic acid and dehydroascorbic acid by reverse phase ion-pair high performance liquid chromatography with pre-column derivatization, *J. Chromatography*, *245*, 249–255.

Ting, S. V. 1967. Nitrogen content of Florida orange juice and Florida orange concentrate, *Proc. Fla. State Hort. Soc.*, *80*, 257–261.

Ting, S. V. and Rouseff, R. L. 1986. *Citrus Fruits and Their Products*. Marcel Dekker, Inc., New York, 128–129.

Tockland, L. B. 1961. Nitrogenous constituents. In *The Orange: Its Biochemistry and Physiology*, W. B. Sinclair, ed. University of California Press, Berkeley, Calif.

USDA. 1955. *Energy Values of Foods—Basis and Derivation*, A. L. Merrill and B. K. Watt, eds. Government Printing Office, Washington, D.C.

U.S. Food and Drug Administration. 1973. Food: nutritional labeling, *Federal Register*, *38*, No. 49, 6959–6961. U.S. Government Printing Office, Washington, D.C.

Chapter 12

Citrus Rheology

"Rheo" comes from the Greek word meaning "to flow," thus rheology is the study of flow or, in the citrus industry, the study of the flow of citrus juices and concentrates. Fluids are divided into two basic types in rheology. Newtonian fluids are chemically pure homogenous fluids whose viscosity, or resistance to deformation, does not change with shear rate. (Shear rate can be described as the flow rate, the stirring rate, or the rate at which one surface passes over another.) Non-Newtonian fluids have different apparent viscosities at different shear rates. The latter description generally characterizes heterogeneous solutions such as citrus juices and concentrates. The non-Newtonian nature of citrus juices is primarily due to pulp and cloud material within the juice. Removal of these components renders the resulting serum Newtonian in nature. The term "consistency" or "apparent viscosity" is generally used with non-Newtonian fluids, while "viscosity" is generally used with Newtonian fluids.

Citrus juices at concentrations above 20°Brix are considered to be pseudo-plastic; that is, their apparent viscosity decreases with an increase in shear rate. In other words, as the concentrate moves more rapidly through a pipe or during mixing in a tank, its apparent viscosity decreases. In order to start flowing, citrus concentrates require an extra amount of energy, which is called the yield stress.

Citrus concentrates are also thixotropic: when flow begins just after the yield stress is applied, there is a breakup of the forces that hold the concentrate together. When the flow is stopped, these forces are restored. In rheology this behavior is called going from a gel to a sol and then back to a gel condition. The opposites of pseudoplastic and thixotropic are dilatent and rheopectic, respectively.

In summary, citrus concentrates are non-Newtonian, pseudoplastic, thixotropic fluids.

Significance of Citrus Rheology

Citrus juice and concentrate consistency plays an important role in the quality of citrus products. It has an important impact on the operation of every piece of equipment used in citrus processing, as well as on consumer acceptance and satisfaction. Even though routine measurements of the apparent viscosity generally are not necessary, consistency becomes an important parameter in designing equipment for new products. Although this usually is a problem for engineers, quality control personnel often are consulted about product consistency, especially in small processing plants.

Equipment Sizing

Citrus processing plants are often referred to as just a bunch of tanks, pipes, and pumps. Even though there may be a little more to processing plants than that characterization implies, juice conveyance and storage are involved in nearly every step of citrus processing. Thus flow rates, which are a function of pumping power and pipe or orifice size, are of vital interest in integrating all the processing steps and in transporting juice products in a timely manner.

Citrus concentrates exist at a wide range of temperatures, concentrations, and compositions during their processing, and fluctuations in their apparent viscosities can be dramatic. A pump, pipe, or other piece of equipment that is too small can become a lethal bottleneck in processing, blending, packaging, and shipping. A pump, pipe, or other piece of equipment that is too large can turn out to be an excessive and costly expenditure or even an electrical tapeworm, consuming large amounts of unnecessary energy. Single-strength juices generally have the same apparent viscosity as water for engineering purposes, or about 18 centipoise. For 42°Brix frozen concentrated orange juice (FCOJ), the apparent viscosities range from 430 to 2749 centipoise at 26.7°C, with grapefruit concentrate ranging from 150 to 1885 centipoise (Ezell 1959). For 60 to 65°Brix concentrates the apparent viscosities range from 400 to 1000 centipoise at 25°C and 3000 to 7000 centipoise at −10°C, all with a shear rate of 115.5 sec^{-1} (Crandall, Chen, and Carter 1982). Bulk storage and shipping viscosities generally are close to the latter extreme.

The apparent viscosity increases dramatically at commercial freezer temperatures, so care should be taken not to drop the temperature of bulk freezers or tank farms much below −10°C (15°F). Quality control personnel generally have to monitor freezer temperatures. As mentioned in Chapter 18, the increase in fluidity that occurs with increased temperatures, compared to drum freezer temperatures of about −20°C or about 0°F, is responsible for the formation of potassium citrate crystals in lower Brix/acid ratio concentrates.

Processing Considerations

During processing, product consistency can play a major role. As described in Chapter 8, failure to deactivate the naturally occurring pectinase enzymes found in the juice can result in gellation and/or a significant increase in the product's consistency; however, most, if not all, modern evaporators heat juice sufficiently to deactivate the pectinase enzymes. High pulp levels also contribute to product viscosity. Juice sac material added to concentrates to give a floating pulp or "just squeezed" appearance adds significantly to product consistency, whereas centrifuged low-pulp concentrates are very fluid. Pulp wash concentrates generally are relatively viscous because of the additional insoluble material contained therein, and can pose serious consistency problems. Enzymes are often used to break up suspended and insoluble material in pulp wash juices. Also, monitoring of the apparent viscosity during pulp wash processing is warranted. Often pulp wash juices are immediately combined with freshly extracted juices to diminish the problem of excessive pulp wash consistency.

In addition to the suspended and insoluble material in citrus juices, the soluble components can significantly contribute to the apparent viscosities. Early-season high-acid concentrates are more fluid than late-season sweeter concentrates. Evaporators employing plate heat exchangers generally can concentrate early-season orange juices to 65°Brix, but late-season juices containing more sugar cannot be concentrated to the same level because of their excessive viscosity. It has been shown that sucrose solutions above 22°Brix involve sucrose-sucrose molecular interactions that strengthen hydrogen bonding (Kimball 1986). These interactions are the major contributors to this type of apparent viscosity. The use of pectinase enzymes or homogenization has been suggested as a way to reduce the apparent viscosity of citrus concentrates so that higher Brix levels can be achieved (Crandall et al. 1988). This will result in a reduction in storage and shipping costs.

Consumer Considerations

Besides cloud loss and gellation, mentioned in Chapter 8, another parameter of citrus juice quality exists that is related to rheology—the mouth-feel. This is especially important in the formulation of juice drinks. Natural juices set the standard for the mouth-feel or consistency desired by consumers of citrus drinks. Various gums and pectins have been used in an attempt to duplicate the mouth-feel or apparent viscosity of natural juices. Also, the use of gel agents is of primary interest in the manufacture of citrus jams and jellies. In such cases, rheological measurements can help quality control personnel to maintain the proper conditions and product components in order to make a successful, good-quality product.

Rheology Measurement

Many instruments and methods are used to measure viscosities and apparent viscosities. To compare results from one method or instrument with another, it must be remembered that for non-Newtonian fluids the apparent viscosity is characterized by the temperature and the shear rate. The apparent viscosity generally is expressed as centipoise (100 g/sec cm) and shear rate (sec^{-1}). Among the methods used are procedures that measure the flow through a capillary tube or orifice, the rate of a falling weight through the sample, the power consumed in stirring, the penetration into a test sample, the rate and degree of spread on a flat surface, ultrasonic vibration, and radioactive density measurements. For non-Newtonian fluids, the most applicable and most commonly used method is the rotation of a spindle or a cylinder within a coaxial cylindrical tube. These instruments generally provide a consistent shear rate or rotational speed for selected spindles, from which the torque needed to maintain that speed is measured and correlated with the apparent viscosity. The shear rate in such measurements can be found by using:

$$\text{shear rate } (sec^{-1}) = \alpha\pi(R_2 + R_1)/60(R_2 - R_1) \qquad (12\text{-}1)$$

where R_1 is the radius of the cylindrical spindle, R_2 is the radius of the cylindrical tube containing the sample, and α is the angular velocity of the spindle (rpm) (Loncin and Merson 1979). The procedure outlined by the manufacturer of each instrument should be followed in making apparent viscosity measurements.

Newtonian methods of measuring apparent viscosities, such as measuring the rate of fall of a heavy object through a sample or the rate of flow through a tube or an orifice, generally involve varying shear rates throughout a single measurement. For Newtonian fluids where the viscosity does not change with shear rate, these methods are acceptable, but with non-Newtonian solutions, such as citrus concentrates, the varying shear rates can give misleading results. Nevertheless, Newtonian methods can have value when applied to non-Newtonian solutions in some cases. For example, if all that is desired is a pass/fail result, Newtonian methods may suffice. Also, analysts developing in-house standards can employ Newtonian methods if the product composition remains sufficiently consistent.

QUESTIONS AND PROBLEMS

1. Define rheology, Newtonian fluid, non-Newtonian fluid, pseudoplastic, dilatent, thixotropic, and rheopectic solutions.
2. How would citrus concentrates be defined in rheology?

3. How do pulp levels, Brix/acid ratio, pectinase enzyme activity, and temperature affect the apparent viscosity of citrus concentrates?
4. Which piece of equipment is *not* generally affected by juice consistency: pipes, evaporators, extractors, chillers, bulk storage, or pumps?
5. What would be the shear rate of a rotary viscometer using a spindle with a radius of 1.0 cm and a sample container with a radius of 1.2 cm with an angular velocity of 100 rpm?

REFERENCES

Crandall, P. G., Chen, C. S., and Carter, R. D. 1982. Models for predicting viscosity of orange juice concentrate, *Food Tech.*, *36(5)*, 245-251.

Crandall, P. G., Davis, K. C., Carter, R. D., and Sadler, G. D. 1988. Viscosity reduction by homogenization of orange juice concentrate in a pilot plant taste evaporator, *J. Food Sci.*, *53(5)*, 1477-1481.

Ezell, G. H. 1959. Viscosity of concentrated orange and grapefruit juices, *Food Tech.*, *13*, 9-13.

Kimball, D. A. 1986. Volumetric variations in sucrose solutions and equations that can be used to replace specific gravity tables, *J. Food Sci.*, *51(2)*, 529-530.

Loncin, M. and Merson, R. L. 1979. *Food Engineering, Principles and Select Applications*. Academic Press, New York, 34.

Chapter 13

Citrus Processing Varieties

Many factors affect the quality of citrus juices, including industrial practices, fruit maturity, climate, soil conditions, rootstock, and fruit variety. Industrial practices and fruit maturity have been discussed in detail in this text. Climate, soil conditions, and rootstock are generally beyond the control of the processing quality control department and/or processing plant. The variety of inbound juices used in blending and/or product manufacture, however, is of interest and is controlled to some degree by the processor, especially in regard to juices received from outside the local area. Therefore, this chapter has been included to give the processor a global picture of citrus varieties. Taxonomy that is related to rootstocks, ornamentals, or other areas not directly relevant to citrus processing has been omitted. Only those taxonomical parameters that concern citrus products are considered.

For many years the diversity of citrus species and varieties was unknown. Citrus species always have been and always will be undergoing natural and manmade changes, which make classification an ongoing and controversial task. Also, the effort to compare varieties grown in distant geographical areas and to separate the effects of local conditions from varietal differences added to the confusion. In addition, the separation of hybrids from true species remains largely a matter of opinion and accepted practice. In 1967 Swingle and Reece proposed 16 species of the *Citrus* subgenus that have generally been accepted by the industry (Swingle and Reece 1967). Tanaka proposed 159 species of the *Citrus* subgenus (Tanaka 1954). Tanaka's species may be of benefit to the horticulturist, but it is considered too complex for industrial use, especially to processors. The author of this text also utilizes extensively the opinions and information supplied by Hodgson (1967), Rangama, Govindarajan, and Raman (1983), and Considine (1982).

Table 13-1 presents a taxonomical description of the commercially important citrus species. The order in which the species appear represents general importance by groups of species that are similar. The commercial importance is a relative and seasonal term that may vary from time to time. Other members of

Table 13-1. Taxonomy of edible citrus fruits.

Kingdom—Plant
Order—Geraniales
Suborder—Geraniineae
Class—Dicotyledoneae
Subclass—Archichalmydeae
Division—Embryophyta
Subdivision—Angiospermeae
Family—Rutaceae
Subfamily—Aurantiodeae
Tribe—Citreae
Subtribe—Citrinae
Genus—*Citus* L.
Subgenus-*Citrus*
Species—*Citrus sinensis* (L.) Osbeck—Sweet Orange
      Citrus aurantium L.—Sour or Seville Orange
      Citrus paradisi Macfadyen—Grapefruit
      Citrus grandis (L.) Osbeck—Pummelo
      Citrus maxima (Burm.) Merrill—Pummelo
      Citrus decumana L.—Pummelo
      Citrus limon L. Burm. f.—Lemon
      Citrus aurantifolia (Christm.) Swing.—Lime
      Citrus reticulata Blanco—common Mandarin/Tangerine
      Citrus unshiu Marc.—Satsuma Mandarin/Tangerine
      Citrus deliciosa Tenore—Mediterranean Mandarin/Tangerine
      Citrus nobilis Lourerio—King Mandarin/Tangerine
      Citrus madurensis (Lourerio)—Calamondin Mandarin/Tangerine
      Citrus medica L.—Citron

Main Varietal Groups

Citrus sinensis (L.) Osbeck—Sweet Orange
  Common Orange
  Acidless Orange
  Pigmented or Blood Orange
  Navel Orange

Citrus paradisi Macfadyen—Grapefruit
  White or Common Grapefruit
  Pigmented or Pink Grapefruit

Citrus grandis (L.) Osbeck, *maxima* (Burm.) Merrill, or *decumana* L.—Pummelos
  White or Common
  Pigmented or Pink
  Acidless

the Rutaceae family are sometimes used for food products, such as the kumquats of the *Fortunella* genus in the Citreae tribe. However, these products are of minor importance compared to those found in the *Citrus* L. genus.

Sweet Oranges (*Citrus sinensis* (L.) Osbeck)

This species constitutes by far the most important class of commercial citrus, with about two-thirds of the citrus produced worldwide falling into this category. Sweet oranges are known by different names around the world, including "naranja" in Spain, "aranico" in Italy, "laranja" in Portugal, "malta" in India, and "kan" in Japan. Sweet oranges rank behind only sour oranges and mandarins for hardiness, except that mandarins are more susceptible to frost injury to the fruit. Sweet oranges are used extensively for fresh fruit, but in the United States and Brazil they are used primarily for juice products. Sweet oranges can be classified into four categories, including the common orange, navel orange, blood orange, and acidless or sugar orange.

Common Oranges

About two-thirds of all sweet oranges fall into the common orange category. Common oranges are referred to as white or blond oranges. In Spain they are called "blanca," in Italy "biondo," and in France "blonde." The predominant variety in this group is the Valencia orange, which constitutes about half of all the oranges grown in the United States. Valencia oranges are also the predominant commercial variety of any type of citrus, one reason for this dominance being their long growing season and their resulting ability to adapt to a wide range of growing conditions. Therefore, this orange is grown in every major orange-producing country in the world. Contrary to popular belief, the Valencia orange came from the Azores Islands and perhaps from Portugal, not from the well-known city of Valencia in southern Spain, which is also known for its citrus production. This variety should not be confused with the Valencia Temprana variety in Spain, which matures earlier than the Valencia and is a smaller fruit.

Table 13-2 lists the major varieties found in the common orange group. Main budlines and related strains are generally classified together under one variety, especially when the literature describes them as indistinguishable. Synonyms, as well as countries where the literature has reported that the variety is of commercial importance, are listed alphabetically, along with subjective and relative descriptions of flavor and color. This relative rating system attempts to provide the processor with a comparison of varieties within a given species according to the literature and should not be construed as absolute. Also, some varieties may be of local use and interest only, and so are not listed here.

Table 13-2. General attributes of the common orange varieties of the _Citrus sinensis_ (L.) Osbeck or sweet orange species.

Variety	Other Names	Location(s)	Flavor	Color
Balta		Pakistan		
Belladonna		Italy	pleasant	
Biondo Commune	Beledi	N. Africa	pleasant	
	Communa-Spain	Egypt		
	Koines-Greece	Greece		
	Liscio	Italy		
	Nostrale	Spain		
Biondo Riccio		Italy	pleasant	
Cadenera	Cadena Fina	Algeria	excellent	
	Cadena sin Jueso	Morocco		
	Precoce de Valence	Spain		
	Precoce des Canaries			
	Valence san Pepins			
Calabrese	Calabrese Ovale	Italy	well flavored	
	Ovale			
Carvalhal		Portugal		
Castellana		Spain	sweet—low acid	pale
Clanor	Clan William	S. Africa	good	
Don Jao		Portugal		
Fukuhara		Japan		
Hamlin	Norris	S. Africa	sweet—low acid	well colored
		Algeria		
		Argentina		
		Brazil		
		Florida		
		India		
		Iran		
		Mexico		
Homosasa		Florida	good	
Joppa		S. Africa		light orange
		Texas		
Khettmali	Katmali	Israel	excellent	
	Hitmali	Lebanon		
Macetera		Spain	unique	
Maltaise Blonde	Maltaise	N. Africa	mild	medium well colored
	Petite Jaffa			
	Portugaise Blonde			

(_Table continued on p. 184_)

Table 13-2. (*Continued*)

Variety	Other Names	Location(s)	Flavor	Color
Maltaise Ovale	Maltaise Oval Garey's or California Mediterranean Sweet	S. Africa California	mild	pale
Marrs	Marrs Early	California Iran Texas	sweet—lacking acid	well colored
Mosambi	Bombay-Deccan Mosambique Musami	India Pakistan	insipid—low acid	straw yellow
Parson Brown		Florida Mexico Turkey	well flavored	dull orange
Pera Rio		Brazil	good	good
Pera Coroa		Brazil	good	good
Pera		Brazil	rich	well colored
Pera Natal		Brazil	rich	deep color
Pineapple		S. Africa Argentina Brazil Florida India Mexico	rich—sweet	light orange
Premier		S. Africa	good	
Queen		S. Africa	rich	red-orange
Salustiana	Salus	Algeria Iran Morocco Spain	rich—sweet	
Sathgudi	Chini of S. India	S. India	fair—sweet	
Seleta	Selecta Siletta	Australia Brazil	acidy	
Shamouti	Chamouti Jaffoui Jaffa Jaffoui Palestine Jaffa	N. Africa S. Africa Egypt China Cyprus Greece India Iran Israel Lebanon Syria Turkey	sweet	light orange

Table 13-2. (*Continued*)

Variety	Other Names	Location(s)	Flavor	Color
Shamouti Masry	Khalily White Egyptian Shamouti	Egypt	rich—sweet	well colored
Tomango		S. Africa		
Valencia	Valencia Late	S. Africa	good—acid	
	Hart's Tardiff	Algeria		
	Hart Late	Argentina		
	same as	Arizona		
	Lou Gim	Australia		
	Gong	Brazil		
	Natal of	California		
	Brazil	China		
	Pope	Egypt		
	Calderon	Florida		
	Delta	India		
	Midknight	Iran		
	Tajamur	Iraq		
		Israel		
		Jamaica		
		Lebonan		
		Mexico		
		Morocco		
		Spain		
		Pakistan		
		Turkey		
Verna	Berna	Algeria	sweet	well colored
	Bernia	Mexico		
	Vernia	Morocco		
	Bedmar	Spain		
Vicieda	Viciedo	Algeria	well flavored	pale
		Morocco		
		Spain		
Westin		Brazil	rich	well colored

The literature cited early in this chapter also includes some information that may be of interest to the processor. One can learn, for example, that the Belladonna variety may be the same as the Shamouti orange. The Calabrese variety is the most important common orange in Italy. The Hamlin orange is the world's principal early-season common orange. The Homosasa is one of Florida's oldest varieties. The Macetera is considered excellent for processing because of its high juice content and unique flavor. The Maltaise blonde closely resembles the Shamouti and Shamouti Masry oranges. The Parson Brown is the earliest-season common orange. The Sathgudi is the principal common orange in India.

Because the Shamouti orange is the most common sweet orange in North Africa and the Near East, it sometimes is simply referred to as the "beledi" or common orange. This designation probably is misleading because other common oranges are grown in the area as well. The Shamouti orange generally is not considered to be a good processing orange because it lacks juice and has a relatively poor color. The development of bitter compounds has also been reported in the Shamouti oranges (Levi et al. 1974), as well as in the Balta, Mosambi, and Pakistan Valencia. The varieties described as early-season include the Balta, Biondo Commune, Hamlin, Marrs, and Mosambi, with the Parson Brown described as very early. Early to midseason fruits include the Premier, Vicieda, and Westin. Midseason fruits include the Khettmali, Macetera, Maltaise Blonde, Pera Rio, Pera Coroa, Pineapple, Queen, and Sathgudi. The late midseason varieties include the Maltaise Ovale and the Seleta. The late-season fruits include the Calabrese, Pera, Pera Natal, and Valencia.

Navel Oranges

Navel oranges are characterized by a rudimentary secondary fruit in the apex of the primary fruit that resembles a navel. This feature is sometimes found in some mandarins but occurs regularly only in navel oranges. Navel oranges are the most popular eating oranges in the world, primarily because they are not very juicy, the flesh has an excellent and rich flavor, and they are crisp, seedless, and easy to peel. However, a delayed limonin bitterness develops upon processing, as described in Chapter 10.

Table 13-3 presents a general description of navel orange varieties in a manner similar to that of Table 13-2. The flavor descriptions in the table do not account for the delayed bitterness mentioned above because those descriptions are intended for fresh fruit products. Debittered or late-season navel juices rival common oranges in juice quality but take a back seat with respect to juice yield and perhaps juice color. Also, navel trees are sensitive to unfavorable conditions and/or neglect. The principal navel variety and the second most popular all-around citrus variety is the Washington navel, which is quite unstable horticulturally and has given rise to all other navel varieties except the Australian navel varieties and their derivatives. The Australian navels, grown to some extent in California and South Africa, are later-season and juicier than Washington navels, with a tart flavor that lingers longer into the season. In Florida, Washington navels produce poorly, leaving Florida to local navel varieties (Summerfield, Glen, and Dream) that are of minor commercial and processing importance. Very early-season varieties include the Bonanza, Leng, and Navelina. Early-season varieties include the Bajaninha Piracicaba, Oberholzer, Rio Grande, and Thomson. Lane's Late is the only variety classified as late-season.

Table 13-3. General attributes of the navel orange varieties of the *Citrus sinensis* (L.) Osbeck or sweet orange species.

Variety	Other Names	Location(s)	Flavor	Color
Baianinha	Bahianinha	Brazil		
Piracicaba		California		
Bonanza		California	sweet	pale yellow
Bourouhaine		Iran		
		Tunisia		
Frost		Arizona		
Washington		California		
		Morocco		
Fukumotto		Japan		
Gillette		California	rich	deep color
Lane's Late		Australia		
		California		
Leng		Australia	fair	well colored
Navelate		Spain	less sprightly	pale
Navelina	Dalmau	Morocco	sweet	deep color
		Portugal		
		Spain		
Oberholzer	Oberholzer	S. Africa		
	Palmer			
Rio Grande		California		
Thomson	Thomson	Algeria	good	well colored
	Improved	California		
		Chile		
		Iran		
		Spain		
Washington	Atwood	S. Africa	rich	deep color
	Bahia	Algeria		
	Baia	Argentina	Florida	Morocco
	Baiana	Australia	Iran	Portugal
	Fisher	Brazil	Iraq	Spain
	Riverside	California	Israel	Turkey
		China	Lebanon	
		Egypt	Mexico	

Blood Oranges

Blood oranges are characterized by their reddish pigmentation, which is due primarily to the presence of anthocyanins in the flesh of the fruit. In Spain, blood oranges are referred to as "sanguin," in Italy as "sanguigna" or "sanguinella," and in French-speaking countries as "sanguine." Blood oranges make up about 40% of the oranges grown in Spain and a slightly higher percentage of those grown in Italy. Blood orange juice has a unique and highly favored flavor and is considered the most delicious of all citrus juices. However, the anthocyanins have a tendency to fade during processing and storage that gives the juice an undesirable muddy color. Activated charcoal has been used commercially in Italy to remove the anthocyanins, restoring a healthy color to the juice. This activated charcoal also removes limonin, which imparts a delayed bitterness to many blood varieties similar to that of navel juice. However, the activated charcoal also removes significant amounts of vitamin C, which cannot be replenished in the United States in 100% juice products without violating federal standards of identity. The color development in blood oranges is directly proportional to the heat applied during the season. High temperatures deepen the pigmentation.

Table 13-4 lists the main blood orange varieties. Blood oranges can be classified into three regional groups. The oldest group is the ordinary blood orange, which is similar to the common orange in every way except the pigmentation. It includes the Sanguinello Commune, Maltaise Sanguine, and other light blood varieties such as the younger Moro and Tarocco varieties. The Doblefina group includes the Doblefina of Spain, Entrefina, Spanish Sanguinelli, and Doblefina Amelioree. The smallest group is the Shamouti or Palestine Jaffa Blood Orange group, including the Shamouti Mawadri and the Mawadri Beledi. The latter is the earliest-season blood orange. The Maltaise Sanguine and the Sanguinello Commune are similar to the Sanguinello Moscato, the latter being the most important commercial blood orange, which is grown primarily in Sicily. The oldest blood orange is the Sanguigno Semplice. The Moro blood orange matures very early, with midseason blood oranges including the Ruby, Sanguigno Semplice, Sanguinello Commune, Sanguinello Moscato, Tarocco, and Tomango. Late midseason blood oranges include the Doblefina, Doblefina Amelioree, Maltaise Sanguine, Mutera, and Spanish Sanguinelli.

Acidless Oranges

Acidless or sugar oranges are characterized by their low or nonexistent acid levels. In France they are called "douceatre" or "douce." In Spain they are referred to as "sucrena," in Italy as "maltese" or "dolce," in North Africa and the Near East as "meski," in Turkey as "lokkum" or "tounsi," in Egypt as "succari," and in Brazil as "lima." Acidless oranges are generally too devoid of acid to be processed without risk of the growth of pathogenic microorganisms. Some varieties, shown in Table 13-5, such as the Succari of Egypt, reach Brix/acid ratios of up to 100, too sweet to process. The sucrena of Spain is the oldest acidless orange. The Vainiglia of Italy is generally of only local use and interest. Acidless oranges are generally early-season fruits.

Table 13-4. General attributes of the blood orange varieties of the *Citrus sinensis* (L.) Osbeck or sweet orange species.

Variety	Other Names	Location(s)	Flavor	Color
Blood Red	Blood Red Malta	India W. Pakistan	very good	variable
Doblefina	Ovale Sangre Sanguina Oval Rojo Oval Sanguine Ovale Morlotte Blood Oval	Algeria Morocco Spain	mild—pleasant	variable
Doblefina Amelioree	Grosse Sanguine Washington Sanguine Washington Sangre Pedro Veyrat	Algeria Morocco Spain	good	variable
Entrefina Malta Blood	Inglesa	Spain Algeria		
Maltaise Sanguine	Portugaise	Algeria Iran Tunisia	excellent	good blood variable
Mawadria Beledi	Damawi Mawadri Beladi	Lebanon Syria		light blood
Moro	Belladonna Sanguigno	California Iraq Sicily	pleasant	deep almost violet red
Mutera		Spain	rich—sweet	lacks red
Ruby	Ruby Blood	California Florida	rich	sometimes streaked red
Sanguigno Semplice		Italy	pleasant	red streaks
Sanguinello Commune		Italy	pleasant	deep red
Sanguinello Moscato		Iran Sicily	well flavored	good blood
Shamouti Mawadri	Shamouti Maouardi	Lebanon Syria		light blood
Spanish Sanguinelli	Sanguinelli Sanguinella Negra	Spain		deep red and constant
Tarocco	Tarocco dal Muso Tarocco di Francofonte	Italy Iraq	rich and sprightly	well pig- mented
Tarocco Liscio	Calabrese Sanguigno Tarocco Ovale	Morocco Sicily Spain	rich and sprightly	well pig- mented
Tomango		S. Africa	excellent	red only in cold winters

Table 13-5. General attirbutes of the acidless orange of the *Citrus sinensis* (L.) Osbeck or sweet orange species.

Variety	Other Names	Location(s)	Flavor	Color
Lima		Brazil	insipid	light yellow
Succari	Sukkari	Egypt	overly sweet	pale
Sucrena	Imperial	Spain	insipid	good
	Grano de Oro			
	Real			
	Canamiel			
Vainiglia	Vaniglia	Italy	sweet—	
	Maltaise		slight	
	Dolce		bitterness	

SOUR ORANGES (*Citrus aurantium* L.)

Sour oranges are also referred to as bitter or Seville oranges. In Spain they are called "naranja agria" or "amarga," in Italy "melangolo" or "arancio amaro," in France "bigarade" or "orange amere," in Israel "khuskhash," in West Pakistan "khatta," and in Japan "daidai." Sour oranges are generally too bitter and sour to be processed into juice, and the oils are generally strong and disagreeable. However, the trees are resistant to excess soil moisture, frost, and neglect and thus are grown in areas prohibitive to other species. Sour oranges are used to make marmalades and, to some extent, essential oils. By British standards, the quality of marmalades made from sour oranges surpasses the quality of marmalades made from sweet oranges.

There are three basic types of sour oranges, the main varieties of which are listed in Table 13-6. Common bitter oranges consist mainly of the Sevillano variety, which is used extensively for the production of marmalades, essential oils, and bitter juices used in drink bases in Spain and Australia, and is a late-season fruit. The bittersweet group contains the essentially synonymous varieties of the Bittersweet of Florida and the Paraguay or Apepu. This group is grown mainly for essential oils. The variant bitter oranges are grown primarily for perfumery and ornamentals.

Hybrids or species that closely resemble the sour orange include the Gunter (*C. maderaspatana* Tan.), Bergamot (*C. bergamia* Risso), Sanbo (*C. sulcata* Takahashi), and Myrtleleaf (*C. myrtifolia* Rafinesque). The Gunter is of importance primarily in South India. Bergamot oil is used in Italy as a base for cologne water. Sanbo remains popular in the Wakayama Prefecture of Japan and the Myrtleleaf (Large Chinotto variety) is used for candied peel (marmalade) or crystallized whole in Italy.

Table 13-6. General attributes of the sour orange varieties of the *Citrus aurantium* L. species. Other names for the sour orange are the common bitter orange or the Seville orange.

Variety	Other Names	Location(s)	Flavor	Color
Bittersweet	Bittersweet of Florida Paraguay Apepu	Florida Paraguay Spain		
Kabusu	Kabusudaidai	Japan		
Sevillano	Agrio de Espana Real Sour Seville	Australia Spain	good for marma- lade	
(Related Hybrids that Closely Resemble the Sour Orange)				
Bergamot		Italy		
Gunter	Kitchli Vadlapudi	India	pleasant with slight bitterness	pale orange
Myrtleleaf	Chinotto Myrtle-leaf orange	China Florida Italy		
Sanbo	Sanbokan	Japan	good	

GRAPEFRUIT AND PUMMELOS

Grapefruit and pummelos (shaddocks) are closely related. In fact, many people believe that the grapefruit is a hybrid or a variety of the pummelo. The grapefruit (so named, it is believed, because the fruit grows in clusters like grapes) is recognized as the separate species *Citrus paradisi* Macfadyen. Grapefruit has a highly distinctive flavor with a lighter color, whereas the flavor and the color of pummelos vary widely and give rise to the three pummelo species—*Citrus grandis* (L.) Osbeck, the most common, *Citrus maxima* (Burm.), and *Citrus decumana* L. In France the pummelos are referred to as "pamplemousse," in Italy as "pompelmo," in Spain as "pampelmus," and in Japan as "buntan" or "zabon." Most pummelos are very thick-skinned and are worthless as a fresh fruit or for juice. However, there are some that are of importance commercially in the Orient. Pummelos are less cold-tolerant than grapefruit, and some of the major varieties do not have a bitter flavor like that of grapefruit.

Tables 13-7 and 13-8 briefly describe commercially important grapefruit and pummelo varieties. In Spain grapefruit are referred to as "toronja." They rank third behind oranges and mandarins in world production. In processed products the glucoside naringin imparts an immediate bitter flavor as well as delayed

Table 13-7. General attributes of grapefruit varieties of the *Citrus paradisi* Macf. species.

Variety	Other Names	Location(s)	Flavor	Color
Common or White Grapefruit				
Duncan		China	excellent	buff-chamois
		Florida		
Marsh	Marsh Seedless	S. Africa	good	buff-chamois
	White Marsh	Argentina		
		Arizona		
		Australia		
		California		
		China		
		Cyprus		
		Egypt		
		Florida		
		Iran		
		Iraq		
		Israel		
		Jamaica		
		Spain		
		Texas		
Triumph	Jackson	S. Africa	lacks bitterness	
Pigmented or Pink Grapefruit				
Flame		Florida		
Henderson	Ruby Red	Florida	slightly	fades to amber with maturity
		Texas	sweet	
Ray Ruby		Texas		
Red Blush	Ruby	Argentina	good	slight pink
	Red Marsh	Arizona		
	Red Seedless	California		
		Iran		
		Iraq		
		Texas		
Rio Red		Texas		deep
Star Ruby		Texas	best flavor	darkest
Thompson		Argentina	good	chaimois or dark buff
		China		

bitterness from limonin. In Florida, a little less than half of the grapefruit that is grown is processed. Grapefruit peel is used in the production of pectin and essential oils. There are two basic types of grapefruit—common or white and pigmented or pink. The late-season Marsh variety is by far the most popular common grapefruit. The mid-early-season Duncan grapefruit is considered to

Table 13-8. General attributes of pummelo varieties of the *Citrus grandis* (L.) Osbeck, *Citrus maxima* (Burm.), and *Citrus decumana* L. species.

Variety	Other Names	Location(s)	Flavor	Color
Common or White Pummelos				
Banpeiyu	Pai You Pai Yau	Japan Taiwan	excellent	pale yellow
Hirado	Hirado Buntan	China Japan	pleasant, trace bitterness	light greenish yellow
Kao Pan	Kao Panne	Thailand	sweet, mildy acid	lemon yellow
Kao Phuang	Siamese	California	acid	
Mato	Mato Buntan	China Japan	lacking acid, sweet, trace of bitterness	light greenish yellow
Tahitian	Moanalua	Hawaii	unique, excel- lent	faintly amber
Pigmented Pummelos				
Ogami	Egami	Florida Japan	good	deep pink al- most red
Pandan Bener		Java	pleasant, sweet, slightly acid	red-fleshed
Pandan Wangi		Java	pleasant, sweet, slightly acid	red-fleshed
Siamese Pink	Siam	California	grapefruit- like, sub- acid, trace bitterness	pink-tinged
Thong Dee		California Florida	good	pink-tinged

be the best flavored, setting the standard for all other varieties. It is a common trend among varieties that the seedier ones (such as the Duncan) have a richer flavor than the less seedy or seedless varieties (such as the Marsh). The Triumph Grapefruit is another early-midseason grapefruit.

Pink grapefruit varieties are characterized by the presence of the pink to red pigment of the carotenoid lycopene. Although attractive in a fresh fruit, this pigmentation has a tendency to fade during processing and storage similar to that of the anthocyanins in blood oranges. For this reason, pigmented grapefruit juice is not popular. However, pink grapefruit juice cocktail contains pink grapefruit juice along with sweeteners, colorants, and other ingredients de-signed to give a pink grapefruit juice appearance without this color fading effect. Consequently grapefruit juice cocktail is somewhat popular. As with the anthocyanins, heat is required during the growing season to develop the pigmentation. In the United States, the Red Blush variety is the most popular, with highly pigmented varieties recently emerging (Star Ruby and Rio Red). The Star Ruby is reported to have the best flavor among the pigmented varieties as well as the deepest color. The Thompson grapefruit is considered a midseason fruit.

Like grapefruit, pummelos are classified as common or white and pigmented or pink. Common pummelos are moderately high in acid, and the pink pummelos range from light pink to deep red. Some pummelo varieties have a highly distinctive and excellent flavor. A third class of pummelos is the acidless pummelo, which resembles the acidless orange mentioned above. Acid levels range from less than 0.1% to about 2.0%, with Brix/acid ratios of up to 150. Some varieties, such as the Tahitian pummelo of Hawaii, are of local use and interest only. Early-season pummelos include the Kao Pan and Mato. Pummelos considered to be medium early-season fruits include the Banpeiyu, Hirado, and Kao Phuang. The Thong Dee is a midseason fruit.

Several tangelo (tangerine × pummelo) hybrids are closely associated with grapefruit and pummelos, along with the orangelo (orange × pummelo) variety "chironja" of Puerto Rico, as shown in Table 13-9. The chironja hybrid is not considered suitable for processing. The more important varieties include the K-early of Florida, the poorman and Wheeny of New Zealand and Australia, and the hassaku and natusdaidai of Japan. Tanaka classified the hassaku hybrid as *C. hassaku* Hort. ex. Tan. (pummelo × mandarin) and the natsudaidai species as *C. natsudaidai* Hayata (pummelo × sour orange or mandarin). Hassaku juice is known to develop severe limonin bitterness with levels as high as 50 ppm. Nomilin, a known metabolic precursor of limonin, has been found to peak about one month before limonin levels reach their maximum (Hashinaga and Itoo 1983). It has also been found that freezing of natsudaidai juice ruptures juice cell membranes in the juice pulp with leakage of additional bitter compounds into the juice (Iwata and Ogata 1976). The melogold variety (common grapefruit × acidless pummelo) is the most recent commercial hybrid.

MANDARINS OR TANGERINES

Mandarins or tangerines are known in Japan as "mikan," in India as "suntura" or "sangtra," in Italy and Spain as "mandarino," and in French-speaking countries as "mandarine." In the Orient mandarins clearly dominate the citrus industry, and they are important in many other parts of the world. Because the characteristics of mandarins vary so widely, they are often regarded as exotics. Mandarins are characterized by their loose and easily peeled rind, an open core, and a deeper orange color than is found in most other types of citrus. The flavor of mandarins also is unique and richer than that of most citrus species. Man-

Table 13-9. General attributes of hybrids that most closely resemble grapefruit and pummelos.

Variety	Other Names	Location(s)	Flavor	Color
Orangelos				
Chironja		Puerto Rico	mild, lacks grapefruit bitterness	yellow orange
Tangelos				
Hassaku (*C. hassaku* Hort. ex. Tanaka)	Hassaku Mikan Hassaku Zubon	Japan	good, turns bitter	light yellow
K-early		Florida	rather acid	yellowish orange
Melogold		California	good	buff-chamois
Natsudaidai (*C. natsudaidai* Hayata)	Natsumikan Natsukan	Japan	too acid for most	yellowish orange
Orobanco		California	good	buff-chamois
Poorman	New Zealand Grapefruit Poorman Orange	Australia New Zealand	pleasantly subacid, trace of bitterness	reddish
Smooth Seville	Smooth Flat Seville	Australia	pleasantly subacid, trace of bitterness	reddish orange
Wheeny	Wheeny Grapefruit	Australia New Zealand	good but acid	straw color

darins rank second behind oranges in global importance and are more cold-resistant than other types of citrus, except that freeze damage to the fruit can be severe because of the loose nature of the rind. Increased heat during the latter part of the season results in lower acid levels and milder juice. In Japan about 15% of the mandarins harvested are processed. Elsewhere, processed mandarin juice is of less importance and is used primarily to enhance color in light-colored orange juices. In the United States, up to 10% mandarin juice can be added to orange juice without declaration or violation of federal standards of identity. There are four basic types of mandarins, which have been assigned their own separate species classification.

Common Mandarin (*C. reticulata* Blanco)

The common mandarins, described in Table 13-10, are the most important group. They are characterized by a tight rind. The Beauty variety is similar to the Dancy, and the Campeona closely resembles the King variety in the King

Table 13-10. General attributes of the common mandarin varieties of the *Citrus reticulata* Blanco species.

Variety	Other Names	Location(s)	Flavor	Color
Bakrai	Rangur type	Iran		
Beauty	Beauty of Glen Retreat	Australia	sprightly	orange
Campeona	Glen	Argentina	rich and sprightly, medium acid	orange
Clementine		N. Africa Algeria Arizona California Iraq Morocco Spain Tunisia	sweet, subacid	deep orange
Coorg Orange		India		
Cravo	Laranja Cravo	Brazil	mild	deep orange
Dancy		Argentina Arizona California China Florida	rich, sprightly, medium high acid	deep orange
Desi of Punjab		India		
Ellendale	Ellendale Beauty	Australia	rich, pleasantly subacid	bright orange
Emperor	Emperor of Canton	S. Africa Australia India	pleasant	light orange
Fewtrell	Fewtrell's Early	Australia	mild	orange
Indian		Iraq		
Imperial		Australia	pleasantly subacid	pale
Kashi Orange of Assam		India		
Kinnow		Arizona California West Pakistan	rich	deep yellowish orange
Lee		Florida	rich and sweet	orange
Malvaiso		Argentina	pleasantly subacid	
Mandalina		Lebanon Syria	pleasantly subacid	
Murcott	Murcott Honey Smith	Argentina Brazil Florida	rich and sprightly	high color orange
Nova		Florida	pleasant	deep orange

Table 13-10. (*Continued*)

Variety	Other Names	Location(s)	Flavor	Color
Ortanique		Cuba	rich	orange
		Jamaica		
		Reunion Island		
Osceola		Florida	rich	deep orange
Ponkan	Nagpur	Brazil	mild and pleasant	orange
	Warnurco	China		
		India		
		Pakistan		
		Philippines		
		Sri Lanka		
		Taiwan		
Robinson		Florida	rich and sweet	deep orange
Sikkim	Darjeeling Orange	India		
Sunburst		Florida		excellent
Tankan		China	rich and sweet	deep orange
		Japan		
		Taiwan		
Wilking		Brazil	rich and sprightly	deep orange
		Morocco		

mandarin group. The Lee, Robinson, Osceola, and Nova varieties are all similar and are actually hybrids of the Clementine mandarin and the Orlando tangelo. The Kinnow and Wilking varieties are hybrids of the King mandarin and the Willowleaf. The Kinnow variety in West Pakistan is known to exhibit delayed limonin bitterness. The mandalina of Lebanon resembles the Dancy variety. In the United States the Dancy tangerine is predominant with the Ponkan variety dominating the Oriental markets, especially in China. In India the Ponkan is referred to as the "nagpur santra" and is the leading mandarin variety. The very early-season fruits include the Cravo, Imperial, and Nova varieties. Early-season fruits include the Clementine, Fewtrell, and Robinson varieties. Early to midseason fruits include the Emperor, Lee, Osceola, Ponkan, and Sunburst. Midseason fruits include the Beauty, Dancy, and Wilking. Late midseason fruits include the Ellendale, Emperor, Ortanique, and Tankan. Late-season fruits include the Campeona and Murcott. Very late-season fruit includes the Malvaiso variety.

Satsuma Mandarin (*C. unshiu* Marc.)

The Satsuma mandarin varieties are listed in Table 13-11, and, as can be seen, dominate the Japanese citrus industry. Satsuma mandarins are also found in

Table 13-11. General attributes of the satsuma tangerine varieties of the *Citrus unshiu* Marc. species.

Variety	Other Names	Location(s)	Flavor	Color
Aoe Wase	Aoe	Japan		
Dobashi Beni		Japan		deep orange-
		Spain		red
Hayashi Unshiu		Japan		
Iseki Wase	Iseki	Japan		
Ishikawa Unshiu	Ishikawa	Japan	excellent, sweet	
Juuman	Juman	Japan	rich	
Kuno Unshiu		Japan		
Matsuyama Wase	Matsuyama	Japan		
Miho Wase	Miho	Japan	sweet	
Mikan	Japanese Mikan	Japan		
Miyagawa Wase	Miyagawa	Japan	excellent	
Nagahashi		Japan		semibright
Nankan-4		Japan		
Okitsu Unshiu		Japan		
Okitsu Wase		Japan		
Owari Unshiu	Owari Satsuma	Japan	rich but subacid	orange
Seto Unshiu		Japan		
Silverhill	Owari Frost Owari	Japan	very sweet	
Sugiyama Unshiu		Japan	excellent, sweet	
Yonezawa		Japan		bright-col- ored

(Satsuma varieties are also grown in California, China, and India but the exact varieties are not clear.)

significant quantities in many other parts of the world, and are considered the hardiest of mandarins as well as the most cold-tolerant of citrus. Satsuma tangerines generally are divided into five categories. The Wase Satsumas include the early-season varieties (September–October) and represent one-fifth of all Japanese mandarins. The oldest variety is the Aoe Wase. The Zairai group includes old native varieties indigenous to Japan. The Owari group, the most important, is grown in the Owari Province and is especially suited for the canning of fruit sections. The Ikeda and Ikiri groups are primarily of local use and interest only. Early-season varieties include the Aoe Wase, Iseki Wase, Matsuyama Wase, Miho Wase, Miyagawa Wase, Okitsu Unshiu, Okitsu Wase, and Silverhill. The medium-early varieties include the Dobashi Beni and the Owari Unshiu. The midseason varieties include the Nankan-4 and the Yonezawa. The late-season varieties include the Ishikawa and the Sugiyama Unshiu.

Mediterranean Mandarin (*C. deliciosa* Tenore)

This group of mandarins, so named because they are primarily grown in the Mediterranean basin, is sometimes referred to as the Willowleaf mandarins (see Table 13-12). This group contains the broadest distribution of varieties, being referred to as: "Ba Ahmed" in Morocco; "Blida," "Boufarik," or "Bougie" in Algeria; "Bodrum" in Turkey; "Paterno," "Palemo," or "Avana Speciale" in Italy; "Nice" or "Provence" in France; "Valencia" in Spain; "Setubal" or "Galego" in Portugal; "Baladi," "Effendi," or "Yousef" in Egypt and the Near East; "Thorny" in Australia; "Mexirica" or "Do Rio" in Brazil; "Chino" or Amarillo" in Mexico; and "Koina" in Greece. This group is the second most important mandarin variety behind the common mandarin but is being rapidly replaced by the Satsuma mandarins as well as the Clementine mandarin. Some Chinese varieties have been reported to resemble Mediterranean mandarins, including the Suhoikan variety. The Mediterranean variety is considered a late-season fruit.

Table 13-12. General attributes of Mediterranean mandarin varieties of the *Citrus deliciosa* Tenore species and the King Mandarin varieties of the *Citrus nobilis* Loureio species.

Variety	Other Names	Location(s)	Flavor	Color
Mediterranean Mandarins				
Bergamota		Argentina		
Malaguina		Argentina		
Mediterranean	Mediterranean	N. Africa	distinctly mild	light
	Common	Algeria	and plea-	orange
	Willowleaf	Argentina	santly aro-	
		Brazil	matic,	
		Egypt	sweet	
		Greece		
		Spain		
Suhoikan		China		
King Mandarins				
King	Indo–China	California	rich	deep orange
	Mandarin	China		
	Camboge	Florida		
	Mandarin	Indo–		
	Kunembo	China		
		Malaysia		
		Okinawa		
		Taiwan		

King Mandarin (*C. nobilus* Loureiro)

This group of mandarins includes a single commercial variety—the King variety shown with the Mediterranean group in Table 13-12. King mandarins are among the least cold-tolerant mandarins and are relatively unimportant as a commercial variety. These mandarins are the largest in fruit size and the latest-maturing among the mandarins as well.

Hybrids

Several important tangor and tangelo varieties that most closely resemble mandarins are listed in Table 13-13. The most important hybrid in the United States is the Temple tangor (*C. sinensis* × *C. reticulata*), which has been assigned the species *C. temple* Hort. ex. Tan. and is considered a medium late-season fruit. The Iyo mikan of Japan has been assigned the species *C. iyo* Hort. ex. Tan. and is a midseason fruit. The Umatilla tangor is a cross between the Sat-

Table 13-13. General attributes of hybrids that most closely resemble the mandarin species.

Variety	Other Names	Location(s)	Flavor	Color
Tangelos (mandarin × pummelo)				
Minneola		Arizona California Florida	rich and tart	orange
Orlando		Arizona California Florida	mildly sweet	orange
Seminole		Florida	sprightly acid, too tart	rich orange
Thorton		Florida Texas	mildly sweet	pale orange
Ugli		Jamaica	rich subacid	orange
Tangors (mandarin × oranges)				
Iyo	Iyo Mikan	Japan	sweet	orange
Sue Linda		Florida	smoother than Temple	orange
Temple		Cuba Florida Jamaica	rich and spicy	orange
Umatilla		Florida	rich, acid	orange
Others				
Fallglo		Florida	pleasant	excellent
Sudachi		Japan		

suma tangerine and the Ruby orange. The Minneola, Seminole, and Orlando tangelos are all crosses of the Duncan grapefruit and the Dancy tangerine. The Iyo tangor is known to exhibit delayed limonin bitterness. Of this group, the Orlando tangelo is considered an early-season fruit, the Thorton tangelo is considered a midseason fruit, the Minneola is a medium late-season fruit, and the Ugli is a late-season fruit. Other mandarin-like hybrids include the Fallgo (Bower × Temple), an early-season fruit, and the Sudachi (*C. ichangensis* × *C. reticulata*), the latter constituting the species *C. sudachi* Hort. Shirai.

LEMONS (*C. limon* (L.) Burm. f.)

Lemons are characterized by their light yellow color and their excessively high acid content. Lemons once were used in the manufacture of citric acid but have been replaced by more inexpensive methods. Lemons are referred to as "limone" in Italy, "limon" in Spain, and "citron" in France, but are different from the Mediterranean limetta (*C. limetta* Risso). They contain the widest divergence of all citrus species. Lemons are primarily processed into juice and lemonade, as well as pectin and lemon oil. Table 13-14 lists the main lemon varieties. The Berna lemon is much like the Lisbon lemon, and the Interdonato is the earliest-season lemon in Italy. The main varieties by far are the Eureka and the Lisbon, with the latter generally replacing the former. The best-processing lemon is said to be the Mesero, with the Monachello being resistant to the mal secco disease. Lemons may have more than one crop per year and sometimes several crops. The Bearss and Mesero are considered winter fruits, with the Interdonato included as an early-season fruit. The Eureka lemon matures in the late winter, spring, and early summer, with the Lisbon lemon maturing in the winter and early spring. The Femminello Ovale is a late winter fruit. The sweet lemon variety dorshapo is of minor importance except in Arabian countries, where it is highly favored. The related species limetta (*C. limetta* Risso) resembles the Dorshapo sweet and includes the single important variety of Millsweet. Some evidence suggests that the elderly prefer the higher-acidity drinks made from lemons.

LIMES

Limes are similar to lemons except that limes have a greener fruit and flesh color than lemons and have a distinct and unique flavor and aroma. In Italy and Spain limes are referred to as "lima," and in French-speaking countries they are called "lime." In Arabian countries both lemons and limes are called "limun," and in the Orient they are referred to as "nimbu" or "limbu." Limes are the most tropical of all citrus and are generally preferred to lemons. India is the largest producer of limes, which in the United States are grown almost

Table 13-14. General attributes of lemon varieties of the *Citrus limon* (L.) Burm. f. species.

Variety	Other Names	Location(s)	Flavor	Color
Bearss	Sicily	Florida		
Berna	Verna	Algeria		
	Bernia	California		
	Vernia	Morocco		
		Spain		
Canton	Rangpur	China		
	Lime	India		
		Japan		
Eureka		S. Africa	highly	greenish
		Argentina	acid	yellow
		Australia		
		California		
		China		
		Greece		
		Israel		
		Mexico		
		Pakistan		
Femminello	Commune	Sicily	high acid	
Ovale	Ruvittaru			
Interdonato	Speciale	Sicily	high acid	greenish
			slight bit-	yellow
			terness	
Karystini		Greece		
Lisbon		Algeria	very acid	pale greenish
		California		yellow
		China		
		Greece		
		Mexico		
		Morocco		
Mesero	Fino	Italy	high acid	
	Primitiori	Spain		
Monachello	Moscatello	Italy	lacking acid	
Polyphoros		Greece		
Sicilian lemon		Brazil		
Vilafranca		Italy	high acid	greenish yel-
				low

entirely in southern Florida. Limes have the highest acid content of all citrus but are lower in vitamin C and other nutrients compared to lemons. Limes are used mainly as additives to alcoholic beverages, in limeade, and to produce lime oil. Table 13-15 lists the main lime varieties.

Limes can be classified into three species. The small-fruited group (*C. aurantifolia* Swing.) consists of a single commercial variety, West Indian, which is the most commercially important of all lime varieties. The large-fruited group

Table 13-15. General attributes of lime varieties, including the small-fruited (*Citrus aurantifolia* Swing.), the large-fruited (*Citrus latifolia* Tan.), and the sweet lime (*Citrus limettoides* Tan.) species. Also, lemon(lime)-like hybrids are listed.

Variety	Other Names	Location(s)	Flavor	Color
Small Fruit Limes				
West Indian	Mexican Key	Brazil California China Egypt Florida India Iran Malaysia Mexico	highly acid, distinct aroma	greenish yellow
Large Fruited				
Bearss	Bearss Seedless Persian	Florida		
Tahiti	Persian	Australia Brazil California Florida	very acid, true lime flavor	greenish yellow
Sweet Lime				
Indian	Palestine	California Egypt India Latin America Near East Pakistan	insipidly sweet, slight bitter after taste	straw yellow
Lemon/Lime-Like Hybrids				
Galgal	Gulgal Kumaon lemon	India	very sour, trace of bitterness	pale yellow
Marrakech limonette	Moroccan limetta Limun Boussera	Morocco	very sour high acid	pale yellow
Mediterranean Sweet Lime	Tunisian Sweet Lime	Italy Tunisia	acidless, insipidly sweet	
Meyer		N. Africa Florida New Zealand Texas	acidy and lemon-like	light orange

(*C. latifolia* Tan.) is second in importance, with two of the varieties, Bearss and Tahiti, closely resembling each other. The Sahesli lime of Tunisian and the Pond lime of Hawaii are very similar to the Bearss lime described in the table. The sweet lime (*C. limettoides* Tan.) is considered too insipidly sweet in the United States but is popular in India and the Near East. The single variety, Indian or Palestine, is referred to as "mitha nimbu" in India and "limun helou" or "sucarri" in Egypt. In California the dessert and coastal fruit of this variety exhibit marked differences. Sweet limes are often used to prevent fevers and liver complaints. The West Indian and Tahiti limes are mainly winter fruits, along with the Meyer hybrid. Some lemon and lime-like hybrids also are described in Table 13-15.

CITRUS GEOGRAPHY AND QUALITY CONTROL

As mentioned early in this chapter, citrus varieties grown in distant localities become important when citrus products are brought in from those areas and used in blending or in the making of various products. Table 13-16 is a listing of citrus varieties grown in the different citrus-producing areas. This table can be used by quality control personnel in explaining characteristics of juice received from these countries, states, or localities. Even though the commercial importance of these varieties changes, Table 13-16 will give the processor a general idea of the citrus industry in these areas. For example, if a lot of concentrate from Brazil has a significantly different flavor from previous lots, investigation may reveal that the difference is due to a varietal change rather than a processing, storage, or shipping error.

Table 13-16. Citrus varieties believed to be of commercial importance by the author, listed by country, state, or region.

Algeria	Common Mandarins	Common Grapefruit
Common Oranges	Clementine	Marsh
Cadenera	Mediterranean Mandarins	Pink Grapefruit
Salustiana	Mediterranean	Red Blush
Valencia	Lemon	Thompson
Verna	Berna	Common Mandarins
Navel Oranges	Lisbon	Campeona
Thomson		Dancy
Washington	*Argentina*	Murcott
Blood Oranges	Common Oranges	Mediterranean Mandarins
Doblefina	Hamlin	Bergamota
Doblefina	Pineapple	Malaguina
Amerlioreo	Valencia	Mediterranean
Maltaise	Navel Oranges	Lemons
Sanguine	Washington	Eureka

Table 13-16. (*Continued*)

Arizona
Common Oranges
  Valencia
Navel Oranges
  Frost
  Washington
Common Grapefruit
  Marsh
Pink Grapefruit
  Red Blush
Common Mandarins
  Clementine
  Dancy
  Kinnow
Mandarin Hybrids
  Minneola
  Orlando

Australia
Common Oranges
  Seleta
  Valencia
Navel Oranges
  Lane's Late
  Leng
  Washington
Sour Oranges
  Sour Seville
Common Grapefruit
  Marsh
Grapefruit-like Hybrids
  Poorman
  Smooth
    Seville
  Wheeny
Common Mandarins
  Beauty
  Ellendale
  Emperor
  Fewtrell
  Imperial
Limes
  Tahiti

Brazil
Common Oranges
  Barao
  Hamlin
  Pera Rio
  Pera Coroa

Pera
Pera Natal
Pineapple
Valencia
Westin
Navel Oranges
  Baianinha
    Piracicaba
  Washington
Acidless Oranges
  Lima
Common Mandarins
  Carvo
  Murcott
  Ponkan
  Wilking
Mediterranean Mandarins
  Mediterranean
Lemons
  Sicilian lemon
Limes
  West Indian
  Tahiti

California
Common Oranges
  Maltaise Ovale
  Valencia
Navel Oranges
  Baianinha
    Piracicaba
  Bonanza
  Frost
    Washington
  Gillette
  Rio Grande
  Thomson
  Washington
Blood Oranges
  Moro
  Ruby
Common Grapefruit
  Marsh
Pink Grapefruit
  Red Blush
Pink Pummelos
  Siamese Pink
  Thong Dee

Grapefruit Hybrids
  Melogold
  Orobanco
Common Mandarins
  Clementine
  Dancy
  Kinnow
King Mandarins
  King
Mandarin Hybrids
  Minneola
  Orlando
Lemons
  Berna
  Eureka
  Lisbon
Limes
  West Indian
  Tahiti
  Indian

China
Common Oranges
  Shamouti
  Valencia
Navel Oranges
  Washington
Common Grapefruit
  Duncan
  Marsh
Common Pummelos
  Hirado
  Mato
Common Mandarins
  Ponkan
  Tankan
King Mandarins
  King
Mediterranean Mandarins
  Suhoikan
Lemons
  Eureka
  Lisbon
Limes
  West Indian

Cuba
Common Mandarins
  Shamouti

Table 13-16. (*Continued*)

Cuba (Cont.)
Common Grapefruit
  Marsh

Egypt
Common Oranges
  Biondo Commune
  Shamouti
  Shamouti Masry
  Valencia
Navel Oranges
  Washington
Acidless Oranges
  Succari
Common Grapefruit
  Marsh
Mediterranean Mandarins
  Mediterranean
Limes
  West Indian
  Indian

Florida
Common Oranges
  Hamlin
  Homosasa
  Parson Brown
  Pineapple
  Valencia
Navel Oranges
  Washington
Blood Oranges
  Ruby
Sour Oranges
  Bittersweet
Common Grapefruit
  Duncan
  Marsh
Pink Grapefruit
  Flame
  Henderson
Pink Pummelos
  Ogami
  Thong Dee
Grapefruit Hybrids
  K-early
Common Mandarins
  Lee
  Murcott
  Nova

  Osceola
  Robinson
  Sunburst
King Mandarins
  King
Mandarin Hybrids
  Minneola
  Orlando
  Seminole
  Thorton
  Sue Linda
  Temple
  Umatilla
  Fallglo
Lemons
  Bearss
Limes
  West Indian
  Bearss
  Tahiti

Greece
Common Oranges
  Biondo
    Commune
  Shamouti
Mediterranean Mandarins
  Mediterranean
Lemons
  Eureka
  Karystini
  Lisbon
  Polyphoros

India
Common Oranges
  Hamlin
  Mosambi
  Pineapple
  Sathgudi
  Shamouti
  Valencia
Blood Oranges
  Blood Red
Common Mandarins
  Coorg Orange
  Desi of Punjab
  Emperor
  Kahsi Orange
  Of Assam

  Kinnow
  Ponkan
  Sikkim
Lemons
  Canton
Limes
  West Indian
  Indian
Lemon Hybrids
  Galgal

Iran
Common Oranges
  Shamouti
  Valencia
Navel Oranges
  Bourouhaine
  Thomson
  Washington
Blood Oranges
  Maltaise
  Sanguine
  Sanguinello
  Moscato
Common Grapefruit
  Marsh
Pink Grapefruit
  Red Blush
Common Mandarins
  Bakrai
Limes
  West Indian

Iraq
Common Oranges
  Valencia
Navel Oranges
  Washington
Blood Oranges
  Moro
  Tarocco
Common Grapefruit
  Marsh
Pink Grapefruit
  Red Blush
Common Mandarins
  Clementine
  Indian

Table 13-16. (*Continued*)

Israel
Common Oranges
  Khettmali
  Shamouti
  Valencia
Navel Oranges
  Washington
Common Grapefruit
  Marsh
Lemons
  Eureka

Italy
Common Oranges
  Belladonna
  Biondo
    Commune
  Biondo Riccio
  Calabrese
Blood Oranges
  Moro
  Sanguigno
    Semplice
  Sanguinello
    Commune
  Sanguinello
    Moscato
  Tarocco
  Tarocco
    Liscio
Sour Oranges
  Bergamot
Acidless Oranges
  Vainiglia
Mediterranean Mandarins
  Mediterranean
Lemons
  Femminello
    Ovale
  Interdonato
  Mesero
  Monachello
  Vilafranca
Lemon Hybrids
  Mediterranean
  Sweet Lime

Jamaica
Common Grapefruit
  Marsh

Common Mandarin
  Ortanique
Mandarin Hybrids
  Ugli
  Temple

Japan
Navel Oranges
  Fukumotto
Sour Oranges
  Sanbo
Common Pummelos
  Banpeiyu
  Hirado
  Mato
Grapefruit Hybrids
  Hassaku
  Natsudaidai
Common Mandarins
  Tankan
Satsuma Mandarins
  Aoe Wase
  Dobashi Beni
  Hayashi
    Unshiu
  Iseki Wase
  Ishikawa
    Unshiu
  Juuman
  Kuno Unshiu
  Matsuyama Wase
  Miho Wase
  Mikan
  Miyagawa Wase
  Nagahashi
  Nankan-4
  Okitsu Unshiu
  Okitsu Wase
  Owari Unshiu
  Seto Unshiu
  Silverhill
  Sugiyama
    Unshiu
  Yonezawa
Mandarin Hybrids
  Iyo
  Sudachi
Lemon
  Canton

Java
Pink Pummelos
  Pandan Bener
  Pandan Wangi

Lebanon
Common Oranges
  Khettmali
  Valencia
Navel Oranges
  Washington
Blood Oranges
  Mawadri Beledi
  Shamouti
  Mawadri
Common Mandarins
  Mandalina

Malaysia
King Mandarins
  King
Limes
  West Indian

Mexico
Common Oranges
  Hamlin
  Parson Brown
  Pineapple
  Valencia
  Verna
Navel Oranges
  Washington
Lemons
  Eureka
  Lisbon
Limes
  West Indian

Morocco
Common Oranges
  Cadenera
  Salustiana
  Valencia
  Verna
  Vicieda
Navel Oranges
  Frost
    Washington
  Navelina
  Washington

Table 13-16. (*Continued*)

Morocco (Cont.)
Blood Oranges
  Doblefina
  Doblefina
    Amelioree
  Tarocco
  Liscio
Common Mandarins
  Clementine
  Wilking
Lemons
  Berna
  Lisbon
Lemon Hybrids
  Marrakech
    limonette

New Zealand
Grapefruit Hybrids
  Poorman
  Wheeny
Lemon Hybrids
  Meyer

North Africa
Common Oranges
  Biondo Commune
  Maltaise
    Blonde
  Shamouti
Common Mandarins
  Clementine
Mediterranean Mandarins
  Mediterranean
Lemon Hybrids
  Meyer

Okinawa
King Mandarin
  King

Pakistan
Common Oranges
  Malta
Blood Oranges
  Blood Red
Common Mandarins
  Kinnow
  Pokan

Philippines
Pink Pummelos
  Siamese Pink

Common Mandarins
  Ponkan

Portugal
Common Oranges
  Carvalhal
  Don Jao
Navel Oranges
  Washington

Puerto Rico
Grapefruit Hybrids
  Chironja

Reunion Island
Common Mandarins
  Ortanique

Sri Lanka
Common Mandarins
  Ponkan

South Africa
Common Oranges
  Clanor
  Hamlin
  Joppa
  Maltaise
    Ovale
  Pineapple
  Premier
  Queen
  Shamouti
  Tomango
  Valencia
Navel Oranges
  Oberholzer
  Washington
Common Grapefruit
  Marsh
  Triumph
Common Mandarins
  Emperor
Lemons
  Eureka

Spain
Common Oranges
  Biondo
    Commune
  Cadenera
  Castellana
  Macetera

Salustiana
Valencia
Verna
Vicieda
Navel Oranges
  Navelate
  Navelina
  Thomson
  Washington
Blood Oranges
  Doblefina
  Doblefina
    Amelioree
  Entrefina
  Mutera
  Spanish
    Sanguinelli
  Tarocco Liscio
Sour Oranges
  Bittersweet
  Sevillano
Acidless Oranges
  Sucrena
Common Grapefruit
  Marsh
Common Mandarins
  Clementine
Satsuma Mandarins
  Dobashi Beni
Mediterranean Mandarins
  Mediterranean
Lemons
  Berna
  Mesero

Syria
Navel Oranges
  Washington
Blood Oranges
  Mawadri
    Beledi
  Shamouti
    Mawadri
Common Mandarins
  Mandalina

Taiwan
Common Pummelos
  Banpeiyu

Table 13-16. (*Continued*)

Common Mandarins	Ray Ruby	Common Mandarins
Ponkan	Red Blush	Clementine
Tankan	Rio Red	Lemon Hybrids
King Mandarins	Star Ruby	Mediterranean
King	*Thailand*	Sweet Lime
Texas	Common Pummelos	
Common Oranges	Kao Pan	*Turkey*
Marrs	Kao Phuang	Common Oranges
Common Grapefruit	*Tunisia*	Parson Brown
Marsh	Blood Oranges	Valencia
Pink Grapefruit	Maltaise	Navel Oranges
Henderson	Sanguine	Washington

QUESTIONS

1. What is the most commercially important citrus species?
2. What are the two most important varieties commercially among all species?
3. What are the main differences between common and navel sweet oranges?
4. What is the main pigment in blood oranges, and what effect does processing have on it?
5. Which citrus juice species is believed to have the best flavor? The best color?
6. Which citrus variety makes the best marmalades?
7. Which of each pair generally has the better flavor: (a) seedy or seedless varieties; (b) light or dark pigmented varieties?
8. What is the main pigment found in grapefruit and pummelos, and what is the effect of processing on this pigment?
9. What are the two main lemon varieties?
10. Which citrus species is the most tropical?

REFERENCES

Considine, G. D. 1982. *Foods and Food Production Encyclopedia.* Van Nostrand Reinhold Company, New York, 410–414.

Hashinaga, F. and Itoo, S. 1983. Seasonal changes in limonoids in Hassaku and pummelo fruits, *J. Jap. Soc. Hort. Sci.*, *51(4)*, 485–492.

Hodgson, R. W. 1967. Horticultural varieties of citrus. In *The Citrus Industry I*, W. Rether, H. J. Webber, and L. D. Batchelor, eds. Division of Ag. Sciences, University of California, Berkeley, Calif., 431–580.

Iwata, T. and Ogata, K. 1976. *J. Jap. Soc. Hort. Sci.*, *45(2)*, 187–191.

Levi, A. et al. 1974. The bitter principle in Shamouti orange juice. 1. Seasonal changes and distribution in different parts of the fruit, *Lebensm.-Wiss. Technol.*, *7*, 234–235.

Rangama, S., Govindarajan, V. S., and Raman, K. V. R. 1983. Citrus fruit varieties, chemistry, technology, quality evaluation, and analytical chemistry II (Review), *Chemical Rubber Company Critical Reviews in Food Science and Nutrition*, *18(4)*, 313–386.

Swingle, W. T. and Reece, P. C. 1967. The botany of citrus and its wild relatives. In *The Citrus Industry*. Division of Ag. Sciences, University of California, Berkeley, Calif., publication 4012.

Tanaka, T. 1954. Species problems in citrus (Revisio Aurantiacearium, IX), *Japan Soc. Prom. Sci.*, Ueno, Tokyo, Japan.

UNIT TWO

CITRUS JUICE SANITATION

Chapter 14

Inspections

Several basic factors affect or contribute to the sanitary condition or standing of a citrus processing or bottling plant. One is geographic location. Rural, suburban, and metropolitan areas each have different sanitation problems. Insect and rodent habitats as well as waste disposal facilities exert an influence on plant sanitation. Weather conditions and climate also play a significant role in sanitation problems. Temperatures, humidity, and precipitation affect insect and rodent fauna and activity, which vary throughout the year as well as from one geographical location to another. Also, sanitation methods and precautions followed during daylight hours differ from those used at night when insects and rodents are most active and attracted to lights.

The engineering of the plant is an important factor. Construction, location, and cleaning facilities for equipment can make a dramatic impact on plant sanitation. Leaks and drainage problems can cause unsightly accumulations. Another important consideration in the effectiveness of a sanitation program is the attitude and capability of plant management regarding plant sanitation. The "sanitation quotient" of a plant must start at the top. Even so, the attitudes of employees must be inbred, along with follow-through, in the observance of company policies and procedures, in order for any sanitation program to be successful. The foundation of any sanitation program is a good system of inspections. These inspections ensure sanitation awareness as well as implementation of the existing program.

There are several reasons why sanitation is important. In the citrus industry, product quality assurance is essential. Contaminated or spoiled products generally do not pose health concerns because of the nonpathogenic characteristics of microorganisms in citrus juices, but broken glass or pieces of metal in the product can be dangerous to the consumer. Also, if flavor and appearance are irrevocably marred, serious financial losses are possible. Because constant surveillance of processing operations generally is not feasible, the concerns of a sanitation program must go beyond visible contamination and spoilage of citrus products. The program must eliminate or minimize "potential" contamination and spoilage. Another objective of a sanitation program is to facilitate a healthy and attractive appearance not only in citrus products, but in citrus processing equipment. Customer and public confidence is enhanced by a plant that looks clean. Also, employees work better in a clean and healthy environment, just as athletes compete better in clean and attractive uniforms. The final reason for a sanitation program is simple: it is a legal requirement.

ROLE OF THE FOOD AND DRUG ADMINISTRATION

The United States Food and Drug Administration (FDA) has the responsibility for regulating and ensuring the sanitation and safety of all foods and drugs produced or imported into the United States according to the Federal Food, Drug, and Cosmetic Act enacted on June 25, 1938. This Act has since been modified and incorporated into the *Code of Federal Regulations* (CFR). In this code, Title 21 part 110 refers to sanitation and defines the "good manufacturing practice" that is required of all food processors. Some of the main features of good manufacturing practice as outlined in the code are the following:

1. No person infected by a communicable disease can work in a food plant.
2. Food plant employees must wear clean clothes, wash their hands, remove jewelry or other items that cannot be sanitized as well as loose items that could fall into the food, keep their gloves sanitary, and wear hair restraints.
3. No eating, drinking, or use of any tobacco products is permitted in food plants or where processing utensils or equipment is stored. Personal belongings should be kept out of food processing areas.
4. Food plant grounds should be kept so as not to contaminate the food product (there should be no dusty roads or yards, poor drainage, or harborages for rodents, insects, or other pests).
5. Food plants should be designed and constructed to allow sufficient space for processing and storage, in order to avoid food contamination.
6. In food plants, extraneous material or chemicals should not be stored or used so as to cause contamination.
7. Food plants should provide adequate lighting for handwashing areas, locker rooms, restrooms, and where food ingredients are stored and used, as well as adequate handwashing facilities. Light bulbs should be covered so that breakage will not contaminate the food.
8. Adequate ventilation should be provided to minimize off odors or fumes that could contaminate the food.
9. There should be adequate screening to protect against birds, animals, and other vermin. No animals of any kind should be allowed into a food plant other than those that must be used as raw material.

10. There should be an ample potable water supply for proper cleaning, of sufficient quality that it will not contaminate the food.
11. Sewage and refuse disposal should be done in an adequate manner to avoid food contamination.
12. Doors to restrooms should be self-closing and not open directly into processing areas. Double doors should be used if necessary.
13. All equipment and utensils used in processing should be cleaned sufficiently and often enough that they will not contaminate the food, and they should be stored in a sanitary fashion.
14. All ingredients or raw materials should be inspected to ensure that they are clean, wholesome, and fit for human consumption.
15. The food processing should be monitored in such a way as to ensure the sanitation of the product, by procedures including, but not limited to, container inspections, temperature monitoring, sanitary inspections, and microbial monitoring.
16. Food products should be so labeled as to allow proper product identification, and records concerning product quality should be retained for at least the shelf life of the product, or no more than two years. The addition of any defective lot renders the final product defective, regardless of the degree of the defect.
17. Storage and shipment of food products should be done in such a way as not to contaminate the food.

FDA Inspections

To enforce the federal regulations regarding good manufacturing practice, the FDA conducts periodic inspections of food plants. These inspections generally are unannounced, and can range from general surveillance to investigation of a complaint or a specific problem. FDA inspectors should always be treated with respect and courtesy, as they are ambassadors of the public, the ultimate consumers of citrus products. Such courteous treatment also produces the best working relationship. Three phases constitute a proper and effective FDA inspection.

Preinspection Phase

When an FDA representative appears at the plant, the first thing one should do is arrange for a preinspection conference. Those persons who should attend include the quality control manager, the chief executive officer of the company (CEO), and the FDA inspector. FDA inspectors are required to inspect plants at reasonable times and not during weekends, holidays, or off shifts. Each company should maintain an FDA file. The inspector should be required to show his or her credentials, and these credentials should be recorded in the file along with the inspector's name and the date and time of the inspection. FDA inspectors do not have unlimited authority to access company records except for shipping documents of interstate commerce under section 703 of the Act and processing records of low acid foods. They also are not granted authority to take photographs. It must be remembered that volunteered information can be used against a processor in a court of law. However, the best policy is to always be prepared for an FDA inspection through a proper sanitation program. Then full cooperation with FDA inspectors can be a pleasant experience for all concerned, perhaps even a time to show off a little.

The preinspection conference is a good time for processors to review the results of the last inspection and clarify any questions they or the inspector may have. The FDA file should include information on all cleaners and additives used in the plant as well as all paperwork from previous inspections. Also, the file may include informative pamphlets and brochures such as "FDA Inspection Authority" (NFPA Bulletin 39-L, NFPA, 1133-20th Street, Northwest, Washington, D.C. 20036) or "Good Manufacturing Practice in Manufacturing, Processing, Packing, or Holding Human Food" (U.S. Department of Health, Education, and Welfare, Public Health Service, Food and Drug Administration, HFF-326, 200 C Street, Southwest, Washington, D.C. 20204). Some companies have a form letter that they give to the inspector on arrival that outlines company policy about what the inspector will or will not be allowed to do beyond his or her inherent authority. The preinspection conference presents a good opportunity for those involved to go over these policies.

Inspection Phase

FDA inspectors should always be accompanied while inspecting the plant, preferably by the quality control manager. The names and titles of those persons who accompany the inspector should be recorded in the file. If a USDA inspector is present, he or she may wish to accompany the inspector as well. The FDA inspector has authority to inspect all areas of the plant without warrant and should be allowed to do so. During the inspection the inspector will fill out form FD-482, shown in Fig. 14-1, a copy of which should be kept with the FDA file. If the inspector takes any samples, he or she should issue a receipt for them, which also is kept in the file. It is recommended that the company take duplicate samples. The form used by the FDA inspector as a sample receipt is shown in Fig. 14-2. If any problems are found, the inspector is required to notify the company promptly by using FD-483, shown in Fig. 14-3.

Postinspection Phase

After the inspection, a postinspection conference should be held by the same people who attended the preinspection conference. If any of the above forms

1. DISTRICT ADDRESS & PHONE NO.

415-556-2062

50 U.N. Plaza

S.F., CA 94102

DEPARTMENT OF HEALTH AND HUMAN SERVICES
PUBLIC HEALTH SERVICE
FOOD AND DRUG ADMINISTRATION

TO

2. NAME AND TITLE OF INDIVIDUAL

Dan A. Kimball, Director of Quality Control

3. DATE *12-4-81*

4. FIRM NAME

California Citrus Producers, Inc.

5. HOUR *1:30* a.m. / p.m.

6. NUMBER AND STREET

525 E. Lindmore Ave.

7. CITY AND STATE

Lindsay, CA 93247

8. ZIP CODE *209 562-5169*

Notice of Inspection is hereby given pursuant to Section 704(a) & (e) of the Federal Food, Drug, and Cosmetic Act [21 U.S.C. 374(a) & (e))[1] and/or Part F or G, Title III of the Public Health Service Act [42 U.S.C. 262-264][2]

9. SIGNATURE (Food and Drug Administration Employee(s))	10. TYPE OR PRINT NAME AND TITLE (FDA Employee(s))
Robert J. Anderson	*Robert J. Anderson Investigator*

Applicable portions of Section 704 of the Federal Food, Drug, and Cosmetic Act (21 U.S.C. 374) are quoted below:

[1]Sec. 704. (a) For purposes of enforcement of this Act, officers or employees duly designated by the Secretary, upon presenting appropriate credentials and written notice to the owner, operator, or agent in charge, are authorized (1) to enter, at reasonable times, any factory, warehouse, or establishment in which food, drugs, devices, or cosmetics are manufactured, processed, packed, or held, for introduction into interstate commerce or after such introduction, or to enter any vehicle being used to transport or hold such food, drugs, devices, or cosmetics in interstate commerce; and (2) to inspect, at reasonable times and within reasonable limits and in a reasonable manner, such factory, warehouse, establishment, or vehicle and all pertinent equipment, finished and unfinished materials, containers, and labeling therein. In the case of any factory, warehouse, establishment, or consulting laboratory in which prescription drugs or restricted devices are manufactured, processed, packed, or held, inspection shall extend to all things therein (including records, files, papers, processes, controls, and facilities) bearing on whether prescription drugs or restricted devices which are adulterated or misbranded within the meaning of this Act, or which may not be manufactured, introduced into interstate commerce, or sold, or offered for sale by reason of any provision of this Act, have been or are being manufactured, processed, packed, transported, or held in any such place, or otherwise bearing on violation of this Act. No inspection authorized by the preceding sentence shall extend to financial data, sales data other than shipment data, pricing data, personnel data (other than data as to qualifications of technical and professional personnel performing functions subject to this Act), and research data (other than data, relating to new drugs and antibiotic drugs and devices and, subject to reporting and inspection under regulations lawfully issued pursuant to section 505(i) or (j), section 507(d) or (g), sections 512(l) or (m), section 519, or 520(g)), and data relating to other drugs or devices which in the case of a new drug would be subject to reporting or inspection under lawful regulations issued pursuant to section 505(j)). A separate notice shall be given for each such inspection, but a notice shall not be required for each entry made during the period covered by the inspection. Each such inspection shall be commenced and completed with reasonable promptness.

Sec. 704(e) Every person required under section 519 or 520(g) to maintain records and every person who is in charge or custody of such records shall, upon request of an officer or employee designated by the Secretary, permit such officer or employee at all reasonable times to have access to and to copy and verify, such records.

[2]Applicable sections of Parts F and G of Title III Public Health Service Act [42 U.S.C. 262-264] are quoted below:

Part F - Licensing — Biological Products and Clinical Laboratories and······

Sec. 351(c) "Any officer, agent, or employee of the Department of Health, Education, and Welfare, authorized by the Secretary for the purpose, may during all reasonable hours enter and inspect any establishment for the propagation or manufacture and preparation of any virus, serum, toxin, antitoxin, vaccine, blood, blood component or derivative, allergenic product or other product aforesaid for sale, barter, or exchange in the District of Columbia, or to be sent, carried, or brought from any State or possession into any other State or possession or into any foreign country, or from any foreign country into any State or possession."

Part F — ······Control of Radiation

Sec. 360 A(a) "If the Secretary finds for good cause that the methods, tests, or programs related to electronic product radiation safety in a particular factory, warehouse, or establishment in which electronic products are manufactured or held, may not be adequate or reliable, officers or employees duly designated by the Secretary, upon presenting appropriate credentials and a written notice to the owner, operator, or agent in charge, are thereafter authorized (1) to enter, at reasonable times any area in such factory, warehouse, or establishment in which the manufacturer's tests (or testing programs) required by section 358(h) are carried out, and (2) to inspect, at reasonable times and within reasonable limits and in a reasonable manner, the facilities and procedures within such area which are related to electronic product radiation safety. Each such inspection shall be commenced and completed with reasonable promptness. In addition to other grounds upon which good cause may be found for purposes of this subsection, good cause will be considered to exist in any case where the manufacturer has introduced into commerce any electronic product which does not comply with an applicable standard prescribed under this subpart and with respect to which no exemption from the notification requirements has been granted by the Secretary under section 359(a)(2) or 359(e)."

(b) "Every manufacturer of electronic products shall establish and maintain such records (including testing records), make such reports, and provide such information, as the Secretary may reasonably require to enable him to determine whether such manufacturer has acted or is acting in compliance with this subpart and standards prescribed pursuant to this subpart and shall, upon request of an officer or employee duly designated by the Secretary, permit such officer or employee to inspect appropriate books, papers, records, and documents relevant to determining whether such manufacturer has acted or is acting in compliance with standards prescribed pursuant to section 359(a)."

······

(f) "The Secretary may by regulation (1) require dealers and distributors of electronic products, to which there are applicable standards prescribed under this subpart and the retail prices of which is not less than $50, to furnish manufacturers of such products such information as may be necessary to identify and locate, for purposes of section 359, the first purchasers of such products for purposes other than resale, and (2) require manufacturers to preserve such information. Any regulation establishing a requirement pursuant to clause (1) of the preceding

FORM FDA 482 (11/80)   PREVIOUS EDITION IS OBSOLETE   NOTICE OF INSPECTION

Fig. 14-1. Form FD-482, filled out by the FDA inspector during an FDA inspection.

415-556-2062

DEPARTMENT OF HEALTH, EDUCATION, AND WELFARE PUBLIC HEALTH SERVICE FOOD AND DRUG ADMINISTRATION	1. DISTRICT ADDRESS 50 Ui Ni Plaza SF, CA 94102		
2. NAME AND TITLE OF INDIVIDUAL Dan A. Kimball, Quality Control Manager	3. DATE 12-4-89		4. SAMPLE NUMBER
5. FIRM NAME California Citrus Producers, Inc.	6. FIRM'S DEA NUMBER	7. FDA'S DEA NUMBER	
8. NUMBER AND STREET 525 E. Lindmore Ave	9. CITY AND STATE (Include Zip Code) Lindsay, CA 93247		

10. SAMPLES COLLECTED (Describe fully. List lot, aerial, model numbers and other positive identification)

The following samples were collected by the Food and Drug Administration and receipt is hereby acknowledged pursuant to Section 704(c) of the Federal Food, Drug, and Cosmetic Act [21 U.S.C. 374(c)] and/or Part F, Sub Part 3, Section 356(b) of The Public Health Service Act [42 U.S.C. 263d] and/or 21 Code of Federal Regulations (CFR) 1307.02. Excerpts of these are quoted on the reverse of this form.

2/1qt jars orange juice.

2/1qt jars debittered orange juice

11. SAMPLES WERE	12. AMOUNT RECEIVED FOR SAMPLE	13. SIGNATURE (Person receiving payment for sample)
☐ PURCHASED N/C ☐ BORROWED (To be returned) N/C	☐ CASH ☐ BILLED ☐ VOUCHER	Dan Kimball
14. COLLECTOR'S NAME (Print or Type) Robert J. Anderson	15. COLLECTOR'S TITLE (Print or Type) Investigator	16. COLLECTOR'S SIGNATURE Robert J. Anderson

FORM FD 484 (4/74)   PREVIOUS EDITION MAY BE USED.   **RECEIPT FOR SAMPLES**

Fig. 14-2. FDA sample receipt, used if samples are taken during an FDA inspection.

DEPARTMENT OF HEALTH, EDUCATION, AND WELFARE PUBLIC HEALTH SERVICE FOOD AND DRUG ADMINISTRATION	DISTRICT ADDRESS 50 United Nations Plaza San Francisco Ca 94102	
NAME OF INDIVIDUAL TO WHOM REPORT ISSUED TO: Philip Libo	DATE OF INSPECTION 12/13/76	C. F. NUMBER
TITLE OF INDIVIDUAL General Manager	TYPE ESTABLISHMENT INSPECTED (i.e. bakery, cannery.)	
FIRM NAME California Citrus Producers	NAME OF FIRM, BRANCH OR UNIT INSPECTED	
STREET ADDRESS 525 East Lindmore Ave	STREET ADDRESS OF PREMISES INSPECTED	
CITY AND STATE Lindsay Calif	CITY AND STATE	

DURING AN INSPECTION OF YOUR FIRM (I) (WE) OBSERVED:

(1) Open and unscreened doors

(2) Standing water outside building beside animal feed operations.

SEE REVERSE OF THIS PAGE	EMPLOYEE(S) SIGNATURE Robert J Anderson Connie Brallier	EMPLOYEE(S) TITLE Investigators

FORM FD 483 (7/75) REPLACES FD-2.76 AND PREVIOUS EDITIONS OF FD-483. INSPECTIONAL OBSERVATIONS

Fig. 14-3. Form FD-483, filled out by the FDA inspector during an FDA inspection.

have not been filled out, this is a good time to complete them. Any problems found during the inspection are discussed, and the FDA inspector outlines a plan for correcting the problems. Notes should be taken on these conversations and any decisions made, and included in the FDA file.

After the inspector leaves, plant management should meet to decide how any requested corrections will be made, in accordance with the FDA inspector's instructions including specific assignments.

Imported Products

Whenever citrus products are imported into the United States, both customs and FDA clearances must be obtained before any product can be further processed or sold. Custom inspectors are primarily concerned with proper representation of the product, contraband, disease, and contamination brought in from foreign countries. FDA officials are concerned with adulteration and proper identification of products. Most citrus importers identify the products they import into the United States by using the U.S. federal standards of identity found in Chapter 19. These standards of identity sufficiently describe imported products for identification by the FDA officials, who then issue form FD-702, shown in Fig. 14-4.

USDA INSPECTIONS

Unlike the FDA inspections, inspections by the U.S. Department of Agriculture (USDA) are a voluntary service to food processors in grading various food commodities. For a food to obtain a USDA grade certification, certain requirements must be met, one of them being sanitation. Failure to meet USDA sanitation requirements can, at worst, lead to a loss of this service.

The USDA is paid by user companies for its service, which can be provided on an occasional basis or continuously, using laboratory facilities provided by the user company. Sanitation requirements are always in force when the USDA issues grades. However, if continuous inspection is desired, a special plant sanitation survey must be performed to determine whether a plant's sanitation program qualifies for USDA grading. Thereafter, routine sanitation inspections are performed.

The plant survey covers plant organization, facilities, and operation. Some of the main concerns of this inspection are as follows:

1. Weeds, trash, rubbish, rodent and insect harborages around the plant, and protection from such in food storage areas.
2. Offensive odors, dust, soot, or inadequate drainage.

Form Approved; OMB No. 0910-0016.
Expiration Date: January 31, 1986.

**DEPARTMENT OF
HEALTH AND HUMAN SERVICES
FOOD AND DRUG ADMINISTRATION**

ENTRY DATA TAKEN FROM

ID Advance Notice	Number 439-0010127-0
Manifest	Date 06/14/88
IT Advance Notice	Commercial invoice attached

439-0010127-0 06/14/88
ENTRY NO. AND DATE

BILL OF LADING NO.	PORT OF LADING Veracruz	COUNTRY OR ORIGIN Mexico	PORT OF UNLOADING Hidalgo, Texas		PORT OF ENTRY Hidalgo, Tx
BROKER'S REF NO. I-4638	C.H. BOX NO. 493	VALUE OF ENTRY IN U.S. $ $100300.00	CONTAINER NO.	IMPORTING VESSEL 2 Trucks	ARRIVAL DATE 06/14/88
FOR THE ACCOUNT OF /Consignee (Name, Address, Zip Code) Tropical Beverages Inc 700 Jackson Ave, Apt 134 McAllen, Texas 78501		IMPORTER OF RECORD (Name, Address, Zip Code) Same		MANUFACTURER/SHIPPER (Name, Address, Zip Code) Juguera Veracruzana SA Km 297 Carrt Mexico-Tuxpan Posa Rica, Ver., Mexico	

	BROKER (If not same as above) Jimmy Santos, Inc	LOCATION OF LOT (For FDA examination) Teyas Citrus Exchange	DATE AVAIL
Number of items sampled from this Entry.	Related Sample Numbers LEAD SAMPLE		same
		PHONE NO. 585 4503	

(FOR BROKER'S USE)

	GENERAL DESCRIPTION OF SHIPMENT		
THIS IMPORTATION	QTY.	PACKAGED	ITEMS (Include IND, NDA, FCE, Antibiotic Cert Nos., etc.)
MAY PROCEED Without FDA Examination	216	DRums	Frozen Orange Juice Concentrate

This notice does not preclude action should the merchandise later be found violative.

VALID ONLY IF SIGNED

SIGNED _____ 6-15-88
FDA Representative   Date

IMPORTANT NOTICE — An import shipment must be held intact locally pending further notice from FDA. With the advance notice, or otherwise, Broker or Importer must inform FDA of the following:

1. Earliest date the shipment will be available for sampling.

2. Location in local area of the shipment on that date.

3. Breakdown as shown on the invoice-number and size of units each lot, and $ value each lot.

FORM FDA 702 (6/85)   PREVIOUS EDITION MAY BE USED.   **MAY PROCEED NOTICE**

Fig. 14-4. Form 702, used by the FDA to clear an imported product.

3. Wall construction and cleanliness, as well as the number of floors and floor materials and their cleanliness.
4. Proper screening of all openings in the plant and adequate shielding of light bulbs.
5. Sanitizing systems, including chlorination facilities and proper cleanup facilities.
6. Separate storage areas for cleaning compounds away from food processing areas.
7. Sufficient lighting for proper operations and cleaning.
8. Sufficient ventilation to prevent excessive heat, steam, condensation, vapors, smoke, or fumes.
9. Leaking roofs, pipes, valves, or other equipment, as well as covered tanks, catwalks, and cross belts.
10. Flaking paint or rust.
11. Material used to make equipment and containers, which should be non-absorbent so as not to contaminate the product, and should not come into contact with unsanitary surfaces.
12. Oil and grease from motors that could contaminate the product.
13. Slime and mold buildup on plant facilities or equipment, as well as the placement of loose items such as lunchboxes or personal gear.
14. Quality of the water used in the plant.
15. Restroom doors, which should not open directly into the food processing area and should be self-closing; the proper number of waste receptacles; handwashing signs; handwashing facilities, including soap, hot water, and hand drying facilities.
16. Adequate lighting for color scoring and sufficient facilities, environment, and equipment for inspection.
17. Protection of food containers from dust and dirt.
18. Cleanliness of employees, including hair restraints, as well as employees with infectious diseases, gum chewing, tobacco use, or other unsanitary personal habits or facilities.

During continuous inspections, the USDA inspector will perform periodic sanitation inspections, using the form shown in Fig. 14-5. When infractions occur, one of three levels of seriousness will be assigned to the problem. A minor infraction (MN) constitutes a problem that it is desirable to correct, which must be corrected within 24 hours. A major infraction (MJ) may result in product contamination and must be corrected by the next shift change. A critical infraction (CR) means that the product is being contaminaated, and the infraction must be corrected immediately. Copies of the sanitary score sheet should be given to plant management promptly. If unsatisfactory or unusual conditions exist, copies must be sent to area, regional, and national USDA offices. If an

FORM FV-416-3
(5-76)

U.S. DEPARTMENT OF AGRICULTURE
AGRICULTURAL MARKETING SERVICE

SANITATION SCORE SHEET FOR
CITRUS PROCESSING PLANTS

NAME OF PLANT	LOCATION	DATE	D.I.R. NO.
Sunnyland	Citrusville	10/5/90	31

RATING SYMBOLS (✓) Satisfactory
MN – Minor   MJ – Major
CR – Critical   U – Unsatisfactory

SIGNATURE OF INSPECTOR(S) *John Doe*

SANITATION DEFICIENCIES -- SHOW ITEM, NO. AND DESCRIBE

		0800	1700	2300	DEFICIENCIES
A	**PREMISES**				
	1. Outside areas	✓	✓	MN	① Doors left open – Notified QC mgr
	2. Waste disposal	MN	✓	✓	② peel trash on ground
	3.				
B	**RECEIVING AREA**				
	1. Unloading pit	✓	✓	✓	
	2. Grading	✓	✓	✓	
	3. Conveyors and chutes	✓	✓	✓	
	4. Bins	MJ	✓	✓	④ flies swarming bins – told foreman
	5.				
	6.				
	7.				
C	**PROCESSING AREAS**				
	1. Washing and grading	✓	MN	✓	① chlorinator not working
	2. Sizers and leads	✓	✓	✓	
	3. Extractors	✓	✓	✓	
	4. Troughs and lines	✓	✓	✓	
	5. Finishers	✓	✓	✓	
	6. Valves and lines	✓	✓	✓	
	7. Blend and storage tanks	✓	✓	✓	
	8. Disposal screw conveyors	✓	✓	✓	
	9. Fillers & closing machines	✓	✓	✓	
	10. Stabilizer	✓	✓	✓	
	11. Floors, gutters & walls	✓	✓	✓	
	12. Waste storage	✓	✓	✓	
	13.				
	14.				
	15.				
	16.				
D	**WAREHOUSE OR COLD STORAGE**				
	1. General housekeeping	✓	✓	✓	
	2. Sugar storage	✓	✓	✓	
	3. Can storage	✓	✓	✓	
	4.				
	5.				
	6.				
E	**REST ROOMS**				
	1. Supplies	✓	MN	✓	① No soap for washing hands
	2. Wash basins	✓	✓	✓	
	3. Toilets & urinals	✓	✓	✓	
	4. Floors & walls	✓	✓	✓	
	5.				
F	**PERSONNEL**				
	1. Cleanliness	✓	✓	✓	
	2. Head covering	✓	✓	✓	
	3. Smoking	✓	✓	✓	
	4. Clean hands	✓	✓	✓	
	5.				

Fig. 14-5.   Inspection form used by USDA inspectors during continuous inspection.

unsatisfactory condition persists for three successive days or three days in one week, the inspector is required to contact his or her supervisor and/or other USDA officials as needed. Again, failure to resolve sanitation problems can result in the withdrawal of grading services.

IN-HOUSE INSPECTIONS

Some good guidelines for in-house inspections can be found in the FDA and USDA inspections. Performing FDA-like and USDA-like inspections is the best way to prepare for such inspections. A good pamphlet that can be used as an additional guide is "Do Your Own Establishment Inspection—A Guide to Self Inspection for the Smaller Food Processor and Warehouse" (U.S. Department of Health and Human Services, Public Health Service, Food and Drug Administration, 200 C Street, Southwest, Washington, D.C. 20204).

Even though the legal requirements of sanitation must be kept in mind, sanitation programs in citrus processing plants should be based on product quality. If proper care is taken to ensure the quality of the products in-house, no serious problems should ever occur with FDA or USDA inspections. A loss of reputation and profits is usually much more devastating than conflicts with regulatory agencies. In-house personnel also are more knowledgeable about equipment and procedures and can implement a more effective sanitation program than outside inspectors. This knowledge of products and equipment should be constantly used in designing and redesigning sanitation programs so that production time and cleaning materials will not be wasted during the production of high-quality products.

There are two types of in-house inspections. One is based on solving a specific problem that has occurred. For example, the presence of a large number of *Drosophila* fruit flies may require the periodic inspection of peel residue accumulations in certain areas of the plant. Once these accumulations are being removed on a regular basis, or when the weather gets cooler and the flies become less active, the procedure can be modified or eliminated. If a particular procedure or inspection system does not solve the problem, new procedures should be tried until the problem is solved. At least on a weekly basis, the entire plant should be shut down for a complete cleanup, and an FDA- or USDA-like inspection should be performed. It is important that those individuals doing the inspecting are not those doing the cleaning. Otherwise, shortcuts will be taken in the inspection. Also, management should always be aware of and supportive of monitoring the sanitation condition of the plant, especially in construction and reconstruction within the plant. Quality control personnel should always be consulted in such modifications.

Inspections form just the foundation for a good sanitation program. The subsequent chapters go into greater detail about the many facets of citrus sanitation.

QUESTIONS

1. What are some of the factors that affect sanitation in processing or bottling plants?
2. What is the foundation of any sanitation program?
3. What are some of the reasons why sanitation is important?
4. What is the law regarding chewing tobacco in processing areas?
5. What is the law regarding restroom doors that open directly into processing areas?
6. Who should attend the preinspection conference when an FDA inspector appears at the plant?
7. What forms will an FDA inspector fill out, and why?
8. What records does an FDA inspector have the authority to access?
9. Can a processor go to jail for failing a USDA inspection?
10. Why is a survey taken prior to USDA continuous inspection?
11. Give examples of minor, major, and critical USDA sanitation infractions.
12. What should be the basis of in-house sanitation programs?
13. What are the two types of basic in-house inspections?
14. How should FDA and USDA inspectors be treated, and why?
15. When is it okay to leave an FDA inspector alone in the plant?

Chapter 15

Citrus Microbiology

One of the major means of contamination of citrus products is by the growth of single-celled microogranisms. Some microorganisms are pathogenic (disease-causing) in humans and are of primary concern to food processors. Other microorganisms, although not pathogenic, destroy the quality of foods and make them unpalatable or undesirable. Microorganisms found in citrus products fall into the latter category. The high acid content of citrus juices prevents the growth of pathogenic microbes while allowing certain acid-tolerant bacteria, yeasts, and molds to flourish. The wastes generated by this biological activity produces undesirable flavors and odors that degrade the quality of citrus products. An understanding of the factors that affect the growth of these microorganisms enables the development of procedures to prevent it. Some of these factors are light, temperature, air availability, pH, moisture, and osmotic pressure.

BACTERIA

Single-cell organisms are divided into two types according to cellular structure. One type is the eucaryotes ("eu" means true; "karyo" is the combining form for nucleus), or microorganisms with a true nucleus. The other category is the procaryotes, which are microbes that lack a true nucleus and usually have their DNA in a single molecule. Procaryotes include bacteria and bluegreen algae, whereas eucaryotes contain all other algae, molds, yeasts, and protozoans. There are three types of bacteria, which differ in their shape or morphology. Bacillus bacteria are elongated or rod-shaped, coccus bacteria are spherical, and spirillus bacteria are spiral or corkscrew-shaped. Bacteria are classified by many parameters, including growth conditions, size, and wastes produced. Even though other acid-producing bacteria have been isolated from citrus juices from the genera of *Aerobacter* and *Xanthomonas* (Faville and Hill 1951a), only lactic acid bacteria from the genera *Lactobacillus* and *Leuconostoc* have been found to significantly affect citrus products.

Lactobacillus

Lactobacillus bacteria (Buchanan and Gibbons 1974) are gram positive rod-shaped bacteria from the Lactobacillaceae family, which derives its name from the fact that at least half of the carbon produced from carbohydrate metabolism is in the form of lactate. Other products from *Lactobacillus* growth are diacetyl, acetate, formate, succinate, carbon dioxide, ethanol, and other one- or two-carbon acids. Diacetyl (2,3-butanedione), which is synthesized by the bacteria from citric acid, imparts an undesirable buttermilk-like flavor and odor to citrus juices and is the major cause of product rejection of single-strength juices due to spoilage.

These nonpathogenic organisms grow at temperatures of 5 to 53°C (41–127°F) with optimal temperatures generally 30 to 40°C (86–104°F). They are very acid-tolerant, with an optimal pH range of usually 5.5 to 5.8 with significant growth below this pH range. *Lactobacillus* bacteria also are very sensitive to osmotic pressure or juice concentration, with slow growth at 35 to 38°Brix and no growth over 45°Brix. The main species found in citrus juices are *Lactobacillus plantarum* and *Lactobacillus brevis* (Hays and Riester 1952). *L. planarum* has an optimal temperature range of 30 to 35°C (86–95°F), with growth at 15°C (59°F) but generally not at 45°C (113°F). *L. brevis* has an optimal growth temperature of about 30°C (86°F), with growth at 15°C (59°F) but no growth at 45°C (113°F). Some strains are pigmeted from orange to red, but most are white in color.

Leuconostoc

Leuconostoc (Buchanan and Gibbons 1974) are gram positive spherical coccoids, usually in pairs or in chains. *Leuconostoc* is a nonpathogenic member of the Streptococcaceae family, which normally produces lactic acid, ethanol, carbon dioxide, and diacetyl with the same buttermilk-like off flavors and odors as produced with *Lactobacillus* bacteria. Optimal temperatures for growth are in the range of 20 to 30°C (68–86°F), which is cooler than the range for *Lactobacillus* bacteria. Most species will grow in media with a pH of 5.5 to 6.5. One species, *Leuconostoc oenos*, will grow in the range of 4.2 to 4.8 and even lower but probably not higher. Generally *Leuconostoc* bacteria grow very slowly at pH levels of around 3.6 compared to *Lactobacillus*, but they grow faster at a pH of 4.0 and above. Most species are facultative anaerobes, which means they can grow in the presence or absence of air. The main species found in citrus are *Leuconostoc mesenteroids* and *Leuconostoc dextranicum* (Hays and Riester 1952). *L. mesenteroids* grows in the 10 to 37°C (50–99°F) range, with an optimal range of 20 to 30°C (68–86°F), but these bacteria cannot withstand

55°C (131°F) for 30 minutes. Some strains, however, may withstand temperatures of 80 to 85°C (176–185°F) for shorter periods of time. *L. mesenteroids* forms a slime of dextran metabolized from sucrose and/or levano polymers from fructose. *L. dextranicum* also produces slimy dextran growth, though not as actively as *L. mesenteroids*. The growth temperature range and the optimal range for *L. dextranicum* are the same as for *L. mesenteroids*.

Bacillus subtilis and *Bacillus pumilus*

A problem that is sometimes encountered in plating is the formation of large colonies that swell with gas formation and then burst, forming a colony that resembles a volcano or has a wrinkled texture. These aerobic colonies spread rapidly on the plate and can obliterate the other colonies, rendering the counting of colonies impossible. The microorganisms responsible for this behavior are *Bacillus subtilis* and/or *Bacillus pumilus*, both aerobic spore formers (York 1988). These bacteria sometimes are found on the surface of citrus fruit and get into the juice during extraction. The pH range for their growth has not been reported. They may be thermophilic or grow at high temperatures, and they can survive the heat treatment of pasteurization or evaporation of citrus juices. The formation of these bacteria on agar plates is enhanced by the presence of moisture. That is why plates are inverted during incubation—inversion keeps the moisture away from the agar surface. These organisms are not pathogenic but can make the counting of colonies on a plate very difficult if not impossible.

YEASTS

Yeasts belong to the eucaryote (true nucleus) category of microorganisms. Taxonomically, yeasts belong to the class Ascomycetes (sac fungi) and to the fungi subclass of the phylum Thallophyta, or plants that have no true roots, stems, or leaves. Yeasts usually can be differentiated under a microscope by the fact that most of them reproduce by budding, as shown below.

Yeast cells are generally larger than bacteria, as illustrated in Fig. 15-1 by bacteria and yeasts that have been isolated from citrus juices. Generally yeasts grow much more slowly than bacteria, with optimum temperatures for growth of 20 to 30°C (68–86°F), the same as for *Leuconostoc* bacteria. Yeasts generally are more tolerant of high temperatures (65–70°C or 149–158°F) than

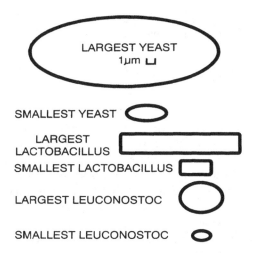

Fig. 15-1. Relative size ranges for bacteria and yeasts that have been isolated from citrus juices.

bacteria or molds. They also have a much higher tolerance for preservatives, cold temperatures, and osmotic pressures than bacteria. In fact, yeasts are sometimes referred to as osmiophilic or osmiotolerant. Within a microorganism there is a delicate balance of aqueous life-giving solutions. If the cell wall comes into contact with a medium that is more concentrated in soluble material than the cell is, the water from these life-giving solutions will migrate out of the cell into the more concentrated solution in an effort to equalize the two concentrations. This osmotic effect is generally lethal to bacteria and is one reason for the use of concentrated syrups in the canning of fruits. However, yeasts are more resistant to osmotic pressures than bacteria and can easily survive in citrus concentrates of 58 to 65°Brix. Yeasts have been shown to remain more viable at 50°Brix than at 12 to 30°Brix after heating at 60°C (140°F) for up to 8 minutes (Juven, Kanner, and Weisslovicz 1978), so they are of primary concern in the microbiology of citrus concentrates. If yeasts are predominant in the juice flora, the juice most likely came from reconstituted concentrate. If bacteria dominate the flora, the juice most likely came from single-strength juice or was contaminated in a single-strength form. Yeasts also grow in media with a pH as low as 1.5, well within the acidity range that is common to citrus juices.

It is interesting to observe that even though yeast can be found in nearly every concentrate that has not been aseptically treated, fermentation occurs inconsistently in lots of citrus concentrates. This phenomenon may be partially due to the number of viable microbes prior to spoilage, but that does not account for such fermentation in samples with lower microbial counts. In the fermentation process, there is oxidation by a microbe in the absence of an electron receptor such as oxygen. This results in the production of more oxidized cabron dioxide gas and more reduced ethanol, as follows:

$$C_6H_{12}O_6 \rightarrow 2CH_2CH_2OH + 2CO_2(g) \qquad \Delta G° = 57 \text{ kilocalories/mole}$$

glucose　　　　　　ethanol

$$(15-1)$$

Not all of the 57 kilocalories thus produced is released as heat. Some of this energy is stored in high-energy phosphate bonds in ATP within the microbe.

One reason for the inconsistent fermentation in citrus concentrates is that not all species that have been found in citrus juices are capable of fermentation. As seen in Table 15-1, the six most commonly found species in one study will not ferment at all (*Rhodotorula* and *Cryptococcus* species). The next eight species ferment only glucose, which comprises only one-fourth of the carbohydrates in citrus juices. The next three species (*Trichosporan*) will not ferment at all, whereas two others that were among the least common yeasts in this study are capable of fermenting both glucose and sucrose. (*Note*: The eight species at the bottom of the table were isolated from citrus concentrates in separate studies; thus their percent of incidence in relation to the others is unknown.) It is evident that the species present may significantly affect whether or not fermentation will occur. A more recent study documents that the most prominent yeast in pasteurized citrus juices is probably *Sacchromyces cerevisiae*, followed by *Rhodotorula* species (commonly called pink yeast) and *Zygosaccharomyces* species (an osmophile) (Parish 1988). *Rhodotorula* species are not thought to be spoilage yeasts and may indicate post-pasteurization contamination.

Mold

Most of the molds of concern to the food industry belong to the same class as yeasts—Ascomycetes (sac fungi). The major contributing factor to mold growth is the availability of moisture. Molds have been known to grow in any location where moisture is found, ranging from air with a relative humidity of 70% to water reservoirs. In addition to moisture, air availability is important. Also, molds generally require an environment free from agitation or disturbance. Idle concentrate tanks exposed to mold spore contamination from the air have been known to accumulate mold growth if not cleaned out or flushed out with fresh concentrate. Mold colonies found in standard plate counts generally indicate that there has been undisturbed concentrate exposed to the air for a period of time up to several days, or that the plate has been exposed in the air. An upper temperature limit for mold growth would be about 60°C (140°F), with a pH tolerance down to as low as zero.

Molds that get into citrus juices from the surfaces of fruit generally are ho-

Table 15-1. Yeast isolated from commercial citrus concentrates (Kobatake, Kurata, and Komagata 1978).

Genus	Species	% of Isolation Out of 24 Samples
Rhodotorula	glutinis	83.3
	minuta	
	rubra	
Cryptococcus	albidus	37.5
	infirm-miniatus	
	laurentii	
Candida	lambica*	33.3
	diddensii*	
	freyschussii*	
	intermedia*	
	parapsilosis*	
	sake*	
	santamariae*	
	valida*	
Trichosporan	cutaneum	25.0
	pullulans	
	variabile	
Debarymoyces	hansenii**	12.5
Kloeckela	apiculata*	4.2
Pichia	etchellsii*	4.2
Hansenula	anomala**	4.2
Cephaloascus	fragrans	4.2
Candida***	magnoliae**	—
Saccharomyces***	chevalieri**	—
Candida****	maltosa**	—
Hanseniaspora****	guilliermondii*	—
Pichiu****	membranaefaciens*	—
Saccharomyces****	cerevisiae*	—
Schwanniomyces****	occidentalis**	—
Torulaspora****	delbrueckii*	—

*Known to ferment glucose only.
**Known to ferment glucose and sucrose.
***Juven, Kanner, and Weisslowicz 1978.
****Found in unpasteurized orange juices (Parish and Higgins 1989).

mogenized and destroyed by the juice acidity, and usually are not likely to affect the quality of the juice, except perhaps for carton juice. Freshly squeezed juice obtained from the processing of moldy fruit may also contain high levels of mold, most of which can be destroyed through proper pasteurization. Black mold growing on the surfaces of moist processing equipment or buildings may break off and create an unsightly cosmetic contamination. However, molds found in citrus juices have not been known to be pathogenic or produce off flavor or odors in citrus products under normal conditions. Some of the molds that have been isolated from citrus juices include, in order of prevalence, *Aureobasidium pullulans*, *Aspergillus niger*, *Botrytis* species, *Phoma* species, *Mucor* species, *Aspergillus fumigatus*, *Cladosporium sphaerospermum*, and *Penicillium* species (Kobatake, Kurata, and Komagata 1978). A more recent study found the following in commercial pasteurized orange juices: *Aureobasidium pullulans* (a black yeast), *Penicillium* species (black yeasts), and *Cladosporium* species. Also the following were found in unpasteurized laboratory samples: *Fusarium* species, *Geotrichum* species, and *Penicillium* species (Parish and Higgins 1989).

DETERMINATIVE TECHNIQUES

Determinations of the number and type of microorganisms are important in any food processing operation. In the citrus industry, several methods are being used, the most rapid one being the direct microscope count. The major disadvantage of this method is that it does not differentiate between living and dead microorganisms. Periodic microscope counts have been used to estimate significant changes in the number of microbes. One must be careful to take a representative sample and perform a uniform count.

Diacetyl

Another method sometimes used to estimate microbial growth is the measurement of diacetyl or 2,3-butanedione levels. As mentioned earlier, diacetyl is produced by lactic acid bacteria, which impart a buttermilk flavor and odor to single-strength juices. Diacetyl reacts with creatine in the presence of naphthol under alkaline conditions to produce a dark brownish-purple dye:

$$CH_3CHCHCH_3 + HN=C \xrightarrow{40\% \text{ KOH}} Dye \quad (15\text{-}2)$$

Diacetyl   +   Creatine

Acetylmethylcarbinol (3-hydroxy, 2-butanone) also is a fermentation product, as well as a derivative of diacetyl in the presence of strong alkali and oxygen. However, it is less volatile than diacetyl and is distilled off at a more constant rate throughout the distillation in the procedure given below. Acetylmethylcarbinol also reacts with the creatine to give the same color as produced with diacetyl; so a correction must be applied to the diacetyl distillate to account for the acetylmethylcarbinol. Reconsituted concentrates result in much less dye formation than freshly extracted juices, because of volatile losses during evaporation.

Diacetyl Test (Hill, Wenzel, and Barreto 1954)

Equipment and Supplies

- Spectrophotometer set at 530 nm.
- Distillation setup shown in Fig. 5-1, as used in the Scott oil determination.
- 1-Naphthol (or alpha naphthol) (5 g in 100 ml of 99% isopropyl alcohol).
- Creatine solution (100 g KOH or 71.3 g NaOH in about 150 ml distilled water in 250 ml volumetric flask. Cool the solution and add 0.75 g creatine, and fill the flask to the mark with distilled water.)
- Timer.
- Three 25 ml graduated cylinders.
- 10 ml pipette.
- 5 ml pipette with bulb or equivalent.
- 2 ml pipette with bulb or equivalent.
- 50 ml beakers or flasks for samples and blank.
- 300 ml graduated cylinder.
- Diacetyl standards in distilled water at 0.5, 1, 2, 3, 5, 7, and 10 ppm.

Procedure

1. Standardize the Brix of the sample of 11.8°Brix.
2. Add 300 ml of the sample to the distillation apparatus shown in Fig. 5-1 (without the alcohol shown in the figure), and distill off three 25 ml portions, discarding the second portion. The first portion contains most of the diacetyl, while the third contains approximately the same amount of acetylmethylcarbinol as in the first portion because it distills off at a steady rate.
3. Pipette 10 ml of the first portion and 10 ml of the third portion into separate 50 ml beakers, and 10 ml of distilled water into a third beaker to serve as a blank.
4. Pipette 5 ml of the naphthol solution into each beaker, using a bulb or pipettor.
5. Pipette 2 ml of the creatine solution into each beaker at 2-minute intervals according to the timer.

6. After 5 minutes of contact time with the creatine solution, measure the % light transmission or absorbance, using a spectrophotometer set at 530 nm (Abs = log (100 / %Trans).

7. Subtract the absorbance of the third 25 ml portion from that of the first, which contains most of the diacetyl. This step corrects for the acetyl-methylcarbinol in the first diacetyl portion. Compare the corrected absorbance with the absorbance of diacetyl standards.

Diacetyl values can be plotted on a control chart, and when those values increase dramatically, a cleanup should be performed immediately. A range of 0.6 to 6 ppm has been suggested for the maximum diacetyl level, depending on the particular operation and desired specifications.

Standard Plate Counts

Standard plate counts comprise the primary method used in monitoring microbial growth in citrus products. Plate counts account only for viable organisms but take 2 to 3 days to get results. The plate count is performed by mixing a juice sample with an agar medium and incubating the resulting gel. Each organism in the sample reproduces to form a visible colony that can be easily counted. The location in the plate, shape, color, and texture of the colonies generally can be used to identify the type of organism that created the colony. *Lactobacillus* organisms form small white to red circular colonies about the size of a pinhead and can be embedded throughout the agar, not on the surface. *Leuconostoc* colonies are large domelike circular colonies, appearing only on the surface of the agar, with a smooth translucent color and texture. One species of *Leuconostoc* produces colonies that appear as a thin film on the bottom of the medium under anaerobic conditions. Yeast colonies generally are recognized by their star shape, formed by two flat circular colonies intersecting at right angles. These colonies are generally between *Lactobacillus* and *Leuconostoc* in size and grow throughout the medium. Surface yeast colonies may resemble *Leuconostoc* colonies but usually are flatter on the surface. Mold colonies are always on the surface and consist of a very fibrous or cottonlike material. Some mold colonies may result from air contamination rather than the sample. A blank plate containing no juice sample aids in the monitoring of sterilization procedures as well as air contamination.

Plate counts usually are expressed as the number of colony forming units (cfu) per milliliters of single-strength juice (11.8°Brix). This allows the comparison of juice products that may vary in concentration. When the number of colonies per plate exceeds 300, a dilution should be performed until the number of colonies per plate is less than 300. Trying to count over 300 colonies may result in counting errors, and too many colonies on a plate may result in the blending together of colonies that are difficult to differentiate. Using a standard dilution, over 300 colonies are usually considered "too many to count" (TMTC) or "too numerous to count" (TNTC).

There is a wide divergence of opinion about what constitutes microbial growth limits or specifications for a contaminated product. The detection of off flavors or fermentative gas production is a clear indicator that microbial growth has gone too far. Maximum microbial levels accepted in the industry range from 500 to 100,000 cfu/ml of single-strength juice (SSJ). Some commercial specifications allow less than 1000 to 5000 bacteria/ml SSJ while limiting yeast and mold levels to less than 10/ml SSJ. Such specifications may be appropriate for pasteurized or "not from concentrate" juices, as bacteria are the main flora in such products. However, for concentrates or products made from concentrates, yeasts are the dominating flora. For such products a more appropriate specification would be the reverse of the above: 1000 to 5000 yeasts/ml SSJ and less than a 10 mold count and bacteria/ml SSJ. Manufacturers of fresh squeezed, lightly pasteurized, or pasteurized juices or those who still use the archaic method of adding cutback juice to concentrates to make FCOJ risk microbial contamination in order to provide a better-tasting product. The shelf life of such products usually is very short, and appropriate limits may be in the neighborhood of 30,000 cfu/ml SSF.

Most processing plants that manufacture citrus concentrates, regardless of the equipment they use or the cleaning procedures employed, should be able to keep microbial plate counts below 5000 cfu/ml SSJ. Higher counts than this generally indicate that a problem exists that, if not corrected, can lead to much higher plate counts and eventual spoilage. It generally is not known what the minimum plate count is that will produce off flavors and odors. Quality deterioration depends again on the temperature, viscosity, and type of microorganism present. However, with citrus concentrates, off flavors and odors probably will not develop until the total viable plate counts reach the tens of thousands or higher.

Standard Plate Count Method

Equipment and Supplies

- Incubator set at about 25°C (80°F).
- Autoclave sterilizer (121–132°C or 250–270°F).
- Sterile 1 ml wide-mouth pipettes (disposable or sterilized glass pipettes).
- Sterile 99 ml dilution bottles containing excess sterile distilled water.
- Sterile 10 ml screw cap test tubes for sample collection (optional if samples are to be taken from retail containers).

- Sterile orange serum agar. (Dissolve 45.5 g of dehydrated orange serum agar in 1 liter of cold water, heat the solution to 28°C (82°F), pour 20 ml in each screw cap test tube of appropriate size, and sterilize in the autoclave. Orange serum agar also can be made by mixing 200 ml of orange juice serum, filtered at 200°F with about 6 g filter aid, with 10 g Tryptone and 3 g yeast extract, 5 g dextrose, 2.5 g K_2HPO_4, 20 g Bacto-agar, and 800 ml water.)
- Sterile lemon serum agar for plate counts of lemon juices. (Mix 50 ml of lemon juice serum, filtered at 200°F with about 6 g filter aid, with 10 g Tryptone, 3 g yeast extract, 10 g dextrose, 4 g K_2HPO_4, 20 g Bacto-agar, and 950 ml water.)
- Sterile petri dishes.
- Quebec Colony Counter.
- Mechanical hand counter.
- Melting bath with thermometer.

Sterilization

1. Dry glassware can be sterilized in ovens at 132°C (250°F) for several hours. Glass pipettes can be sterilized in canisters, a practice that allows them to remain in a sterile environment until use.
2. Presterilized pipettes and petri dishes can be purchased. Care should be taken to avoid contamination in partially used packages.
3. Dilution bottles containing dilution water and test tubes containing orange or lemon serum agar or for use to take samples can be sterilized in the autoclave at 15 lb pressure for 15 minutes. It is important that the caps are loosened during sterilization to avoid pressure breakage of the glassware. Also, gradual cooling is necessary after sterilization to avoid boiling of the liquids in the glassware, causing spillage and desterilization.
4. Before use, you should adjust the volume of water in the 99 ml dilution bottles to exactly 99 ml taking care to keep the openings covered as much as possible to avoid air contamination.
5. Plating should be done in an area that is free of extraneous drafts or other contamination.

Procedure

1. Samples can be collected from bulk containers and tanks in sterile 10 ml screw cap test tubes or directly from retail containers. Samples should be used as is, without dilution, to avoid contamination. A factor can be applied to give all results on an 11.8°Brix basis.
2. One milliliter of the sample is pipetted into the 99 ml dilution bottle, with the bottle opening kept covered as much as possible. It is important that

the 99 ml dilution bottles contain exactly 99 ml of sterile water. A common practice is to sterilize the bottles with slightly less than 99 ml water. A separate bottle containing sterile water then can be used to adjust the water level in these bottles. You should be careful not to contaminate the water. Do not flame the neck of the bottle as it draws nonsterile air into the bottle. For concentrate samples, rinse the concentrate from the inside of the pipette using the sterile water in the bottle, by repetitively withdrawing and expelling the water, into and out of the pipette.

3. All equipment and solutions should be at room temperature to avoid killing the flora in the sample.

4. The dilution bottles containing the sample should be sealed and shaken well until fully mixed. Using another sterile 1 ml pipette, pipette 1 ml of the dilution into a petri dish, keeping the dish covered as much as possible.

5. The presterilized agar test tubes, containing about 20 ml of agar, should be melted in a hot water bath and cooled to at least 45°C (113°F) before the agar is poured into the petri dish, with care again taken to keep the dish covered as much as possible.

6. The dish must be swirled gently so that the agar and sample are mixed and just cover the bottom of the dish. Allow the dish to cool until the agar sets up or gels.

7. The petri dish is labeled and placed in the incubator (25°C or 75°F) in an inverted (gel side up) position in the dark for 48 hours. Lights in the incubator during incubation will inhibit growth.

8. After incubation, the top cover of the dish is removed and the bottom part of the dish placed on the Quebec Colony Counter. The colonies are counted by using the mechanical hand counter or its equivalent. Separate counts for *Lactobacillus*, *Leuconostoc*, yeast, and mold can be done if desired. The number of colonies per 1 ml of 11.8°Brix equivalent can be calculated from:

$$C = \frac{PD(1.029 \text{ lb soluble solids/gal } @11.8°\text{Brix})}{S} \qquad (15\text{-}3)$$

where C is the count/ml SSJ or the number of colony forming units (cfu), P is the number of colonies counted on the plate, D is the dilution or 100 in the procedure above, and S is the pounds soluble solids/gallon according to the Brix of the original sample, using Equation 2-8 or 2-10. If greater dilution is required in order to keep the number of colonies counted below 300, then 1 ml from the first dilution bottle can be added to a second 99 ml dilution bottle to give a dilution of 10,000 instead of 100. This value then can be used for D above.

Automated Methods

Recently developed computerized methods of rapid microbial analysis, involving measurement of the change of conductance of samples with microbial activity, are being used in some parts of the industry. These methods provide rapid results, sometimes in hours, as well as the ability to measure many samples at once. The high cost of the equipment can be justified only if numerous microbial determinations are required. For this reason generally only larger processing plants with special applications, such as aseptic processing, are employing these methods. These automated systems are the closest thing available to timely and accurate microbial determinations. Standard plate counts often are used to calibrate and check results from these methods.

PROCESSING CONSIDERATIONS

Fruit

The peel of citrus fruit is a natural barrier to most microbial contamination. Surface molds eventually will cause fruit to decompose, but once a fruit is broken, microbial decay can occur rapidly. For this reason, under USDA inspection, only 10% or less of the fruit leaving the grading table is allowed to be broken, and only 2% or less can show visible evidence of decay (off color or spongy texture). Also, inbound fruit traditionally undergoes some form of trash elimination and washing prior to grading. In California, where most of the fruit has undergone washing and waxing in a packing house prior to its arrival at the processing plant, fruit washing is done more to enhance the conveyance of the fruit than to clean it. Such washing removes waxes applied at packing houses, which cause the fruit to stick to belts and other conveyance equipment. Surface microbial counts of the fruit generally cannot be correlated to the plate counts of the resulting juice (Faville and Hill 1951b). One study involving over 300 commercial lots of orange juice gave a pure random correlation between the percent broken and decayed fruit and the final plate counts of the juice (Kimball 1982).

Because most juices undergo pasteurization in modern evaporators or pasteurizers, the microbial flora of freshly extracted juice rarely correlates with the flora in the processed juice. However, overly decayed fruit can easily produce a juice with off flavors and odors developed prior to processing. This becomes a greater problem in areas dominated by fresh fruit markets, where processing plants are considered to have top priority for getting rid of fruit regardless of its condition. Many processors, however, set standards regarding the degree of broken and decayed fruit that is permissible before fruit is processed.

Water containing 15 to 20 ppm chlorine is often used for fruit rinsing or for flushing of equipment that may be vulnerable to microbial growth or accumulation. Chlorine levels are easily determined by using locally available swimming pool test kits, which involve the yellow color development of chlorine with o-tolidine. Dilution of water samples may be required in order to keep within the chlorine concentration range on the color comparison chart included with such a kit. Fruit grading, washing, and the use of chlorinated water is generally more important in warmer humid climates or seasons. It should be remembered that fruit that has undergone spoilage prior to chlorinated washings may still impart off flavors and odors to the resultant juice.

Pasteurization

Heat treatment is the most common method of reducing microbial activity in foods. Bacteria, the fastest-growing microorganisms in freshly extracted citrus juices, undergo cell division about once very 30 minutes, depending on the temperature and growing conditions. This means that in 6 hours over 4000 microbes can be produced from one bacterium. Unpasteurized juice can easily contain thousands or tens of thousands of bacteria per milliliter, which can result in significant spoilage if the juice is held for several hours without pasteurization. Pasteurization temperatures as low as 65.6°C (150°F) have been shown to be adequate in the control of microbial growth (Berry and Veldhuis 1977). However, temperatures of at least 91°C (195°F) are required to deactivate pectinase enzymes, which can cause cloud loss and/or gelation. Pasteurization time can be as long as 40 seconds, but modern evaporators elevate juice temperature to pasteurization levels for only about 10 to 15 seconds. Aseptic systems sterilize the juice with similar pasteurizers, and use hydrogen peroxide to sterilize juice containers prior to filling. Aseptic packaging for citrus juices is not as popular as with other juices because of the higher amino acid content of citrus juices, which results in the development of browning and off flavors. The main objective of aseptic packaging is to minimize or eliminate the need for refrigeration. Even though aseptically treated juices have no microbial activity, refrigeration still is required for citrus juices in order to prevent the development of off flavors and colors from the oxidation of amino acids, ascorbic acid, and sugars.

Cooling Effects

Immediate chilling of pasteurized juices or concentrates not only inhibits the growth of surviving microbes, but reduces the rate of oxidative reactions as well. Single-strength juices should be kept close to 0°C (32°F), and concentrates should be packaged and stored at −4°C (25°F) or lower. Concentrations stored at these temperatures consistently show a reduction of microbial flora with time. Concentrates with plate counts in the tens of thousands can experience a reduction of contamination with prolonged storage in freezers, usually in a matter of months. This is due not only to cold temperatures but also to the osmotic effects mentioned earlier. Single-strength juices usually have a limited shelf life (weeks) regardless of their storage temperature, but concentrates can often be stored for years.

Cleanup and Water Effects

Both bacteria and yeasts have difficulty growing in concentrates of 60°Brix and higher; so the careful control of water addition to concentrate lines and tanks has been found to be of great importance in controlling the growth of microorganisms in citrus concentrates. Partial rinsing of a concentrate line or tank is one of the major causes of high microbial plate counts in concentrate lots because the microbes grow much more rapidly under the more dilute conditions. Also, a constant flow of fresh concentrate through the lines and tanks minimizes the time available for growth and reduces contamination. The only growth that can occur is on the inner surface of the pipes where yeast may adhere and feed on the flow of juice passing by, and this is why light yellowish yeast slime can be found on the inner surfaces of contaminated lines. This buildup should be cleaned immediately. During busy times of the year when fresh concentrate constantly flows through the lines and the weather is cooler, the number of cleanups can be reduced. Hot summer temperatures and infrequent processing usually require a frequent-cleanup program. Cleanups may occur at various intervals, from daily to monthly. Incomplete cleanups run the risk of inducing microbial growth through dilution of the concentrate; therefore, partial cleaning is worse than no cleaning at all for citrus concentrates.

Contamination of the lines from other sources also may cause spoilage. For example, condensate may form inside the concentrate tank, diluting the concentrate and then spoiling it when left for a long period of time such as over the weekend. Another source is blending with contaminated product, which contaminates the entire blend. Another is blockage in plate heat exchangers or lines, which provides a site for the accumulation of proudct that is not removed during cleanups, and then spoils when left for long periods of time in what is mistakenly thought to be clean equipment. Entrapped debris also provides a site for microbial growth, and dead spaces in product lines present another hazard where microbial growth may thrive. Such dead spaces should be taken apart and checked regularly or removed altogether. A buildup of pressure from carbon dioxide production during fermentation literally can cause product hoses, lines, containers, and even steel drums to explode, creating a safety as well as a quality hazard.

Juice lines, tanks, evaporators, extractors, fillers, and packagers can all be cleaned by using automated CIP (Cleaning In Place) systems. They also can be taken apart manually and cleaned. Manual cleaning is the easiest and surest method to inspect, but permanent lines are difficult or impossible to take apart and inspect completely. However, most processing plants have enough equipment to justify the use of an automated CIP system for at least part of it. Properly programmed CIP systems provide a more efficient and consistent cleanup even though they are more difficult to inspect than manual cleaning systems.

Because citrus juices are acidic, 2% caustic solutions are often used for cleanups. After the juice is removed, it is common practice to use a chlorine or iodophor sanitizer rinse to reduce surface contamination of the equipment. However, a rule of thumb may be that if a piece of equipment can be totally inspected (all surfaces can be seen), such as a tank, and it looks clean to the eye and there are no off odors, it can be considered clean. Equipment that cannot be totally inspected, such as process lines, must be inspected by indirect methods such as observance of clear water rinses and the presence of off odors. Laboratory plate counts, however, are the surest means of determining the level of product microbial contamination.

QUESTIONS

1. What is the main by-product of lactic acid bacteria that degrades citrus juices?
2. What other genera of bacteria have been found in citrus juices besides *Lactobacillus* and *Leuconostoc*?
3. What microorganisms are considered pathogenic to humans that are found in citrus juices?
4. How can you recognize yeast cells under the microscope?
5. Why do yeasts survive better in citrus concentrates than bacteria?
6. To what eucaryotic class of microorganisms do the yeasts and molds commonly found in citrus juices belong?
7. How can mold contaminate citrus juices?
8. What can be deduced from each of the following statements about samples?
   - "A" has a high level of bacteria.
   - "B" has a high level of yeasts.
   - "C" has a high level of mold.
9. What is the minimum plate count before off flavors and odors will develop from microbial spoilage?
10. Cite evidence from the chapter that the growth of microorganisms found in citrus products is inhibited by light.
11. What is the maximum allowable % broken and % decayed fruit leaving the grading table under USDA inspection?

12. How does the condition of fruit correlate with the microbial plate counts of the resulting processed juice?
13. What is the minimum temperature that has been shown to adequately control microbial growth?
14. About how long does a concentrate juice lot with a high microbial count need to be stored in a commercial freezer before the counts are reduced to acceptable levels?
15. What are some of the most likely causes of high plate counts in citrus concentrates?

PROBLEMS

1. What is the cfu of a 65°Brix sample that resulted in 316 colonies appearing on the plate?
2. What is the cfu of a 60°Brix sample yielding 146 colonies of yeasts, 4 colonies of *Lactobacillus*, and 1 colony of *Leuconostoc*?
3. If a 12.3°Brix sample yielded 25 colonies when plated, would it be more or less contaminated than the concentrate sample in problem 2 using the same 100 dilution factor?
4. Suppose that a linear relation was found between absorbance and the concentration of diacetyl, using the procedure in the chapter, which was:

$$\text{Absorbance} = (\text{Concentration in ppm})(0.0623) \qquad (15\text{-}4)$$

What would be the diacetyl value of a sample with an absorbance of 0.7 for the sample and 0.5 for the blank?
5. Identify the following plates:

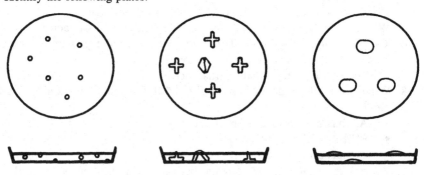

REFERENCES

Berry, R. E. and Veldhuis, M. K. 1977. Processing of oranges, grapefruit and tangerines. In *Citrus Science and Technology II*, S. Nagy, P. E. Shaw, and M. K. Velduis, eds. The AVI Publishing Company, Inc., Westport, Conn., 202.

Buchanan, R. E. and Gibbons, N. E., eds. 1974. *Bergey's Manual of Determinative Bacteriology.* The Williams & Wilkins Company, Baltimore, Md., 510-513, 576-593.

Faville, L. W. and Hill, E. C. 1951a. Recent investigations pertaining to acid-tolerant bacteria in citrus juices. *Second Annual Citrus Processors Meeting,* Citrus Experiment Station, Lake Alfred, Fla., October 9, 1951.

Faville, L. W. and Hill, E. C. 1951b. Incidence and significance of microorganisms in citrus juices, *Food Tech., 10,* 423-425.

Hays, G. L., Riester, D. W. 1052. The control of "off odor" spoilage in frozen concentrated orange juice, *Food Tech., 6(8),* 386-389.

Hill, E. C., Wenzel, F. W., and Barreto, A. 1954. Colormetric method for detection of microbiological spoilage in citrus juices, *Food Tech., 3,* 168-171.

Juven, B. J., Kanner, J., and Weisslowicz, H. 1978. Influence of orange juice composition on the thermal resistance of spoilage yeasts, *J. Food Sci., 43,* 1074-1080.

Kimball, D. A. 1982. Unpublished.

Kobatake, M., Kurata, H., and Komagata, K. 1978. Microbiological studies of fruit juices I & II, *J. Food Hygenic Sci. of Japan (Shokuin Eiseigaku Zasshi), 19,* 449-461.

Parish, M. E. 1988. Microbiological aspects of fresh squeezed citrus juice. In *Ready to Serve Citrus Juices and Juice Added Beverages,* R. F. Matthews, ed. Univ. of Florida, Gainsville, Fla., 79-96.

Parish, M. E. and Higgins, D. P. 1989. Yeasts and molds isolated from spoilage citrus products and by-products, *J. of Food Protection, 52(4),* 261-263.

York, G. K. 1988. Private communication with microbiologist, University of California, Davis, Calif.

Chapter 16

Insects, Rodents, and Birds

Citrus processing plants generally are located in areas with mild climates, which offer ideal living and growing conditions for many insects, rodents, and birds. However, citrus juices may be processed and packaged in nearly every climate, each with its own problems in regard to animal infestation. Regardless of the processing conditions, the same basic principles of pest control are generally applicable and important in producing a good-quality citrus product.

INSECTS

Even though the insects are a specific taxonomical class, the term "insect" is generally applied to any creature that one would colloquially refer to as a "bug." Insects are known for having six legs and three separate body sections. Such creatures as spiders, mites, ticks, chiggers, millipedes, centipedes, snails, slugs, and so on, are not insects, but for the purposes of food sanitation they can be considered as such because of the methods used for their containment and control. Insects are the most abundant animal form on earth, with three to four times as many species as the rest of the animal kingdom has put together. As many as 700,000 species are known, with about two new species being added each year. There is much that we do not know about them, but one thing we do know is that insects cost the agricultural industry billions of dollars each year. With respect to agricultural growth, insects are a concern because of their effects on the quality and yield of certain crops. In the citrus processing industry, however, the concern is with insects being embedded in fruit or being attracted to the processing areas and entering the product during processing or storage. Insect fragments in citrus products do not pose a health danger because the acids of the juices kill any pathogenic microbes, but visible insect fragments constitute a serious quality infraction, rendering the juice unpalatable. Also, insect fragments and eggs that are visible only under a microscope can result in lower defects scores upon USDA inspection; and citrus products contamination by insects can lead to FDA action, up to costly fines or worse.

Insects are able to readily adapt to a wide range of environments. It can be safely assumed that anywhere humans can live, so can insects. It may be next to impossible to completely purge a processing plant of all insects, but, by using basic techniques, processors can control insects sufficiently to produce a good-quality product. The war on insects is too vast for processors to rely on merely one or two battle plans. They need an integrated system, using all available resources and tactics, to stay on top of an insect invasion. This requires a knowledge of insects, including what factors affect them and how.

Food and Water

Hunger and thirst are the greatest motivators on earth, and insects are constantly seeking food and water. Processing plants have a tendency not only to exist by large bodies of war, but to have a large amount of water scattered around the premises. Water, clean or dirty, has a tendency to draw insects, expecially crawlers, some of which can detect water from long distances; and crickets and cockroaches are especially attracted to wet areas. The field cricket (*Gryllus assimilis*) (Fabricius)) is found in many parts of the world and across the United States. The most common types of cockroaches are the Croton bug or the *Blattella germanica* (German cockroach) (Fig. 16-1) and the *Blatta orientalis* (oriental cockroach); these cockroaches can hatch up to 50 offspring at one time. These insects congregate around sinks, ditches, and wherever water collects. They are most active at night or in the dark, but they often can be seen during daylight hours. Crickets are generally easy to locate because of the noise they make with their legs. Puddles of water not only attract insects, but they look messy, and if they are in walkways or roadways, they become unpleasant obstacles for people to wade through. Water should be drained or removed by squeegee in such areas, and long ditches or waterways should be covered or shortened to minimize insect attraction.

Insects feed on a variety of materials, especially those that humans consider to be trash, such as paper. Trash receptacles should be well placed and used, and should be emptied regularly. Floors in processing and eating areas should be swept and/or mopped often. Fruit residue common around fruit conveyers should be routinely removed, and juice spillage should not be left very long before cleanup. Fruit waste that must be stored should be stored away from the plant, and, if an insect infestation is noticed, the waste should be sprayed. Weeds should not be allowed to grow near the plant, and landscaping should be well cared for and sprayed if necessary. Unnecessary piles of parts, equipment, or pallets provide excellent insect breeding and hiding areas, and should be eliminated or removed from the plant. If the food and water are taken away, insects will go somewhere else, not into your plant or your products.

The natural oils found in the outer peel of citrus fruit act as a natural barrier

Mediterranean Fruit Fly
(Epochra canadenis)

Mexican Fruit Fly
(Anastrepha ludens)

Oriental Fruit Fly
(Dacus dorsalis)

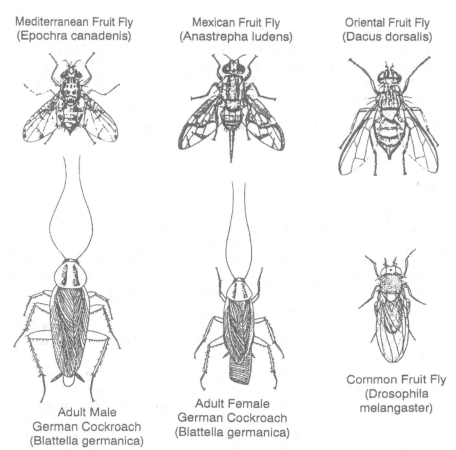

Common Fruit Fly
(Drosophila
melangaster)

Adult Male
German Cockroach
(Blattella germanica)

Adult Female
German Cockroach
(Blattella germanica)

Fig. 16-1. Insects of common concern to citrus processing plants (USDA drawings). The *Drosophila* is much smaller than the other insects in the figure.

to most insects. Also, the acidity of the juice inside the fruit is unpalatable to insects, as a rule. However, there are some insects that thrive in citrus fruit, namely, the Mediterranean fruit fly, the Mexican fruit fly, the Oriental fruit fly, and a few others that are less common. These pests have been essentially eradicated from the continental United States but flourish in citrus processing areas in other parts of the world. These insects enter the fruit and lay their eggs within the fruit itself. This not only damages the fruit but provides unsightly contamination when it is processed. Grade-out of infested fruit is not always 100% efficient, and centrifuging of contaminated citrus juices is commonly used to remove insect fragments. Imported juices often will have lower pulp levels because centrifuging to remove insect fragments also removes pulp. Other insects grow only on the surface of citrus fruit and generally do not contaminate juice extracted from the fruit. Lemons are not considered a host for the Mediterranean fruit fly.

Drosophila

Even though the natrual aromas and essential oils of citrus repel most insects, the common fruit fly or *Drosophila* (Insecta, diptera—also known as vinegar flies or pomace flies) seems attracted by them. The genus *Drosophila* has up to 50 different species, with the predominant species being *Drosophila melanogaster*, which thrives in mild weather during the late summer and early fall (see Fig. 16-1). *Drosophila simulans* also becomes abundant in areas such as central California but does not thrive in hot weather. *Drosophila pseudobscura* predominates in cooler weather.

Drosophila melanogaster are 1.5 to 2.5 mm long with red eyes and a black abdomen, the first three segments having a yellow band. They convey spoilage organisms from fruit to fruit and are strong fliers, flying over 6 miles in one day, although they become immobilized in wind over 5 miles per hour and a light intensity greater than 150 foot-candles. They are day fliers with very rapid reproductive cycles. Although the flies spend only 24 hours in the egg, the egg may remain in the female to within an hour of hatching. The larva and pupa stages take three days each, giving a reproductive cycle of about seven days. Females may lay up to 2000 eggs at one time but average only about 1000. Adults live from 40 to 70 days, depending on the temperature. Thus, if two *Drosophila* mate on June 1, assuming an average life span and two days between reproductive cycles, by mid-July, when the original parents die, there will have been over 30 quadrillion (3.125×10^{16}) fruit flies produced. If this many fruit flies were placed end to end (assuming an average 2 mm length), they would extend out into space nearly $2\frac{1}{2}$ light days or the distance that light traveling at 186,000 miles per second would go in $2\frac{1}{2}$ days. Within a few days after this, the average rate of production from the time of original mating would have exceeded the speed of light in placing the flies end to end as they were born. If the parents were to remate during the larva and pupa stages of the last generation, the reproduction rate would be two to three times faster. These staggering statistics suggest the importance of insect control and the need for action early in the season before the numbers get out of control. Vigorous reproduction also produces sex odors that drift in the wind and attract even more insects—another reason why breeding areas should be removed and cleaned.

Temperature

Insects are cold-blooded animals; that is, they require an outside source for body heat. The ideal temperature for insects is generally 28.0°C (82.4°F), which is very close to humans' ideal temperature. Therefore, insects will always try to invade the human environment. The comfort range for insects is also very narrow, generally 21 to 32°C (70–90°F); temperature outside this range seriously diminish insect activity. Solar heat stored in cement and metals provides a heat source for insects after sunset. The abundance of cement surfaces, brick or stone buildings, metal buildings, vehicles, equipment, or railroad rails commonly found at processing plants provide a comfortable habitat for insects. Once activated by temperature, they begin their search for food and water, which are generally plentiful around processing plants. When the climate is cooler or during the winter, insects like to move indoors, where they can vacation for the winter. Some like it so much that they never bother to go home, especially if temperature controls are left on all the time. Turning off the temperature controls over the weekend, especially during cold weather, inhibits insect activity and growth. Insects also like to inhabit electrical control boxes and equipment because of the warmth of the electronics. As no processing plant can operate without these devices, the sealing of cracks and crevices and the use of chemicals (see below) are needed to control infestations in these areas. Insects often cause damage to such equipment, especially sensitive laboratory equipment.

Light

All insects are attracted to light to some extent, a characteristic that is of great importance to night operations where nocturnal (night-active) insects are attracted to processing lighting. Some insects, such as moths, can see for hundreds of yards, whereas others, such as *Drosophila*, can see only a few yards. The compound eye structure that is common to insects prevents them from seeing linearly, so they usually approach lighting in a wandering fashion.

Insect are particularly sensitive to the ultraviolet light range, just outside humans' visible region. Mercury vapor lights attract over 100 times as many insects as softer yellow sodium vapor lights, which emit a smaller amount of ultraviolet light. For this reason, the use of mercury vapor lights should always be discouraged at food processing plants, especially in rural areas. When outside lighting is necessary, it should be placed away from entries into the plant or locations where product is exposed to the air, such as tanker loading areas. Tankers always need air access to the tanker during filling or emptying, which usually means that insects have access as well. If these areas are well lit at night, insects usually will congregate there. Lights within the plant are best placed where they cannot be viewed directly from outside the plant. Blinds and curtains should be closed at night to avoid attracting insects, and screens should be used to cover windows that must remain open. In rural areas where a processing plant is the only light source for miles around, insects especially can be a problem, but in a city where there are many lights, the problem may be less severe. Using lights with a lower wattage will also diminish insect attraction. Again, once they are inside the plant, they may decide to stay there and reproduce.

The attraction insects have for light can also be used to the advantage of the quality control department. Ultraviolet light traps, consisting of an ultraviolet light surrounded by an electrified screen, attracts insects. When they land on the screen, they are electrocuted and fall into a collecting tray at the bottom of the trap. These trays, which should be changed often, will give a good indication of what kind of insects are infesting an area. These lights should be placed in areas that draw the insects away from processing areas, and where they cannot be seen from the outside—you would not want a light trap to lure insects in from the outside. Most flying insects fly only about 5 feet from the floor or less, so the light traps are best used at these heights. However, higher processing areas that have attractive debris will lure insects higher off the ground. Ultraviolet bulbs lose their ability to emit ultraviolet light with time and should be changed periodically (at least annually), even if no visible change is observed in their light.

Insecticides

Due to their obvious toxicity, most insecticides are not permitted for use in food processing plants. The federal code (Title 21 110.37(b)) states: ''The use of insecticides or rodenticides is permitted only under such precautions and restrictions as will prevent the contamination of food or packing materials with illegal residues.'' The pesticide tolerances that define ''illegal residues'' are in a constant state of change, so that only by subscribing to the *Federal Register* can one be assured of staying on top of changes in the law, but this generally is not required of citrus processors. The use of pyrethrin insecticides is permitted by the Food and Drug Administration and has been shown to be effective against *Drosophila* and many crawling insects. Pyrethrins have been used orally for the treatment of intestinal worms in humans and thus pose a smaller toxic threat than other chemicals used to control pests. Extracted from pyrethrin plants in Africa and South America, pyrethrins consist basically of four compounds, known as pyrethrin I and II and cinerin I (allethrin) and II; and they are effective against a wide variety of insects including *Drosophila*, aphids, bettles, cabbage worm, housefly, leaf hooper, louse, mealybug, mosquito, sod webworm, thrip, and mare. These insecticides also work quickly, with a quick kill rate. Sunlight quickly breaks pyrethrins down, making them a contact insecticide that must be sprayed directly on the insect rather than a residual insecticide that will kill the insect at a later time. This breakdown of pyrethrins prevents the buildup of insecticides in the plant, but it also necessitates cyclic application. Pyrethrins should never be used with alkaline materials, as they would thereby lose their effectiveness. Similar pyrethrin-containing compounds are Drione, Foliafume, Prentox Pyronyl, Pyrenone, and Tri-Excel DS (Considine 1982). In particularly difficult situations, professional exterminators with experience in food processing plants can be consulted and retained.

Whenever insect buildup is observed, the areas should be sprayed as soon as possible, both inside the plant and outside. Foggers are very effective in killing flying insects such as *Drosophila*. Care should be taken that no one is in the plant during fogging. The use of timers will allow the use of foggers at night or over the weekend. Doors to laboratories or lunchrooms should be left open so that the insecticide can reach those areas. Care should be taken that all areas of the plant be fogged. If one room is neglected, the flies will resort to that room until the air clears, and then will reinfest the plant. Processors should also be careful to close all product tanks and vessels and to cover any surface that may come into contact with the food product. Carrier gases associated with the insecticides can cause illness and can contaminate citrus products. The fumigated rooms should be left vacant for several hours after fogging, again preferably overnight or over the weekend. If a particular area is found to contain insects, hand sparying also can be done.

Even though many crawling insects may be killed in fogging operations, many will escape because crawlers are well adapted to hiding in places that the fog cannot reach. Roach powders made of borax or boric acid, sulfur, or pyrethrum can be used in electrical boxes, in cracks and crevices, under or behind equipment or fixtures, or wherever there is not much water. Care should be taken not to place these powders where they may fall or get into the product. Roach traps consisting of sticky surfaces also can be used, not only to kill roaches but to monitor their activity. Again, in particularly difficult situations, a professional exterminator can be employed.

Outside the plant, harsher chemicals can be used. A chemically treated perimeter around the plant will prevent many crawlers from entering the plant. All insecticides should be stored in isolated areas away from food ingredients in order to avoid the accidental incorporation of these dangerous chemicals as food additives.

Physical Barriers

An open door or product tank is an invitation to dinner for an insect that is best canceled. All doors to processing areas should be closed when not in use and equipped with automatic closing arms if necessary. Few insects can fly through a closed door. Those doors that must remain open for long periods of time should be equipped with properly functioning air screens. As shown in Fig. 16-2, air screens should not blow insects into the plant. Air flow should be uniform across the entire door area without quiet "holes" where insects can

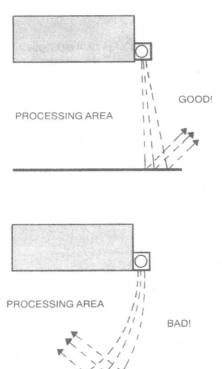

Fig. 16-2. Proper and improper air screens over doorways that must be open for long periods of time, showing good air flow and bad air flow. Care should be taken not to blow the insects into the processing area.

slip through. Also, ventilation systems used to remove peel oil volatiles that burn the eyes in extraction areas can disrupt air curtain flow, a situation that must be avoided. Because fruit must be conveyed into the plant and the wastes removed, there usually are openings in plant walls that can allow the entry of insects. These should be as small as possible and covered with plastic strips or other suitable barriers. Windows that open should always be screened. Also, windows should be caulked, thresholds sealed, and other holes and cracks repaired. Even a solid wall a few feet high around the plant can be a significant barrier to a crawling insect infestation.

Product tanks should always be closed except when it is absolutely necessary that they be open. Some tanks require air vents during filling and emptying, which should be screened to avoid the entrance of insects. Drum, can, or bottle filling stations should be entirely free of insects. Filling is usually the most vulnerable time in the process for product contamination by insects.

Education

Often, in order for the previous methods to work, plant personnel need to be educated on what the plant is doing for the control of insect contamination. Purchased or home-made slide or video presentations can be used as a part of new-employee orientations. Unlike microorganisms, insect problems are something that everyone can see and do something about, not just the quality control department.

The important thing to remember is that all of these methds need to be used. No one or two methods are sufficient. In fact, using all of these methods does not guarantee that you will not have insect problems. As with many quality control parameters, however, the consumer demands a sanitary product regardless of the problems involved.

RODENTS AND BIRDS

It is contrary to the *Code of Federal Regulations* (CFR, Title 21 110.37(b)) to permit any animals or birds into any area of a food processing plant except those that are essential as a raw material. As with insecticides, the chemicals used to control animals in food processing plants are limited to those that will not leave illegal residues on any food products. Rats, mice, and squirrels are the most common types of rodents that infest citrus processing plants. Squirrels are less bold than rats or mice and generally live in outlying areas away from the bulk of plant activity. Rats and mice readily move inside the plant and live in ceilings, walls, control panels, equipment, offices, laboratories, lunchrooms, and so on. Rodents are generally nocturnal in nature and thus are most often seen at night. They are more visible than insects, and sightings should be reported right away. The exact location and direction of movement should be noted so that breeding areas can be located. Evidence of chewing or feces, as well as grease streaks along walls that rats may rub against, also are important. Rat urine will fluoresce under long ultraviolet light, emitting a greenish color. Poisoned bait should be set out in areas suspected as hiding or breeding areas but never where it can contaminate food products or containers. Each day the bait should be checked for feeding. When the feeding has stopped, the rodents usually have been killed. Pallet and drum storage areas and old abandonded equipment or structures are ideal habitats for rats and mice, and usually a change in season will drive rodents into plant areas. All holes, cracks, or other openings that rodents can use to enter the plant should be sealed off as much as possible.

Birds are attracted to high structures where they can nest out of reach of predators. They generally like quiet places free from vibration and noise. Thus they generally shy away from tall evaporators that are constantly in use, but idle tall structures away from processing noise are vulnerable to nesting. Commercial ultrasonic or chemical repellants or avicides can be used to discourage infestation, according to good manufacturing practice.

DETECTION OF CONTAMINATION

Detection of an insect or an animal presence in the plant is discussed above. However, the detection of animal contamination in the product requires different methods from those previously described. The best way to detect this type of filth in citrus products is to alert production and quality control personnel who routinely inspect the product. Any type of off-colored specks observed against the bright colors of citrus juices should be reported immediately. Finding one insect fragment in a product must mean that the rest of the insect is somewhere else. Filth of any kind should be removed from the product and separated by using sieves. The filth must be identified visually or under a microscope as an insect fragment, hair, and so forth, so that steps can be taken to prevent further contamination. Imported or purchased products should be inspected for evidence of insect contamination upon their arrival at the plant. Severe problems usually can be seen through plastic liners without their being opened. Lots severely contaminated with such filth are nearly impossible to salvage and should be discarded or rejected. The economic loss of so doing should be sufficient incentive to impose a proper sanitation program on the process. Accepting and/or selling contaminated products can cost processors their reputations and a portion, if not all, of their customers and business.

QUESTIONS

1. How many species of insects are known?
2. What can attract insects to a processing plant?
3. What is the main concern of processing plants regarding insects?
4. Why is it that more insects do not infest the interior of citrus fruit?
5. What insect is attracted to the citrus aromas common in citrus processing plants?
6. What can be done to prevent *Drosophila* infestation in a processing plant?
7. How many *Drosophila* can be produced from two flies during an average life span?
8. What is the comfort temperature range for insects?
9. What is the difference in insect attraction for mercury vapor lights and sodium vapor lights?
10. Why are pyrethrins just contact insecticides?

REFERENCES

Considine, D. M. 1982. *Foods and Food Production Encyclopedia*. Van Nostrand Reinhold Company, New York, 1626.

Chapter 17

Physical and Chemical Contamination

The sanitary production of foods as outlined in the federal code for good manufacturing practice includes restrictions that prevent contamination by physical
or nonliving objects or chemicals. The physical objects range from dust particles to hammers, and the chemicals range from toxic lubricants to juices of
another variety. Most of these objects and chemicals are necessary for the operation of a citrus processing plant. However, regardless of their value, they all
have the potential for misuse and can damage the quality of the plant's products.
Such misuse can range from accidental or careless contamination to intentional
adulteration, and, unlike the contamination discussed in previous chapters, can
be lethal to product consumers. For this reason, prevention of such contamination should be a vital part of any sanitation or quality control system in citrus
processing plants.

PHYSICAL CONTAMINATION

Product containers, whether they be cans, bottles, cartons, drums, tanks, or
tankers, usually are protected from physical contamination by lids or covers.
However, during processing, there is always a time when this protection is
removed or not in place, so that there is some risk of objects other than the
intended product getting into the container. Common sense dictates that these
containers should be clean and sanitary before use. Also, during processing,
the containers should be protected from falling objects or airborne contamination. Production personnel, who are usually the first line of defense, should be
trained and held responsible for this type of contamination. Loose items, such
as tools, bolts, nuts, lunch boxes, clothing, jewelry, pens, and so on, should
not be placed near open product containers where they may accidentally get
into the product. Loose items in shirt pockets pose a serious risk when one
bends over an open product tank. Hair protection is required by federal law,
including beard covers as well as hair nets or hats. Hair longer than shoulder
length should be kept up inside a hair cover because long hair poses a safety as

wall as a sanitary hazard. Even visitors should wear hair protection when they tour proudct areas. Also, good personal hygiene should be practiced by food handlers, including clean clothes and hands.

Insulation near product containers, commonly used to insulate product and refrigeration lines, may become worn and break off, causing unsightly or even dangerous contamination. Dust and dirt inside the plant should be removed as much as possible, and cleaned equipment should not be placed on the ground or in any location where it may pick up dust or dirt. Dusty roads or travelways should not be allowed near or inside the plant, especially in tanker and tank loading and unloading areas. After a tanker has been cleaned, it should never travel any distance without all openings closed tightly. Condensate water that commonly accumulates on cold surfaces, such as refrigerated lines and tanks, should not fall in or near product containers, as this water picks up dust and dirt from the air and the exterior surfaces of processing equipment.

Citrus products should never come into contact with unsanitary surfaces such as wood or rusty steel. Drums and drum lids used to package citrus concentrates should contain an inner baked enamel finish that is free from rust, dirt, or chipping paint. Plastic containers should be free from plastic dust and residue. Paperboard containers should be checked for dust and dirt accumulation prior to use. During barreling, two plastic liners should be used, and the innermost liner should not be opened until just before filling and should be closed just after filling. Before barrel dumpers are used, all extraneous material, such as labels, liner fasteners, and so on, should be removed so that it will not fall into the product as it empties.

If glass objects are dropped into a tank full of product, there is no sanitary way to get them out, and the potential for breakage could result in the total rejection of the entire lot of juice. Plastic or metal containers should be used to take samples, and bimetal thermometers should be used to take temperature readings. Similarly, the Food and Drug Administration requires all glass lighting over product vessels or containers to have plastic covers or sleeves in order to avoid contamination if the glass should break. Because debris from eating, drinking, smoking, or chewing gum or tabacco often finds its way into product containers, the FDA prohibits these activities in processing areas. Food machines should be placed in lunchrooms and not in processing areas, and employees should be provided a time and place for meals and breaks.

CHEMICAL CONTAMINATION

The processing of citrus juices always requires the use of chemicals in one form or another. Water itself is a chemical, and is used not only for cleaning but to adjust the concentration of concentrates and juices from concentrate. It should go without saying that only potable water should be used in citrus products and for cleaning surfaces that come into contact with citrus products. Recycled or reuse water should never be used for such purposes. Even though the water so used in potable, that does not guarantee that the minerals, salts, or other chemicals (such as chlorine) in the water will not affect the quality of the citrus product. Dilution water should be checked for off flavors and odors prior to use or whenever contamination is suspected. Water used for cleanups also should be potable and should not contain any extraneous debris. Many beverage manufacturers go to great lengths to treat the water they use in the making of their products.

Only food-grade caustics and sanitizers should be used for cleanups, and FDA inspectors generally require information on the cleaners used in processing. After cleanups, equipment should be thoroughly rinsed and drained in order to avoid contamination the next time that it is used. Some mixing of the FDA-approved cleaning agents may not pose much of a health hazard for consumers; however, off flavors or odors may arise with such mixing, the acid nature of the juice may change, and/or the federal standards of identity will most likely be violated.

Citrus processing sometimes also includes the use of food additives, pesticides, lubricants, and other chemicals. Nonfood and food chemicals should be stored separately so that toxic materials will not be added accidentally to food products. Care should be taken not to add the wrong food additive because so doing not only could affect the quality of the product, but it could also violate standards of identity. If polychlorinated biphenyls (PCBs) are used in any transformers, in capacitors, as heat transfer or hydraulic fluids, or in lubricants, coatings, inks, and so on, the federal code should be consulted (Title 21 110.40). These compounds are considered toxic, and their use is regulated in food plants. The many moving parts of citrus machinery require lubricants, and care should be taken that these chemicals not drip into product containers or anyplace where they could result in product contamination. Also, refrigerants used to chill citrus juices and concentrates have the potential of leaking into the product, which would diminish their chilling capacity in addition to contaminating the product. Containers, such as drums, that have been previously used for nonfood purposes, such as petroleum, should never be used for food products.

A good rule of thumb regarding physical or chemical contamination is sometimes referred to as Murphy's law: If anything has the possibility of getting into a food product, most likely it will. Eliminate the possibility, and you will eliminate the contamination.

QUESTIONS

1. Who should be responsible for it if containers are contaminated during processing?

2. Who should be responsible for inspecting containers prior to their use?
3. What loose items are allowed to be placed near open product vessels or containers?
4. Why is it that glass objects cannot be used to take samples from product vessels?
5. Why is eating, drinking, smoking, or chewing prohibited in processing areas?
6. Why should food and nonfood chemicals or materials be stored in separate areas?

Chapter 18

Processing Contamination

Besides microbial, insect, animal, physical, and chemical contamination, there is also contamination due to processing errors. As with many types of contamination mentioned before, this type of contamination does not produce any kind of health dangers; only the product quality is affected. The detrimental effects range from cosmetic or appearance changes to texture and flavor changes. Because processing errors are the source of this type of problem, proper processing can be the cure. A good quality control program can virtually eliminate all of the problems discussed in this chapter.

HESPERIDIN

Hesperidin belongs to the group of compounds called flavonoids, like naringin, which was discussed in Chapter 10. Unlike naringin, hesperidin imparts no flavor to citrus juices, and it exists as the predominant flavonoid in orange and tangerine juices. Its chemical structure is:

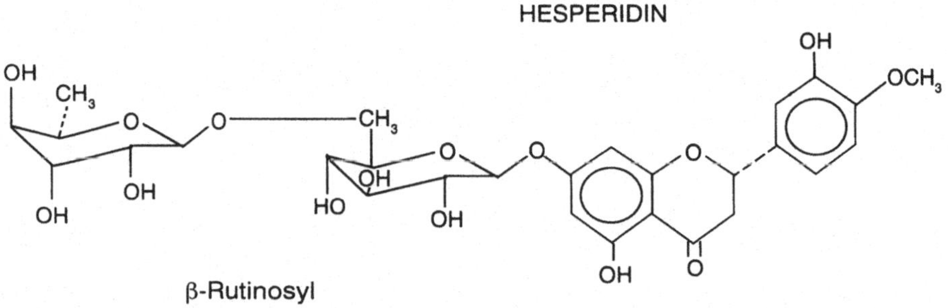

HESPERIDIN

β-Rutinosyl

Like other flavonoids, hesperidin is found primarily in the membrane and peel material of the fruit. It is not soluble in neutral aqueous solutions and is only slightly soluble in acidic solutions such as citrus juices. Upon extraction of the juice from the fruit, hesperidin comes into contact with the acidic juice and will begin to form crystals. Under the microscope, these crystals are long and needlelike. On processing equipment, they appear as a white scale or film that can build up until it breaks off as white flakes. Hesperidin also contributes to about 10 to 20% of the juice cloud. Its crystals can clog finishing screens, thereby decreasing juice yield. Also, during evaporation concentration of the juice accelerates hesperidin crystallization, producing white flakes that appear in the concentrate or reconstituted juice.

Hesperidin levels within the fruit generally decrease with maturity because as the fruit matures, it accumulates moisture, which dilutes the hesperidin concentration. In spite of this, the appearance of hesperidin flakes in citrus juices increases with fruit maturity and may become an acute problem in the late season, especially in Valencia juice. This is probably due to lower acid levels, which reduce the solubility of the hesperidin. Also, greater breakdown of the peel in late-season fruit may cause increased amounts of hesperidin to get into the juice. The USDA has set an arbitrary standard for the number of hesperidin flakes permitted for various defects scores, as shown in Fig. 18-1. Hesperidin flakes can be observed (USDA method) using 710 ml of 11.8°Brix reconstituted juice that has sat undisturbed for 5 minutes in a 1000 ml beaker with a 4-inch-diameter bottom. White hesperidin flakes are apparent on the bottom of the beaker, as observed from below with the aid of a flashlight.

Even though hesperidin formation cannot be prevented, routine hot caustic cleanups can keep finisher screens and processing equipment free from hesperidin buildup. Again, late-season processing requires more frequent hot caustic cleanups.

SCORING GUIDE

FOR HESPERIDIN
FROZEN CONCENTRATED ORANGE JUICE
AND CONCENTRATED ORANGE FOR MFG.

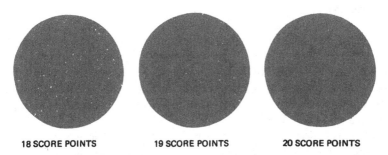

| 18 SCORE POINTS | 19 SCORE POINTS | 20 SCORE POINTS |

Fig. 18-1. Arbitrary defects scoring guide for hesperidin flakes used by the USDA.

BLACK FLAKES

Whenever heat is used during processing, as in evaporation or pasteurization, there is risk of product overheating, scorching, or burn. (Off flavors and colors produced from heat abuse are discussed in the next section.) If the flow of the juice through such heating equipment is obstructed, excessive heat can be applied, which may result in product burn. Plate heat exchangers used in evaporation or pasteurization are especially susceptible because they have very small orifices that are easily blocked. The main purpose of finishing operations is to remove pulp and other particles that may block the openings and passageways in the evaporators or adhere to heated surfaces. Once a pulp or hesperidin particle is trapped and begins to burn, other particles will become attached to it, and the burn area can spread until the entire flow of juice is blocked. Superconcentration of the juice results in solidified sugars or ''candy'' formation, so that the evaporator must be shut down and cleaned. The first evidence of this problem is the appearance of black flakes in processed concentrate. Evaporators with plate heat exchangers should be shut down and taken apart immediately if black flakes appear in the concentrate because the burn area will only spread and increase. TASTE evaporators should be cleaned right away if more than a few black flakes appear. Small amounts of very tiny black flakes are commonly found in concentrates processed in TASTE evaporators and generally are of no concern. If juice sac material is to be added, the addition should be made after heat treatment, especially if plate heat exchangers are used.

Black flake formation usually can be attributed to holes worn into finisher screens or excessive hesperidin buildup due to insufficient cleanups. Because black flakes are highly visible against the bright colors of citrus juices, they can be easily recognized and perhaps even removed by using a sanitary ladle. If the contamination is not too severe, the problem can be diminished by blending with uncontaminated lots. However, it should be remembered that, unlike hesperidin flakes, most specifications require a zero level of black flakes or specks because of their high visibility.

JUICE OXIDATION

Any food that contains highly oxidizable components, such as carbohydrates and ascorbic acid, is vulnerable to oxidation. Oxidation can occur in many ways, but catalyzed oxidation is usually the most rapid and therefore of greatest concern to food processors. Citrus juice oxidation produces both off colors and off flavors, but generally off flavors develop before off colors and are of greater concern than the off colors.

In flavor oxidation, there are three basic catalysts. One is the naturally occurring enzymes—nature's catalysts—that cause chemical changes in all living things. These enzymes are generally deactivated through heat processing and cause no further changes in the juice. The organic acids present in citrus juices also are catalysts and can assist in breaking up carbohydrate chains through acid hydrolysis. However, the amine-assisted breakdown of carbohydrates has been shown to be the major cause of the development of cooked off flavors (Handwerk and Coleman 1988). The mechanism for the amine catalysis is similar to that of acid catalysis, but has been shown to occur under milder conditions of heat and acidity than acid catalysis. Also, the oxidation of ascorbic acid or vitamin C has been shown to be a significant factor in the development of off flavors in citrus juices.

Citrus juices contain greater amounts of amino acids than other juices such as apple or grape, which make citrus juices more vulnerable than other juices to oxidation involving amino acids and the plentiful carbohydrates. These sugar–amino acid reactions were first studied by Louis-Camille Maillard (1878–1935) and have since taken his name as nonenzymatic Maillard reactions. The Maillard reactions involve the reaction of the aldehyde group of the sugar with the amino group of the amino acid, according to the following condensation reaction:

$$\underset{\text{sugar}}{R'-CHO} + \underset{\substack{\text{amino}\\\text{acid}}}{R-NH_2} \rightarrow \underset{\text{Schiff base}}{\underset{\overset{|}{R'}}{HC=N-R}} \qquad (18\text{-}1)$$

Further reactions can occur from the Schiff base, including various condensation reactions and Strecker degradations. At least 20 products of oxidation of citrus juices have been isolated, and many more exist. However, only six have been shown to contribute significantly to the quality of heat-abused citrus juices (Handwerk and Coleman 1988).

1-Ethyl-2-formylpyrrole

This compound, also known as N-ethylpyrrole-2-carbaldehyde, is formed from hexose–amino acid reactions followed by a Strecker degradation. It has a taste threshold of only 2 ppm compared to a typical level of 0.5 ppm in dehydrated orange juice, which generally contains larger amounts of oxidative products due to the additional heat required in processing. No data are available for other citrus juices, but the levels would be expected to be much less. Excess amounts of this compound impart a piney stale odor and flavor to heat-abused juices. Of the six heat-abuse products that affect citrus juice quality, this is the only one that actually incorporates the nitrogen from the amino acid into its structure, illustrative of the role of amino acids in the oxidative process. Its boiling point is 48 to 53°C.

2-Hydroxy-3-methyl-2-cyclopenten-1-one

Other common names associated with this compound are methylcyclopentenolene, hydroxymethylcyclopentenone, and cyclotene. Cyclotene is considered a major contributor to the off flavor of citrus juices because of its low taste threshold of 5 ppm, compared to levels of 1 ppm commonly found in dehydrated orange juice. This compound most likely results from the acidic degradation of ascorbic acid because the chemical structures are similar. However, acid-catalyzed hexose degradation also is possible. Excessive amounts of cyclotene produce a maple or caramel-like odor and flavor.

5-Methyl-2-furaldehyde

Also known as 5-methylfurfural, this compound is formed from amino acid–catalyzed oxidation of hexoses such as glucose or fructose. It has a boiling point of 187°C, a specific gravity of 1.1072_4^{18}, and an index of refraction of 1.5263^{20}, and is soluble in water, very soluble in alcohol, and infinitely soluble in ether.

It can be detected at 10 ppm, compared to less than 0.5 ppm found in dehydrated orange juice. Adding a hydroxy group to the methyl group to give 5-hydroxymethyl-2-furaldehyde results in a compound with a taste threshold in canned orange juice of over 200 ppm with a typical level of 14 ppm. Grapefruit juice held at 50°C for 12 weeks can exceed this taste threshold. This hydroxy compound is virtually absent in fresh juices and may serve as a general measure of product oxidation. It has a boiling point of 114 to 116°C, a melting point of 32 to 35°C, and an index of refraction of 1.5627^{20}.

2,5-Dimethyl-4-hydroxy-3(2H)-furanone

This compound is also known as furaneol, and is also produced from hexoses and amino acids. It has a taste threshold of 0.05 ppm in orange juice and imparts a pineapple-like flavor when it occurs in excessive amounts.

2-Methoxy-4-etheneylphenol

This compound is also known as 4-vinyl guaiacol or guaiacol ethylene ether. It is one of the only two compounds from citrus juice oxidation that produce detectable quality changes but do not come from sugar–amino acid reactions or ascorbic acid oxidation. It is formed from ferulic acid, which occurs in small amounts in most plants. Ferulic acid has been found to occur in single-strength orange juice at levels of 0.18 ppm, which increase to 0.30 ppm upon pasteurization. When exposed to air, 4-vinyl guaiacol is converted to vanillin (Naim et al. 1988). It has a melting point of 138 to 139°C and is slightly soluble in water and soluble in ethanol. It has a taste threshold of only 0.05 ppm, with a typical canned orange juice level of 1 ppm (which is over the taste threshold). Excessive amounts impart an old fruit or rotten flavor and odor.

α-Terpineol

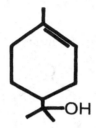

This compound is a derivative of *d*-limonene, the primary constituent in citrus oils, and *d*-limonene is commonly found in citrus juices. α-Terpineol is a colorless liquid with a lilac odor, which has a boiling point of 210 to 218°C and a density of 0.935 g/ml. It is volatile with steam but is insoluble in water and soluble in alcohol or ether. It has a taste threshold of 2.0 ppm, with a typical level of 4 ppm in canned orange juice, and in excessive amounts it imparts a stale, musty, or piney aroma to heat-abused juice.

General Observations

All these compounds have very low taste thresholds, and some of them have been shown to impart off flavors below their thresholds when combined with other heat-abuse products. For example, 5-hydroxymethyl-2-furaldehyde, hydroxymethylcyclopentenone, and N-ethylpyrrole-2-carboxaldehyde will impart an off flavor even though each exists at only half of its taste threshold. Also, 2,5-dimethyl-4-hydroxy-3(2H)-furanone, α-terpineol, and 4-vinyl guaiacol impart off flavors even though they occur at only half of their individual taste thresholds.

   Additional amounts of citric acid also have been shown to enhance Maillard oxidation, an effect attributed to the buffering action of the citrate ion because the rates of these oxidative reactions are pH-sensitive. Also, the methyl groups of citric acid assist in catalyzing oxidation to some extent, and phosphate addition accelerates oxidation regardless of buffering effects.

   Restricting the amount of available oxygen obviously reduces the rate of oxidation of citrus products. Steam injection or vacuum filling minimizes off flavor development. Deaeration of citrus juices has not been shown to significantly reduce oxidation because there is always enough air in the juice to facilitate oxidation. Increasing the dissolved oxygen increased the rate of browning in lemon juice but did not affect the rate of ascorbic acid degradation (Robertson and Samaniego 1986). In another study, raising the initial levels of dissolved oxygen increased browning and ascorbic acid degradation in orange juice, but reducing the initial dissolved oxygen levels did not extend the taste-panel-determined shelf life of pasteurized orange juice (Trammell, Dalsis, and Malone 1986).

Even though large amounts of heat are used during most citrus processing, the majority of heat abuse occurs during storage. Shelf-life studies of 66°Brix orange concentrate have shown that a reduction of storage temperature to 5.5°C (4.4°C to −1.1°C) can increase the shelf life by 4 months (5 months to 9 months) (Marcy et al. 1984). With storage temperatures of −6.6°C (20°F) or below, orange concentrate can be stored for at least a year. These results are comparable to what is found industrially. Juices being evaporated are generally held at 90 to 95°C (195–205°F) for about 10 to 20 seconds, with a total retention time in the evaporator of about 10 minutes, depending on the evaporator. Most of this time the temperature of the juice is around 35 to 70°C (100–150°F). Processed concentrate can withstand maximum evaporator temperatures for about 30 minutes. Aseptic juices and concentrates are known for their burnt-off flavors, most of which occur during unrefrigerated or prolonged storage.

Analytical Techniques

Citrus juice oxidation products, like most chemicals, can be measured by using sophisticated or time-consuming analytical techniques, including mass spectrometry, gas chromatography, and so forth. These methods are generally too expensive and involved for routine quality control, and are generally not found in quality control laboratories. For quality control purposes, the general degree of oxidation and the shelf-life determination are the main factors of interest. The simplest method for measuring the general heat abuse of citrus juices is the formol or formaldehyde test, which measures the level of primary amino acids. Most of the amino acids in citrus juices are primary amino acids, the exception being proline, which is a secondary amino acid. Proline occurs in significant amounts in citrus juices and even constitutes the major amino acid in many varieties. Because amino acids are consumed during Maillard oxidation, the loss of amino acids can be used as an indicator of heat abuse. The formol test involves a Strecker degradation of the amino acids, generating acidity according to the following equation:

$$R-\underset{\substack{\text{amino acid}\\\text{(except proline)}}}{\overset{\overset{\displaystyle NH_3^+}{|}}{C}}-\overset{\overset{\displaystyle O}{\|}}{C}O^- + \underset{\text{formol}}{\overset{\overset{\displaystyle O}{\|}}{\underset{H\ H}{C}}} \rightarrow R-\overset{\overset{\displaystyle \underset{N}{\overset{H\ H}{\diagdown C \diagup}}}{\|}}{C}-\overset{\overset{\displaystyle O}{\|}}{C}O^- + H_2O + H^+$$

$$(18\text{-}2)$$

The acidity generated can then be titrated with a standard base.

The advantage of this method is its simplicity, speed, and convenience, as it can be done at the same time as a routine acid titration. Its disadvantages include a lack of specificity as to which amino acid is being measured and the fact that oxidation can occur without consuming any amino acids. Also, because the decrease of amino acids is what is important, two tests must be done, one before and one after heat abuse or storage. Still, the formol test can provide a quick estimate of heat abuse, and studies in the author's laboratories have shown that a decrease of about 0.10 meq/100 ml of a single-strength juice equivalent constitutes a detectable flavor change in California orange juices. Formol values generally range from 1 to 3 meq/100 ml of single-strength juice. The following procedure can be used.

Formol Test

Equipment and Supplies

- Same as for acid titration (see Chapter 3).
- 37% formaldehyde neutralized to a pH of 8.4 within the hour.
- 10 ml graduated cylinder.

Procedure

1. Proceed to the endpoint (8.2 pH) as described in the section on acid titration (see Chapter 3), and continue titrating to a pH of 8.4.
2. Add 10 ml of the neutralized formaldehyde solution. The pH will drop. Rezero the buret containing the NaOH solution.
3. Again titrate to a pH of 8.4, and read the ml of NaOH needed to do so. The total amino acids (except proline) can be calculated by using:

$$\text{meq AA}/100 \text{ ml} = \frac{(\text{ml titrated})\,(N)\,(100 \text{ ml juice})\,(S_j)}{(\text{ml sample})\,(S_s)} \quad (18\text{-}3)$$

where N is the normality of the base ($0.1562N$) and S_s and S_j are the lb solids/gal of the sample and $11.8°$Brix juice. S_s and S_j are the same if the sample has been reconstituted or adjusted to $11.8°$Brix. S_j is equal to 1.029 lb solids/gal. The amino acid level is usually expressed as milliequivalents amino acids per 100 ml of juice. Equation 18-3 can be condensed to:

$$\text{meq AA}/100 \text{ ml} = 16.07(\text{ml titrated})/(\text{ml sample})S_s \quad (18\text{-}4)$$

Another method suggested for the monitoring of heat abuse in citrus juice products is the measurement of the furfural level. Furfural is a product of ascorbic acid degradation, but it does not contribute directly to off flavor development in the juice. Even though some oxidation reactions occur at different rates, the rate of furfural development is believed to be an adequate indicator of heat abuse for quality control purposes. The advantages of this method include the measurement of a specific oxidative product, and the fact that because furfural is essentially nonexistent in fresh juice, only one test is needed to estimate the heat abuse of the juice. Also, furfural is easily distilled from the juice by using an apparatus common in citrus quality control laboratories and is concentrated sixfold in so doing. The main disadvantage is that it involves the use of aniline, a toxic chemical that can cause intoxication when inhaled, ingested, or absorbed through the skin. Consumption of 0.25 ml of aniline can cause serious poisoning. Acute symptoms include cyanosis, methemoglobinemia, vertigo, headache, and mental confusion. Chronic symptoms include anemia, anorexia, weight loss, skin lesions, and bladder tumors. The maximum safe level is only 5 ppm. Analysis using aniline should be done under expert supervision.

The furfural procedure involves the following reaction:

FURFURAL   ANILINE

This is a form of the Strecker degradation that results in a highly colored conjugated product that can be detected with a spectrophotometer. The procedure uses $SnCl_2$ to stabilize the color formation as well as to decrease interference from codistilled compounds.

Furfural Test (Ting and Rouseff 1986)

Equipment and Supplies

- Spectrophotometer.
- Oil distillation apparatus used in oil determination plus boiling chips or antifoaming agent.
- 10 ml ground glass stoppered graduated cylinder.
- 2 ml pipette.
- One 13 × 150 mm test tube.
- 1 ml pipette.
- Furfural standards (0.5, 1.0, 2.0 μg/ml in distilled water; make a 1 : 100 dilution of a 1 mg/ml stock solution, and use 5, 10, and 20 ml of this solution, diluting each to 100 ml).
- 95% ethanol.
- Aniline/acetic acid solution. (Mix 2 ml of purified aniline with 1 ml 20% $SnCl_2 \cdot 2H_2O$ in concentrated HCl. Mix and add glacial acetic acid to 20 ml.) Observe necessary safety precautions in working with aniline.

Procedure

1. Add 200 ml of reconstituted (11.8°Brix) or single-strength juice to the 500 ml boiling flask in the distillation setup, along with a drop of anti-foaming agent or boiling chips.
2. Collect 10 ml of distillate in the graduated cylinder, and stopper it tightly.
3. Mix the distillate and pipette 2 ml of it into the test tube; then add 2 ml of ethanol and 1 ml of aniline/acetic acid solution to it. If analyzing grapefruit juices, add 10 ml of distilled water.
4. Allow the tube to stand for 10 minutes, and then read the absorbance at 515 nm on the spectrophotometer.
5. Compare to standards, and calculate the μg/liter of the furfural accordingly. Standards should be added to fresh juice and distilled in order to account for the distillation efficiency. The following equation can be used:

$$\mu g/l = \frac{(10 \text{ ml distillate})(1000 \text{ ml/liter})(2 \text{ ml std})(\mu g/ml \text{ std})}{(200 \text{ ml juice})(2 \text{ ml distillate})}$$

(18-6)

or:

$$\mu g/\text{liter furfural} = 50/(\mu g/ml \text{ standard})$$

(18-7)

For grapefruit juices, multiply the furfural level by 3. Off flavor detection correlates with furfural levels of about 50 to 70 μg/liter in canned and glass-packed orange juice and 150 to 175 μg/liter in canned or glass-packed grapefruit juice. Grapefruit juice has a greater tendency to mask the off flavors formed from product oxidation and thus has a higher taste threshold.

If the equipment is available, HPLC can be used to measure the amount of furfural, which would eliminate the need to handle the toxic aniline.

HPLC Furfural Test (Marcy and Rouseff 1984)

Equipment and Supplies

- Isocratic HPLC system with a 4.6 mm × 25 mm Zorbax ODS column and a UV detector set at 280 nm with a solvent system of 35:65 methanol and water.
- Furfural standards prepared in the same way as in the previous procedure.
- Distillation apparatus used in the previous procedure with antifoam agent and boiling chips.
- 10 ml graduated cylinder.

Procedure

1. Add 200 ml of reconstituted (11.8°Brix) or single-strength juice to the 500 ml flask in the distillation setup, and add 1 drop antifoam and boiling chips. The standards used in the previous procedure should be used with fresh juice here as well, in order to account for distillation efficiency.
2. Collect 10 ml of the distillate in the graduated cylinder and inject 15 μl into the HPLC system with a flow rate of 1.0 ml/min.
3. The elution time is about 8.5 minutes. Use the same calculations as in the previous procedure.

If a gas chromatograph is available, some of the oxidative products that directly affect off flavors can be measured. The main drawback to the determination of some of these compounds is that they cannot be purchased in pure form for standard preparation and must be synthesized in the laboratory. The synthesis of these compounds is usually the difficult or prohibitive step, rather than the analysis itself.

GLC Oxidation Products Test (Tatum, Nagy, and Berry 1975)

Equipment and Supplies

- Rotary evaporator.
- 100 g NaCl.

- 1500 ml beaker.
- Two 1-liter separatory funnels.
- 1 liter methylene chloride.
- Na_2SO_4 desiccant.
- 10 ml volumetric flask or graduated cylinder.
- GLC with a 9' × 0.25" Carbowax 20M pack stainless steel column (20% on 60–80 mesh Gas Chrom P) with a helium flow rate of 200 ml/min. The temperature is programmed at 80°C for 6 min., and increased to 130°C for 6 min., 135°C for 14 min., 140°C for 24 min., 155°C for 30 min., 180°C for 46 min., 190°C for 56 min., 200°C for 64 min., and 220°C for 76 min.
- 100 μl syringe for injection into GLC system.

Procedure

1. Add 100 g NaCl to 1 liter of reconstituted (11.8°Brix) or single-strength juice in the 1500 ml beaker, and stir it for 5 minutes to dissolve the salt.
2. Separate the sample into two 1-liter separatory funnels, and extract each portion 5 times with 200 ml of methylene chloride.
3. Dry the extracts over Na_2SO_4, combine them, and evaporate them in a rotary evaporator at 30°C.
4. Transfer the concentrate and washings into a 10 ml volumetric flask, and fill it to the mark with methylene chloride.
5. Inject 100 μl of the concentrated extract into the GLC system, and compare the peaks to standards. The retention time and source of synthesis have been reported for the relevant compounds shown in the table:

Compound	Standard Source	GLC Ret. Time (min.)
α-Terpineol	c	21.0
2,5-Dimethyl-4-hydroxy-3(2H)-furanone	x	43.0
4-Vinyl guaiacol	a	54.5
5-Hydroxymethyl furfural	c	75.0
5-Methyl-2-furaldehyde	c	*
1-Ethyl-2-formylpyrrole	b	*
2-hydroxy-3-methyl-2-cyclopenten-1-one	c	*
Furfural	c	12.5

a—Synthesized as described in references (Tatum et al. 1975).
b—Synthesized as described in references (Tatum et al. 1967).
c—Commercially available.
x—Not commercially available and no reported synthesis.
*—Not reported using this method.

Brown Color

As mentioned earlier, during the oxidation of citrus juices browning follows off flavor development. Browning with heat is common in nearly all foods. The development of brown compounds in crude plantation white sugars has been attributed to the development of humic acids, caramel, 5-hydroxymethyl furfural, and melanoidins (Cheng, Lin, and Wang 1983). 5-Hydroxymethyl furfural has already been mentioned as a remote participator in off flavor development in oxidized citrus juices. The dark pigment in caramel is probably related to the melanoidins, as is the pigments found in humic acids (Benzing-Purdie, Ripmeester, and Preston 1983). Melanoidins have long been associated with Maillard reactions (Spark 1969) and are the result of a slow polymerization of reducing sugars along with inorganic salts. Melanoidins also are what constitutes the brown pigment in skin. Even though the chemical structures of melanoidins have not been determined, the related compound melanin has been shown to be a complex quinoidal compound consisting of the following units:

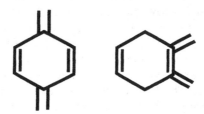

Melanin is formed from the oxidation of tyrosine, which occurs at levels of up to 3 mg/100 ml of single-strength juice and below 1 mg/100 ml in single-strength grapefruit and lemon juices. As with many chemical reactions that occur in citrus juices, the formation of melanin is commonly found in nature. This oxidation product is found in feathers, hair, eyes, and skin, with albinos lacking the enzyme needed to convert tyrosine to melanin.

The browning of citrus juices has been linked to ascorbic acid degradation, which is at least partially responsible for the off color development in citrus juices. Increased citrate concentrations enhance browning reactions by increasing the buffering of the pH. Increased phosphate levels enhance browning regardless of any buffering effect (Spark 1969).

Chemical measurement of the off colors found in heat-abused juices generally is not done. Visual observations, along with color scoring and taste confirmation, are usually sufficient for one to conclude that the shelf life of the juice has expired. The actual amount of browning usually is not important, especially to the pass/fail shelf-life tests. Heat-abused juice can be carefully reblended to dilute the browning effect to below detection levels. Otherwise, browned juice should be discarded as unmarketable. Even though sulfites, thiols, mercaptans, and other sulfur-containing amino acids have been shown to inhibit browning, addition of these compounds might pose a health risk to consumers and most likely would violate most federal standards of identity. The most viable means of preventing both off flavor and off color formation in citrus juices is to minimize oxygen contact and reduce the temperature as much as possible during processing and storage. In other words, do not heat-abuse the juice.

POTASSIUM CITRATE CRYSTALLIZATION

When high Brix (58–65°Brix), low Brix/acid ratio citrus concentrates are stored in commercial freezers, crystallization of some of their organic salts may occur. These crystals generally have a negative effect on the citrus juice, resulting in lower USDA defects scores as well as clogging plate heat exchangers and other processing equipment that have restricted flow areas. These crystals also do not reconstitute as easily as citrus concentrates. Larger crystals usually are laced with insoluble hesperidin, which further hampers reconstitution attempts.

The crystals are hard and have a characteristic acid taste. They range in size from fine sand up to 3 mm in diameter, depending on how long the crystallization has been allowed to proceed. The composition of the crystals from California and Brazilian concentrates is illustrated in Table 18-1. As can be seen, the Brazil concentrate and crystals have higher levels of potassium than the California counterparts. The higher potassium levels resulted in higher acid levels in the Brazilian crystals as well. This predominance of potassium and citric acid in the crystals agrees with other studies (Hils 1973; Koch 1980;

Table 18-1. Composition of MPC crystals and the composition typical of orange juice concentrate for comparison (Kimball 1985).

| | 60 Brix Concentrate | | MPC Crystals | |
w/w % of:	Kimball	Bielig	Kimball	Koch
Total acid (as citric)	4–7	4–5	30–40	68.7
Reducing sugars	20–30	31.1–36.2	<0.5	0.09
Sucrose	—	16.1–27.1	—	0.06
Glucose	—	13.2–17.1	—	0.03
Fructose	—	13.5–18.1	—	0.06
Potassium	0.51–0.71	1.09–1.46	9.15	13.78
Sodium	0.03–0.04	0.002–0.006	0.133	—
Ascorbic acid	1.23–1.54	1.01–1.78	0.02	—
Hesperidin	—	0.37–0.46	—	0.38
Pulp	8–12 (v/v)	6–8 (v/v)	—	4.75
Water	40	40	—	13.00

Bielig et al. 1983). The stoichiometric relationship between potassium and citric acid suggests that monopotassium citrate (MPC) is the primary species of salt in the crystals. The citrate ion has so many sites for internal rotations that it takes on a variety of shapes, leading to an amorphous or irregular crystal lattice that gives rise to amorphous crystals. Also, other juice components, especially cloud material, pulp, and hesperidin, become entrapped in or adhere to the crystals, adding to their amorphous nature.

There are three main factors that affect the appearance of these crystals, one being the thermodynamic favorability of the crystallization. Solubility tests have shown that the Ksp of MPC in sucrose solutions follows this relationship:

$$(CA)(K^+) = Ksp \, (moles/liter)^2 = 9.47 \times 10^{10} \, e^{-8060/T} \quad (18\text{-}8)$$

where (CA) and (K^+) are the concentrations of the citric acid and potassium ions in moles/liter, and T is the absolute temperature between 253°K (-20°C or -4°F) and 293°K ($+20$°C or 68°F) (Kimball 1985). If the product of the molar concentrations of citric acid and potassium ion exceeds the Ksp value, crystallization is thermodynamically favored. It should be remembered that citrus juices, like most other fruit juices, contain excess amounts of citrate beyond that measured by acid titrations—up to 20% more citrate (Shaw, Buslig, and Wilson 1983). Measurements of the citric acid and potassium levels for navel and Valencia concentrate throughout a season in California are shown in Fig. 18-2 (Kimball 1985). The dashed line at the bottom is the Ksp value calculated from Equation 18-8 at 0°C. As can be seen, throughout the entire season for both varieties of orange concentrate the Ksp value was exceeded, a situation that would strongly favor MPC crystallization. This suggests that the high-Brix

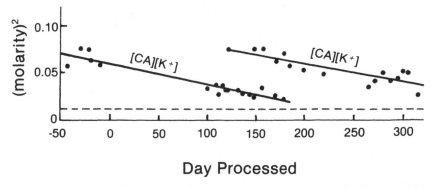

Day Processed

Fig. 18-2. The product of the citric acid and potassium concentrations for California navel and Valencia concentrates during a processing season. The dashed line represents the Ksp value calculated from Equation 18-8 for 0°C (Kimball 1985). (Reprint from *Food Technology* 1985, *39(9)*, 79-81, copyright © by the Institute of Food Technologists.)

citrus concentrates are always susceptible to MPC crystallization and can be considered to be supersaturated solutions of MPC. Using the relationships in Fig. 18-2 and the Ksp relationship in Equation 18-8, Fig. 18-3, can be constructed, giving the estimated MPC saturation curves in orange juice as a function of Brix and temperature. Early-season and midseason navel and Valencia curves were essentially the same; separate late-season navel and Valencia curves are given. In Fig. 18-3, if concentrated orange juice is diluted and warmed, it passes to the right and down in the figure, passing through the saturation curve until it reaches the unsaturated area. In this area, MPC crystallization is no longer thermodynamically favorable, and the crystals will begin to dissolve.

Another factor that affects this crystallization significantly is the juice pH. In aqueous solutions, the highly polarized water molecule solubilizes the ionic species—here potassium ions, citrate ions, and hydrogen ions. As water is removed through evaporation, the ions begin to compete for available water molecules. The small size of the hydrogen ion and its ability to hydrogen-bond make it an easy winner. This means that increased hydrogen ion or acid levels exclude the potassium ions from the solubilizing water molecules, leading to crystallization between potassium and citrate ions as shown in Fig. 18-4. This

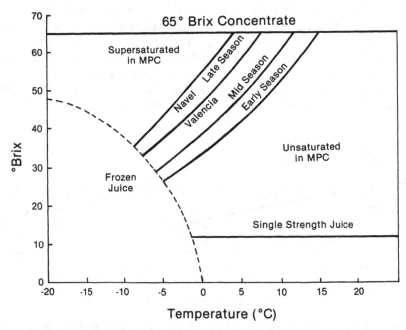

Fig. 18-3. Saturation curves for California navel and Valencia orange juices for MPC crystallization as a function of Brix and temperature (Kimball 1985). (Reprint from *Food Technology* 1985, *39(9)*, 79–81, copyright © by the Institute of Food Technologists.)

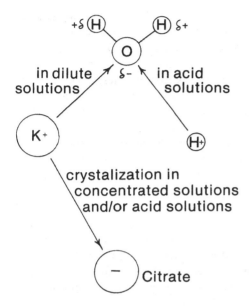

Fig. 18-4. The effect of acid on MPC crystallization. The solubilizing water molecules favor the hydrogen ion, leaving the potassium to crystallize with citrate (Kimball 1985). (Reprint from *Food Technology* 1985, *39(9)*, 79–81, copyright © by the Institute of Food Technologists.)

explains why early-season high-acid concentrates produce more crystals than their sweeter late-season counterparts. High-acid tangerine and tangelo concentrates are especially susceptible to MPC crystallization. Lemon juices usually are not so concentrated, and MPC crystallization thus has not been so much of a concern with them.

The final factor affecting this crystallization is the rate of crystal formation. Even if crystallization is thermodynamically favorable, if it does not occur within a significant period of time, the thermodynamic favorability becomes less important. A general equation that describes crystal formation is:

$$\text{rate} = \sqrt{(2DV(C - S)t)/q} \qquad (18\text{-}9)$$

where D is the diffusivity factor, V is the ionic volume, C is the ionic concentration, S is the ionic solubility, t is the time, and q is the charge of the ionic species. The diffusivity factor is generally constant in dilute solutions, but here this factor is significantly affected by the Brix, temperature, and acid content of the concentrate. The acid levels can be reduced by ion exchange or blending. It is desirable to keep the Brix as high as possible in order to minimize storage and shipping costs. The temperature can be controlled somewhat. Bulk storage tanks cannot be kept much below $-10°C$ (about $15°F$) because if the temperature is too low, the concentrate becomes too viscous to be pumped in and out of the storage tank, but drum freezers are kept much colder ($-20°C$ or $0°F$). The higher heat needed by bulk storage systems increases the ion mobility in the concentrate, permitting thermodynamically favorable MPC crystallization. Some bulk freezers are allowed to warm up to $-4°C$ ($25°F$), which would favor MPC crystallization even more. Reduction of bulk freezer temperatures as low as possible would minimize MPC crystallization. Also, increased agitation, which is common in the handling of bulk concentrates, increases the formation of seed crystals. The result is the initial formation of many small crystals that later aggregate into larger crystals.

Because the exact relationship of the D factor in Equation 18-9 to the Brix, acid level, and temperature is unknown, it is not useful in predicting the actual crystal growth. A regression equation has been found that does a better job of this when compared to experimental data:

$$t = 648d^{0.334} \qquad (r^2 = 0.91) \qquad (18\text{-}10)$$

where t is the time in days of storage in an industrial freezer at about $-10°C$ (about $15°F$), and d is the diameter of the crystals in millimeters. Usually a Brix/acid ratio of about 15.0 or less and a Brix of at least 60°Brix are required for Equation 18-10 to be effective in crystallization prediction. Concentrates with higher Brix levels will form crystals more slowly.

As large MPC crystals can cause processing quality problems, care should be taken not to store low Brix/acid ratio concentrates in bulk freezers for more than a few months. Such early-season concentrates should be barreled or blended soon after production to avoid crystallization. This precaution and minimizing bulk freezer temperatures are generally the best ways to control MPC crystal formation.

QUESTIONS

1. What quality problem does hesperidin produce in citrus juices, and how can this problem be eliminated?
2. During what part of the season is hesperidin contamination the biggest problem, and why?
3. What causes black flakes to form, and how can this contamination be eliminated?
4. How many black flakes are generally acceptable in citrus concentrates?
5. What compounds in citrus juices are mostly responsible for off flavor development during excessive heating?
6. Which of the six heat-abuse products mentioned in the chapter imparts a pineapple-like flavor?
7. Which two of the six heat-abuse products mentioned in the chapter that affect flavor are not produced by Maillard reactions between sugars and amino acids or from ascorbic acid degradation?

8. When is the most heat abuse generally applied to citrus juices?
9. What are some of the advantages and disadvantages of the formol test and the furfural test as mentioned in the chapter?
10. Which occurs first, browning or flavor change during juice oxidation?
11. What causes the brown coloring in heat-abused citrus juices?
12. Does the fact that citrus concentrates are supersaturated in MPC mean that MPC crystals will definitely form during normal industrial storage? Why?
13. How does the acid level affect MPC crystallization?
14. What is the most feasible way to prevent MPC crystallization?
15. What varieties of concentrate are the most vulnerable to MPC crystallization?

PROBLEMS

1. In the formol test, using 9.83 ml of 60.2°Brix concentrate for both tests, would the concentrate be expected to have developed off flavors if the initial test titrated 9.82 ml of the base and the test after heat abuse titrated 9.51 ml of the base?
2. In the furfural test, suppose that the absorbance matched the 1.02 μg/ml standard. Would this orange juice be expected to have off flavors from heat abuse?
3. What would be the Ksp value for MPC at 0°C, the freezing point of water? (0°C = 273°K)
4. Would a concentrate containing 0.15M citric acid and 0.10M potassium ion be favored to form MPC crystals at the same temperature mentioned in problem 3?
5. About how long would it take to form 2 mm crystals in 60°Brix concentrate with a Brix/acid ratio of less than 15.0?

REFERENCES

Benzing-Purdie, L., Ripmeester, J. A., and Preston, C. M. 1983. Elucidation of the nitrogen forms in melanoidins and humic acid by nitrogen-15 cross polarization–magic angle spinning nuclear magnetic resonance spectroscopy, *J. Agric. Food Chem.*, *31*, 913–915.

Bielig, H. J., Hofsommer, H. J., Fischer-Ayloff-Cook, K. P., and Balcke, K. J. 1983. Crystalline precipitation in frozen orange juice concentrates, *Flussiges Obst, 50(3)*, 105.

Cheng, H. T., Lin, W. F., and Wang, C. P. 1983. Studies on the color development in stored plantation white sugars. In *The Maillard Reaction in Foods and Nutrition*, G. R., Waller and M. S. Feather, eds. ACS Symposium Series 215, American Chemical Society, Washington, DC., 91–102.

Handwerk, R. L. and Coleman, R. L. 1988. Approaches to the citrus browning problem, *J. Agric. Food Chem.*, *36*, 231–236.

Hils, A. 1973. Identification of a crystalline substance (K-citrate) in orange concentrate, *Flussiges Obst, 40(12)*, 496.

Kimball, D. A. 1985. Crystallization of potassium citrate salts in citrus concentrates, *Food Tech.*, *39(9)*, 76–81, 97.

Koch, J. 1980. Formation of monopotassium citrate (MPC) in frozen concentrated orange juice (FCOJ) at 65°Brix. Presented at the 31st Annual Citrus Processors' Meeting, Lake Alfred, Fla., September 10, 1980.

Marcy, J. E. and Rouseff, R. L. 1984. HPLC determination of furfural in orange juice, *J. Agric. Food Chem.*, *32*, 979–981.

Marcy, J. E., Graumlich, T. R., Crandall, P. G., and Marshall, M. R. 1984. Factors affecting storage of orange concentrate, *J. Food Sci.*, *49*, 1628–1629.

Naim, M., Striem, B. J., Kanner, J., and Peleg, H. 1988. Potential of ferulic acid as a precurser to off-flavors in stored orange juice, *J. Food Sci.*, *53(2)*, 500–504.

Robertson, G. L. and Samaniego, C. M. L. 1986. Effect of initial dissolved oxygen levels on degradation of ascorbic acid and the browning of lemon juice during storage, *J. Food Sci.*, *51(1)*, 184–187, 192.

Shaw, P. E., Buslig, B. S., and Wilson, C. W. III. 1983. Total citrate content of orange and grapefruit juices, *J. Agric. Food Chem.*, *31(1)*, 182–184.

Spark, A. A. 1969. Role of amino acids in non-enzymatic browning, *J. Sci. Food Agric.*, *20*, 308–316.

Tatum, J. H., Nagy, S., and Berry, R. E. 1975. Degradation products formed in canned single strength orange juice during storage, *J. Food Sci.*, *40*, 707–709.

Tatum, J. H., Shaw, P. E., and Berry, R. E. 1967. Some compounds formed during nonenzymatic browning of orange powder. *J. Agric. Food Chem.*, *15*, 773.

Ting, S. V. and Rouseff, R. L. 1986. *Citrus Fruits and Their Products*. Marcel Dekker, Inc., New York, 177—178.

Trammell, D. J., Dalsis, D. E., and Malone, C. T. 1986. Effect of oxygen of the taste, ascorbic acid loss, and browning for high-temperature-short-time pasteurized single strength orange juice, *J. Food Sci.*, *51(4)*, 1021–1023.

Chapter 19

Juice Adulteration

Fraudulent representation of foods and medicines has a long history. In the days of the wild west, snake oils and medicinal tonics were sold to gullible passersby at a price that was hard to turn down. These concoctions were consumed in the hope of there being at least a remote possibility that some of their claims would be valid but many of these ''cure-alls'' caused more damage than they remedied.

In 1938, the United States Congress enacted the Federal Food, Drug, and Cosmetic Act. Even though there was an earlier law, and the Act has been modified somewhat since, it remains essentially intact today. This law gave definition to the term ''adulteration,'' which is described in section 402 of the Act. The Food and Drug Administration (FDA) has the responsibility for enforcing this law, which requires different emphases and methods depending on which type of adulteration is involved. The four basic types of adulteration are: (1) filthy, putrid, decomposed, or harmful; (2) unsafe or unsanitary; (3) statutorily unsafe; and (4) economic (Stutsman 1988).

The first kind of adulteration is a concern for the manufacturers of any product. Chapter 17 covers a portion of this type of contamination. There is no product made that cannot be dangerous in the wrong situation. The production process for any consumable food, such as citrus juices, is not totally invulnerable to the possible addition of poisons or harmful materials. In recent times a great deal of attention has been paid to packaging that prevents the addition of dangerous substances by persons other than the manufacturer. Citrus juices often are packaged in cans, plastic containers, or cartons that are difficult to tamper with in retail stores. Industrial processing and storage areas should maintain a security system to prevent outsiders from roaming around the plant. Processing records should be detailed enough that any problems can be traced back to specific employees. This way employees can be held accountable for their own actions, in addition to serving as watch dogs for the product during processing.

The second type of adulteration involves processing in such a manner that the food *may* become adulterated. Because the FDA cannot continually inspect a processing area, it needs the flexibility of preventing adulteration by preventing the conditions that would cause adulteration as much as possible. For example, if there are many insects in the plant, their presence would constitute an unsanitary condition because the insects would be likely to cause contamination even if they were not actually found in the product.

The third type of adulteration involves substances that are declared by law to be unsafe, which usually are excluded from the group referred to as "food additives." Food additives must undergo thorough screening in order to determine if they can be safely added or used in food products. In order for a substance to be used in a food, it must be "generally recognized as safe" or achieve GRAS status. Food additives that are considered safe because of common usage prior to January 1, 1958 have GRAS status (Title 21 CFR 170.30(c)) without extensive additional testing. Food additives developed since then must undergo a rigorous screening process. Detailed descriptions of those substances that can be used in foods or come into contact with foods can be found in the *Code of Federal Regulations*, Title 21 170-189. This list continues to be augmented with the development of new food additives and processes.

The last type of adulteration is economic adulteration. Laws relating to this type of adulteration do not permit the use of materials that are not described or implied on the label. Because a full description of how a product, such as orange juice, was manufactured would be lengthy and confusing, standards of identity have been invoked to simplify the identification of foods and beverages. As a result, the federal standards of identity for citrus juices have evolved. These standards of identity go far beyond descriptions of the other forms of adulteration in stating exactly how a citrus juice can be made and/or formulated. Many acceptable food additives that have achieved GRAS status and are acceptable with respect to the first three forms of adulteration are not acceptable in regard to the standards of identity. It should be remembered that the standards of identity and the Food, Drug, and Cosmetic Act are designed to prevent contamination, fraud, or a decrease in the value of food products. New technologies that legitimately enhance the quality of food products, such as debittering, generally do not come under restriction according to the intent of the Act.

STANDARDS OF IDENTITY

The standards of identity for various citrus products are contained in the *Code of Federal Regulations* and the *Federal Register*. These standards and their interpretation may change from time to time, and the current status of such standards should be monitored by quality control management with the aid of legal counsel if necessary. With this charge in mind, the following constitutes a brief description of the most current standards of identity for citrus products available:

Orange Juice (CFR Title 21 146.135). This product contains juice from unfermented fruit of sweet orange varieties (*C. sinensis*) with seeds and excess pulp removed. The product may be chilled but not frozen.

Frozen Orange Juice (CFR Title 21 146.137). This product is the same as orange juice except that it is frozen.

Pasteurized Orange Juice (CFR Title 21 146.140). This product is the same as orange juice except that pasteurized orange juice can contain up to 10% of tangerine juice (*C. reticulata*) or its hybrids. Also, oil and pulp (not washed or spent pulp) can be added. Concentrated orange juice, sugar, invert sugar, dextrose dried corn sirup, or dried glucose sirup can be added to raise the Brix to a normal range for orange juice, but such addition must be declared on the label. The juice can be heat-treated to reduce enzyme activity and the number of viable microorganisms. This juice can be frozen and must have a Brix of at least 10.5, not including added sweeteners. The B/A ratio must be at least 10.

Canned Orange Juice (CFR Title 21 146.141). This product is the same as pasteurized orange juice with certain additional characteristics: the condensate water from deoiling may be added, the minimum Brix is 10, and the minimum B/A ratio is 9. Also, if the juice is not refrigerated, the word "canned" can be omitted from the label.

Orange Juice from Concentrate (CFR Title 21 146.145). Orange juice from concentrate contains reconstituted frozen concentrated orange juice or orange concentrate for manufacturing and may contain orange juice, pasteurized orange juice, orange juice for manufacturing that has been preserved by freezing and not canning, orange oil, unwashed pulp, and water. It may contain the same declared sweeteners as pasteurized orange juice, and may be heated or pasteurized. It must contain at least 11.8°Brix exclusive of any added sweeteners.

Frozen Concentrated Orange Juice (CFR Title 21 146.146). This product may contain 10% tangerine juice or its hybrids (*C. reticulata*) and/or 5% sour orange juice (*C. aurantium*). It may contain water, orange oil, orange essence, unwashed pulp, and the same declared sweeteners as above. The product can be heat-treated. In order to reconstitute this product to not less than 11.8°Brix, the dilution ratio should be not less than 3 parts water to 1 part concentrate.

Reduced Acid Orange Juice (CFR Title 21 146.148). This product is the same as frozen concentrate orange juice except that it must employ the use of anionic ion exchange as permitted in 173.25 of Title 21 in reducing the acid levels, and the B/A ratio must be between 21 and 26.

Canned Concentrated Orange Juice (CFR Title 21 146.150). This product is the same as frozen concentrated orange juice except that it is not frozen, and it is canned and heat-treated so as to prevent spoilage.

Orange Juice for Manufacturing (CFR Title 21 146.151). This product is the same as orange juice except that it can have a lower Brix and B/A ratio than required for orange juice and can contain up to 10% tangerine juice or its hybrids (*C. reticulata*). Unwashed pulp and orange oil also may be added, and it may be heat-treated, chilled, and/or frozen.

Orange Juice with a Preservative (CFR Title 21 146.152). This product is the same as orange juice for manufacturing except that a declared preservative has been added.

Concentrated Orange Juice for Manufacturing (CFR Title 21 146.153). This product is the same as frozen concentrated orange juice except that the Brix and B/A ratio may be less than that required for frozen concentrated orange juice. However, this product must be concentrated to at least 20°Brix.

Concentrated Orange Juice with a Preservative (CFR Title 21 146.154). This product is the same as concentrated orange juice for manufacturing except that it contains a declared preservative.

Grapefruit Juice (CFR Title 21 146.132). This standard is similar to the orange juice standard except that the use of a mechanical means of juice extraction is mentioned, and it must come from fruit of the *C. paradisi* species. Only 10% by volume of this product can come from grapefruit hybrids, and no more than 15% can come from grapefruit concentrate. Grapefruit pulp, oil, and essence can be added, as well as the same declared sweeteners mentioned in the pasteurized orange juice standard. A minimum of 10°Brix exclusive of added sweeteners is required, with the Brix corrected for acid by adding 0.012 + $0.193x - 0.0004x^2$, where x is the % acid. The product may be heat-treated, canned, or frozen.

Lemon Juice (CFR Title 21 146.114). This product is made from fruit of the *C. limon* (L.) Burm. f. species and has a similar standard to those for orange and grapefruit juices. Acceptable additives include reconstituted lemon concentrate so as not to increase the acidity by more than 15% of the finished food, water, lemon oil, and preservatives. The juice can be manufactured from lemon concentrate, but it must have a temperature-, but not acid-, corrected Brix of at least 6, and a titratable acidity of 4.5% if this is done. Lemon juice can be heat-treated, canned, chilled, or frozen. It also must fill 90% of a standard container unfrozen.

Frozen Concentrate for Lemonade (CFR Title 21 146.120). This product is made from lemon juice along with any combination of suitable nutritive carbohydrate sweeteners. It must have a corrected Brix of at least 48.0, with the acidity in the final product being at least 0.70 g/100 ml with a concentration of at least 10.5°Brix. Acceptable additives include lemon pulp and lemon oil.

Frozen Concentrate for Artificially Sweetened Lemonade (CFR Title 21 146.121). This standard is the same as that for frozen concentrate for lemonade except that in lieu of nutritive sweeteners it is sweetened with one or more of the artificial sweeteners listed in sections 172, 180, or 184 of Title 21, such as Aspartame. If such artificial sweeteners are used, the Brix requirement mentioned in the previous standard does not apply. Dispersing agents can be used to disperse the lemon oil, as well as thickening agents, as long as they are not food additives as defined in section 102(s) of Title 21. If the thickening agents are defined as food additives, they can be used, provided that they comply with section 409 of Title 21.

Economic Fraud

As mentioned in Chapter 15, citrus juices pose no health risk to consumers because no pathogenic organisms can survive the juice acidity. However, economic fraud is a very real concern to citrus processors. Economic fraud basically consists of the use of a cheaper material to formulate a juice without declaring it on the label, and selling it as if it contained the more expensive but authentic juice. Greed among manufacturers of citrus products has led to economic fraud in the industry, and such fraud drives prices down and hurts the honest citrus juice manufacturer besides deceiving consumers. Some consumers have allergic reactions when certain changes are made in juice composition, and if these changes are not declared on the label, a serious health problem may arise in those rare but very real circumstances. If a processor purchases adulterated juice, even in ignorance, and uses it to formulate its own products, both of the companies involved become legally liable for the adulteration. For this reason, it behooves every citrus processor to be vigilant in the detection of juice adulteration. Failure to do so can be very costly. Those companies convicted of adulteration face millions of dollars in fines, including the costs of the investigations, costly recalls, and years of prison terms for management personnel so involved (see "3 Charged in Sale of Phony Orange Juice," *Los Angeles Times*, Thursday, July 27, 1989, page 14 of part I).

Besides the specific descriptions found in the standards of identity, section 402(b) of the Food, Drug, and Cosmetic Act gives four charges that can be used to describe adulteration (Stutsman 1988):

1. If any valuable constituent has been in whole or in part omitted or abstracted therefrom.
2. If any substance has been substituted wholly or in part therefor.
3. If damage or inferiority has been concealed in any manner.
4. If any substance has been added thereto or mixed or packed therewith so as to increase its bulk or weight, or reduce its quality or strength, or make it appear better or of greater value than it is.

An example of violation of the first charge would be the removal of vitamin C using activated charcoal. Activated charcoal has long been known to remove limonin bitterness from citrus juices as well as sensitive anthocyanin pigments that may turn brown and produce off flavors. However, activated charcoal also removes significant amounts of vitamin C. The addition of undeclared vitamin C to make up for this would constitute a violation of the second charge. Adding artificial colors to mask poor juice color would be a violation of the third charge. The addition of cheaper carbohydrates would be a clear violation of the fourth charge.

TYPES OF ECONOMIC ADULTERATION

There are many types of economic adulteration. However, nearly every type of adulteration that has been known to occur in the industry falls into one of four different categories: water addition, carbohydrate addition, cover-up, and blending of unauthorized juices or juice products.

Water Addition

Some of the standards of identity exclude the addition of water. These standards involve single-strength juices that are neither concentrated nor made from concentrate. These juices include lemon juice, grapefruit juice, orange juice, frozen orange juice, pasteurized orange juice, canned orange juice, orange juice for manufacturing, orange juice with preservative, and canned tangerine juice. The last is defined under USDA grade standards rather than as a separate standard of identity. On the other hand, the citrus juice products whose standard of identity allows for the addition of water include orange juice from concentrate, frozen concentrated orange juice, reduced acid frozen concentrated orange juice, concentrated orange juice for manufacturing, and concentrated orange juice with a preservative.

Water comprises over 80% of the composition of natural juices, so detection of its illicit addition can be difficult; but it can be done in one of two ways. One method is to monitor the mineral content of the juice. Natural juices have a certain range of mineral levels as follows:

	Orange	*Grapefruit*
Sodium	< 50 ppm	< 50 ppm
Potassium	> 1400 ppm	300–500 ppm
Calcium	65–120 ppm	100–150 ppm
Magnesium	95–170 ppm	90–140 ppm
Phosphorous	120–310 ppm	—

Potable water from the city, which is used in many plants, may contain much higher levels of these minerals, and high mineral levels would indicate illicit water addition. Because pulp washing utilizes plant water, high mineral levels also may be indicative of pulp wash addition although legal water addition can occur in citrus concentrates without the addition of pulp wash. Minerals can be determined by using a variety of methods, which usually involve instrumentation beyond the resources of most quality control laboratories. Professional laboratories provide an economic means of determining mineral levels.

Treated water naturally low in minerals could not be detected through mineral analyses. However, a method has been developed that can distinguish between water that is natural to the juice and water added from other sources (Winters et al. 1988; Brause et al. 1984). This method involves the measurement of the oxygen isotope ratios. When water is absorbed into the plant through the root system, it has a tendency to lose the lighter ^1H and ^{16}O isotopes through evapotranspiration and to keep the heavier deuterium (^2H) and ^{18}O isotopes. The oxygen isotope ratios are determined by using a complex procedure, also recommended for professional laboratories, and the results are reported as $\delta^{18}O$ by use of the following equation:

$$\delta^{18}O\,(\text{ppt}) = (R_{\text{sample}} - R_{\text{standard}})\,(1000)/R_{\text{standard}} \qquad (19\text{-}1)$$

In this equation the R values are the $^{18}O/^{16}O$ isotope ratios. For concentrated orange juice above 60°Brix, the $\delta^{18}O$ value should be above $+10$. For authentic single-strength orange juice the $\delta^{18}O$ value should be positive. Anything less than these values is indicative of water addition.

Carbohydrate Addition

Because citrus juices are comprised primarily of carbohydrates and water, the addition of cheaper carbohydrates to authentic juices has been a lucrative way to falsify citrus products. Citrus juices are sold by the weight of the soluble solid equivalent of carbohydrates. When commercial sugars are sold as low as 5¢/lb, and the equivalent orange juice price is up to $2.00/lb soluble solids, one can see how much the profit margin can be increased with substitution of commercial sugars or similar products.

Several methods can be used to detect carbohydrate adulteration. One of the simplest is based on the sugar composition of the juice itself. Citrus juices contain sucrose, glucose, and fructose, in a ratio of about 2 : 1 : 1. Glucose and fructose are known as reducing sugars because they can be oxidized by using mild oxidizing agents. Sucrose, however, cannot be oxidized by mild oxidizing agents because its anomeric carbon forms the bond between its glucose and fructose monomers. Thus a measure of the reducing sugars, a relatively easy determination, should result in the determination of about half of the carbohydrates, which in turn constitutes about 80 to 90% of the total soluble solids of the Brix value. The use of high-fructose corn syrup or cane sugars (predominately sucrose) can be detected quickly by the following method. However, invert beet sugars or special combinations of reducing and nonreducing sugars cannot be detected from a reducing sugar test; so this method should constitute a preliminary screening and not serve as a verification of authenticity.

The following reducing sugar test is based on the reduction of Cu^{2+} to Cu^{1+} by the reducing sugar. The Cu^{2+} ions usually form a light blue color in aqueous solution, when they are reduced to Cu^{1+}, an orange-colored oxide precipitate is formed, according to the following equation:

$$2Cu^{2+} + H_2O + 2C_6H_{12}O_6 \rightarrow Cu_2O(s) + 6H^+ + 2C_6H_{10}O_6^- \quad (19\text{-}2)$$
$$\text{blue} \qquad\qquad \text{sugar} \qquad \text{orange}$$

The reaction is catalyzed by tartrate, and methylene blue dye is used as an indicator. When the copper has all been reduced, the dye indicator is quickly reduced, resulting in a dramatic color change from the deep navy blue of the indicator to the bright orange of the copper oxide.

Reducing Sugars Test (AOAC 1984)

Equipment and Supplies

- 10 ml buret, 125 ml Erlenmeyer flask or beaker, and magnetic stirrer/ heater for titration.
- Copper solution. (Dissolve exactly 69.278 g of $CuSO_4 \cdot H_2O$ in 1 liter of distilled water.)
- Tartrate solution. (Dissolve 346 g of potassium sodium tartrate (Rochelle salt) and 100 g of NaOH in 1 liter of distilled water.)
- 1% methylene blue indicator. (Dissolve about 1 g in 100 ml of distilled water.)
- Two 2 ml pipettes.
- Lab centrifuge.
- Boiling chips.
- 10 ml graduated cylinder.

Procedure

1. Centrifuge reconstituted or single-strength juice (11.8°Brix) for 10 minutes as outlined in Chapter 7. Add 1 ml of the serum to the 10 ml graduated cylinder, and fill it to 10 ml with water.
2. Pipette 2 ml of the copper solution and 2 ml of the tartrate solution into the 125 ml Erlenmeyer flask, and fill the flask to about 50 ml with distilled water. Add a few boiling chips, and heat this solution to boiling.
3. Pour the centrifuged juice sample serum into the 10 ml buret, and zero it.

4. Titrate the copper solution while stirring with the sample serum until a dark orange color is observed. Add 3 drops of the indicator, and continue titrating until the bright orange of the copper oxide is fully restored.

5. The % reducing sugars (RS) can be calculated from the following:

$$\%RS = \frac{(0.002 \text{ liter Cu})\,(69.278\text{g}/1 \text{ Cu})\,(180.16\text{g}/\text{mole sugar})}{(249.68\text{g}/\text{mole Cu})(4 \text{ moles Cu}/\text{mole sugar})} \quad (19\text{-}3)$$

or:

$$\%RS = 23.93/(\text{ml titrated}) \quad (19\text{-}4)$$

For example:

$$23.93/4.77 \text{ ml titrated} = 5.02\% \text{ reducing sugars}$$

Levels below 4.5 or above 5.5 might be suspect in regards to carbohydrate adulteration.

A surer way of determining the amount of nonjuice carbohydrates in citrus juices involves the use of $^{13}C/^{12}C$ ratios (Bricout and Koziet 1987). These ratios usually are expressed in the same manner as the $\delta^{18}O$, as follows:

$$\delta^{13}C(\text{ppt}) = (R_{\text{sample}} - R_{\text{standard}})\,(1000)/R_{\text{standard}} \quad (19\text{-}5)$$

It has been found that there is a difference in this ratio for different plants in nature. Simpler C_4 plants, such as corn and sugar cane, have $\delta^{13}C$ values around -10, whereas higher C_3 plants, such as flowers and trees (citrus), have a $\delta^{13}C$ value below -22. Therefore, if sugar from cane or corn is used in citrus juices, a higher $\delta^{13}C$ value will be observed. The analytical methods required for such a determination are, again, expensive and complex and best left to professional laboratories.

Another procedure has been developed (Low 1989) to detect invert beet sugar down as low as 1%, and perhaps even as low as 0.1%. This method employs the use of HPLC with a pulse amperometric detector that is several orders of magnitude more sensitive than refractive index detectors. The method detects the adulterating carbohydrates by determining the oligosaccharide pattern. With this new technique, any illicit carbohydrate adulteration can be detected.

Cover-Up

When juice has been adulterated, it often becomes necessary to add something else in order to cover up the adulteration. For example, one of the simplest tests used in detecting the illicit addition of pulp wash has been the formol or formaldehyde test, which measures the amount of primary amines. Pulp wash has a lower concentration of amino acids than authentic juice has, and its illicit addition may be detected by this very fast and simple technique. However, it is possible to get around the test with the addition of very cheap amino salts—the addition of the amino salts being a second adulteration used to cover up the first adulteration. Because it is so easily circumvented, the formol test has become a weak test for juice adulteration. However, it is useful in other parts of citrus quality control, as mentioned in the preceding chapter, and as a preliminary adulteration screening.

Another common cover-up technique is the addition of colors. When a juice is adulterated with something other than juice, it must become more dilute in authentic components including colors. The addition of turmeric, annatto, or other dyes (FD&C dyes), which makes the juice a darker orange, has been used in adulteration cover-ups. However, these dyes can be easily tested for by any quality control laboratory using the following techniques.

Turmeric is prepared from the dried roots of the tuberous rhizomes of *Curcuma longa* L., a member of the ginger family. It is native to southern and southeastern Asia and is commercially cultivated in Malaysia. The color component of turmeric is curcumin, also known as turmeric yellow, with the following chemical structure:

Annatto dyes are natural carotenoid colorants derived from the seed of the tropical annatto tree or *Bixa orellana*. The primary color component is bixin, a carotenoid with the following chemical structure:

The water-soluble form of bixin is norbixin, which is formed from the saponication of the methyl ester group. A single test can be used to detect either annatto or turmeric. Even though you can continue the test to determine which dye has been added, the presence of either dye indicates adulteration.

Turmeric and Annatto Determination (AOAC 1980)

Equipment and Supplies

- 1-liter Erlenmeyer flask.
- 100 ml graduated cylinder.
- 500 ml graduated cylinder.
- Magnetic stirrer.
- Rotary evaporator or equivalent.
- 500 ml separatory funnel.
- Two 150 ml Erlenmeyer flasks or beakers.
- 600 ml of 95% ethanol.
- Vacuum filter apparatus.
- Ammonium hydroxide (reagent grade).
- About 400 ml hexane.
- Glacial acetic acid.
- Curcumin and annatto for color comparison.

Procedure

1. Place 100 ml of reconstituted or single-strength orange juice (11.8°Brix) in the 1-liter Erlenmeyer flask.
2. Add 600 ml of 95% ethanol, stir the mixture occasionally, and let it stand for several hours. Then filter it.
3. Evaporate the filtrate to 100 ml using a rotary evaporator or the equivalent, and pour it into a 500 ml separatory funnel.
4. Add 25 ml of 25% NaCl and enough ammonium hydroxide to make the solution alkaline.
5. Extract the lipids and oily pigments from the solution using 200 ml of hexane.
6. Immediately acidify the bottom aqueous layer with acetic, and extract it with hexane. If the hexane layer is colored, then turmeric or annatto is present. To see the difference between the presence and the absence of turmeric or annatto, add enough curcumin or annatto to the juice sample to see a detectable color change. Then repeat the test. The presence of turmeric or annatto should be readily discernible by comparison with samples with no dyes present.

This test is a pass/fail test. Any discernible deepening of the color in the last hexane layer indicates the presence of these dyes.

The Food and Drug Act of 1906 certified 7 dyes that could be used in foods. By 1938, when the Food, Drug, and Cosmetic Act was passed, 11 more dyes were included, and thereafter these dyes were referred to as FD&C dyes by number. The azo dyes, FD&C Yellow No. 5 and FD&C Yellow No. 6, have been used to restore diluted orange juice color. FD&C No. 5 is also called tartrazine or hydrazine yellow. It has the following chemical structure:

In 1856 dyes from coal tar were discovered, and for the next 80 years these dyes were used extensively in food products. More modern synthetic techniques generally have replaced the extraction of these dyes from coal tar. The presence of coal tar dyes and of FD&C dyes can be detected simultaneously using the following simple test. Again this is a pass/fail test. Any detection of these dyes would constitute adulteration in 100% orange juice products.

FD&C and Coal Tar Dyes (Ting and Rouseff 1986)

Equipment and Supplies

- 1 g of calcium carbonate.
- Acetone and petroleum ether in the ratio of 1:3.
- 500 ml separatory funnel.

Procedure

1. Add 1 g of calcium carbonate to 5 ml of reconstituted or single-strength orange juice (11.8°Brix).
2. Add 30 ml of the acetone/petroleum ether solution, shake the mixture for 3 minutes, and allow the layers to separate.
3. If the aqueous layer is orange-yellow, it has FD&C or coal tar dyes added. A slight yellow color is acceptable. Just enough FD&C colors should be added to a juice sample to cause a noticeable color change, which should be compared to the color of authentic samples.

Other colorants and cover-up compounds can be and have been used to cover up citrus juice adulteration. Because any adulteration involves juice dilution, the measurement of authentic juice components that are too expensive or impossible to falsify can be used to detect juice dilution rather than adulterant addition. One such component is isocitric acid, for which there is no commercial source. If the isocitric acid content falls below 44 ppm, one can be assured that the authentic juice has been diluted. Affordable test kits are available for the determination of isocitric acid, which can be used routinely in any quality control laboratory (Boehringer Mannheim Biochemicals, Indianapolis, Ind.).

Addition to Unauthorized Juices or Juice Products

Most orange juice products are allowed to contain up to 10% tangerine or tangerine hybrids while still being called 100% orange juice. The high color of tangerine juice is used to enhance the appearance of orange juices of weaker color, primarily in Florida. However, other fruit juices usually are less expensive than tangerine juice and have been illegally added to orange juices in direct violation of the standards of identity. Apple, pear, white grape, orange pulp wash, and grapefruit juices usually are sold at lower prices than that of orange juice, and when they are blended into orange juice products, their economical returns are augmented greatly. Apple, pear, and white grape can be detected by using isocitric tests or other more complex testing schemes based on compositional differences. Grapefruit juice contains the characteristic compound naringin, so naringin determinations can be used to detect grapefruit adulteration in orange juice (see Chapter 10 for analytical methods).

Federal law allows pulp wash to be added to citrus juices on-line in an amount that constitutes about 5 to 10% of the juice, but if the pulp wash is separated and packaged separately, it can no longer be legally added to other citrus products. Florida law prohibits any addition of pulp wash to any 100% citrus juices, and in order to detect illicit pulp wash addition, Florida requires the addition of sodium benzoate to all pulp wash juices to act as a tracer. If sodium benzoate is found in 100% orange juice products, it is assumed that Florida pulp wash was used as an adulterant. However, failure to find the benzoate tracer is no guarantee that the product does not contain illicit pulp wash. Pulp wash is generally used in drink bases that add preservatives, such as benzoates, and the pulp wash product itself is not damaged by the benzoate addition.

ADULTERATION SCREENING PROCEDURES

Several adulteration screening procedures are used around the world, ranging from special computerized packages available from a single manufacturer to specialized techniques developed by individual laboratories to government-regulated procedures and criteria. These procedures are constantly in a state of development and change, and new methods and technologies are constantly appearing (Nagy, Attaway, and Rhodes 1988; Fruit Juice Adulteration Workshop 1989). However, most of these techniques are beyond the scope and resources of most industrial quality control laboratories; so professional laboratories usually are consulted regarding the analytical work. However, a knowledge of the standard methods can be helpful to any citrus quality control facility.

Europe

Northern Europe has seen the establishment of some of the most sophisticated adulteration procedures in the world, and presently three standards are being used in Europe: the German RSK (Hofsommer 1989; Association of the German Fruit Juice Industry 1987), the French AFNOR (Hofsommer 1989; Union Nationale des Producteurs de Jus de Fruits 1989), and a criterion used by the Netherlands and Belgium, referred to here as the Dutch standard (Hofsommer 1989; Warenwet P.B.O.-voorschriften (E-17a) 1977). The European Community is working toward a single EC standard but has encountered some difficulty. One problem, for example, is that many different products are of concern to these countries. The following listing shows the juice products that each standard covers:

AFNOR (French)	RSK (German)	Dutch
Apple	Apple	Apple
Grape	Apricot	Grape
Grapefruit	Black currant	Grapefruit
Lemon	Grape	Orange
Orange	Grapefruit	
Pineapple	Lemon	
Tomato	Orange	
	Passion fruit	
	Pear	
	Raspberry	
	Sour Cherry	

With so many different products, it is difficult to establish a single unified standard. Also, the use of standards does not guarantee the absence of adulteration, as standards can be circumvented. However, properly used standards can constitute a significant deterrent to juice falsification. Tables 19-1 through 19-3 show the ranges of juice components that characterize each of the three European standards. Juices entering the affected countries must meet these standards.

United States

As yet, the United States does not have an official standard for authentic juices that compares with the European standards. The standards of identity simply declare that the juice must be authentic and describe the natural components permitted in it, such as pulp, citrus oils, and so on. When citrus juices are imported into the United States, or when a commercial product is suspected of adulteration, any and all means can be used to prove such adulterations, including employee depositions, chemical tests, and/or inspection of company records. Imported juices usually go through a screening process by customs and FDA officials that is somewhat easy to monitor, even though it is impossible to thoroughly analyze every imported lot. However, the inspection of domestic juice is not required, and the monitoring of adulteration is much more difficult domestically. The most likely person to detect domestic adulteration is a competitor or a customer of the juice adulterator. Adulteration may be suspected if a product is selling for too low a price, or an in-house screening procedure may unearth suspect material. A competitor may wish to notify the FDA and pursue criminal prosecution of the adulterator, while a customer may wish just to cease doing business with a suspected supplier. The latter does not require as much rigorous proof.

Even a rumor that a company is adulterating its juices can cause it difficulty in marketing its products. Even if a company purchases blend components from a suspected company, the purchaser's reputation may be on the line. Not only are reputations of companies at stake, but also the reputations of management personnel, especially quality control management. It would be difficult to convince a judge that quality control personnel were entirely ignorant of adulteration that was going on within the company because it is their specific job to monitor the composition of the company's products.

The adulteration tests are quite numerous and complex, including sophisticated computerized systems that are now on the market. However, the resources of most quality control facilities are usually quite limited and exclude the bulk of these techniques. Professional laboratories have tried to find an economic and thorough method of adulteration detection that can be of use to any citrus quality control laboratory. One such method is the matrix method of detecting juice adulteration (Brause et al. 1984), which consists of the measurement of

Table 19-1. French orange juice authenticity standards (AFNOR) (Hofsommer 1989; Union Nationale des Producteurs de Jus de Fruits 1989).

Parameter	Acceptable Range
Density at 20°C	≥ 1.040
Total solids (g/100 g)	≥ 10.5
Titratable acidity (mmol/L)	100–250
Refractometer dry extract (%)	≥ 10.2
Ash (g/L)	2.8–5.5
Ethanol (g/L)	≤ 3
Sulfur dioxide (mg/L)	≤ 10
d-Limonene number (ml/L)	≤ 0.3
Copper (mg/L)	≤ 5
Zinc (mg/L)	≤ 5
Iron (mg/L)	≤ 15
Lead (mg/L)	≤ 0.3
Arsenic (mg/L)	≤ 0.2
Cadmium (mg/L)	≤ 0.05
Tin (mg/L)	≤ 100
Mercury (mg/L)	≤ 0.01
Sodium (mg/L)	≤ 40
Potassium (mg/L)	1100–2500
Calcium (mg/L)	50–240
Magnesium (mg/L)	70–170
Phosphorus (mg/L P)	100–225
Pulp (% v/v)	≤ 15
Total carotenoids (mg/L)	3–26
β-carotene (mg/L)	≤ 1.8
Naringin (mg/L)	≤ 10
Hesperidin (mg/L)	150–1000
Sucrose (g/L)	5–50
Glucose (g/L)	20–50
Fructose (g/L)	20–50
Maltose (g/L)	0
Glucose/fructose ratio	0.8–1.2
Total sugars (g/L)	80–115%
L-Malic (g/L)	0.6–4
Tartaric acid	0
Citric acid (g/L)	6–16
Isocitric acid (mg/L)	60–200
Ascorbic acid and dehydroascorbic acid (mg/L)	≥ 200
Aspartic acid (mg/L)	100–500
Threonine (mg/L)	10–60
Serine (mg/L)	80–250
Asparagine (mg/L)	180–1000
Glutamic acid (mg/L)	50–200
Proline (mg/L)	400–2500
Glycine (mg/L)	10–40
Alanine (mg/L)	45–200
Valine (mg/L)	10–40

Table 19-1. (*Continued*)

Parameter	Acceptable Range
Methionine (mg/L)	1–5
Isoleucine (mg/L)	2–15
Leucine (mg/L)	2–15
Tyrosine (mg/L)	15–20
Phenylalanine (mg/L)	15–50
γ-Aminobutryric acid (mg/L)	150–800
Ornithine (mg/L)	5–20
Lysine (mg/L)	20–80
Histidine (mg/L)	5–30
Arginine (mg/L)	400–1400
Total free amino acids (mg/L)	2000–5000

21 parameters, as shown in Table 19-4. Many of these tests can be performed in a citrus quality control laboratory, but, as with any set of standards, analysts should be careful in interpreting the results. The FDA has established a pattern recognition method, involving many juice parameters and using specialized

Table 19-2. Dutch orange juice standards (Hofsommer 1989; Warenwet P.B.O.-voorschriften (E-17a) 1976).

Parameter	Acceptable Range
(Sugar?) extract (g/L)	≥ 25
(Acid?) (g/L)	≥ 3.3
Potassium (mg/L)	≥ 1500
Phosphorous (mg/L)	≥ 120
Hesperidin (mg/L)	≤ 1000
Proline (mg/L)	≥ 575
Pectin (mg/L)	≤ 700
D-Isocitric acid (mg/L)	≥ 70
Ratio citric acid/D-isocitric acid	≤ 130
Ratio glucose/fructose	≤ 1.00
Pulp (%)	≤ 10
(Formol number?) (ml/100 ml)	≥ 15.0
Aspartic acid (mmol/L)	≥ 1.7
Serine (mmol/L)	≥ 1.0
Asparagine (mmol/L)	≥ 1.7
Glutamic acid (mmol/L)	≥ 0.5
Alanine (mmol/L)	≥ 0.7
Aminobutyric acid (mmol/L)	≥ 1.7
Arginine (mmol/L)	≥ 2.5
Glycine (mmol/L)	≤ 0.3
Glutamine (mmol/L)	≤ 0.5
Acid-corrected sugar extract	≥ 11.2°Brix

Table 19-3.   German orange juice standards (RSK) (Hofsommer 1989; Association of the German Fruit Juice Industry 1987).

Parameter	Acceptable Range
Color (points)	≥ 3
Aroma (points)	≥ 3
Flavor (points)	≥ 5
Relative density (g/ml 20°/20°C)	≥ 1.0450
Corrected Brix	≥ 11.18
Soluble solids (g/L)	≥ 116.8
Titratable acids (pH 7.0)	
as tartaric acid (g/L)	≥ 8.0
Sulfurous acid (mg/L)	≤ 10
Ethanol (g/L)	≤ 3.0
Volatile acids as acetic acid (g/L)	≤ 0.4
Lactic acid (g/L)	≤ 0.5
L-Malic acid (g/L)	≤ 2.5
Citric acid (g/L)	≥ 8.0
D-Isocitric acid (mg/L)	≥ 70
Citric acid/D-isocitric acid ratio	≤ 130
Tartaric acid (g/L)	0
Glucose (g/L)	≥ 22
Fructose (g/L)	≥ 24
Glucose/fructose ratio	≤ 1.0
Sucrose (g/L)	≤ 45
Sucrose (% in total sugar)	≤ 50
Reduction free extract (g/L)	≥ 26
Ash (g/L)	≥ 3.5
Alkalinity number	≥ 11.0
Potassium (mg/L)	≥ 1700
Sodium (mg/L)	≤ 30
Magnesium (mg/L)	≥ 90
Calcium (mg/L)	≤ 110
Chloride (mg/L)	≤ 60
Nitrate (mg/L)	≤ 10.0
Phosphate (mg/L)	≥ 400
Sulfate (mg/L)	≤ 150
Formol number	
(ml 0.1 mol NaOH/100 ml)	≥ 18
Corrected formol number	≥ 90% of formol no.
Proline (mg/L)	≥ 575
L-Ascorbic acid (mg/L)	≥ 200
Flavonoid glycosides	
as hesperidin (mg/L)	≤ 1000
Water-soluble pectins expressed as	
galacturonic acid anhydride (mg/L)	≤ 500
Aspartic acid (mmol/L)	1.7–3.0
Threonine (mmol/L)	0.10–0.30
Serine (mmol/L)	1.0–1.8
Asparagine (mmol/L)	1.7–4.5

Table 19-3. (*Continued*)

Parameter	Acceptable Range
Glutamic acid (mmol/L)	0.5–1.1
Glutamine (mmol/L)	≤ 0.5
Proline (mmol/L)	3.9–11.3
Glycine (mmol/L)	0.15–0.30
Alanine (mmol/L)	0.7–1.5
Valine (mmol/L)	0.07–0.23
Methionine (mmol/L)	trace–0.03
Iso-leucine (mmol/L)	0.025--0.065
Leucine (mmol/L)	0.020–0.060
Tyrosine (mmol/L)	0.025–0.10
Phenylalanine (mmol/L)	0.8–0.30
γ-Aminobutyric acid (mmol/L)	1.7–3.5
Ornithine (mmol/L)	0.025–0.10
Lysine (mmol/L)	0.15–0.40
Histidine (mmol/L)	0.03–0.12
Arginine (mmol/L)	2.5–6.0
Ammonia (mmol/L)	≤ 1.5
Ethanolamine (mmol/L)	≤ 0.6
Total carotenoids (mg/L)	≤ 15
β-Carotene as % total carotenoids	≤ 5
Cryptoxanthinester as % total carotenoids	≤ 15
Pectic substances as galacturonic acid anhydride	
oxalate-soluble pectin (mg/L)	≤ 200
alkali-soluble pectin (mg/L)	≤ 300

computer software called ARTHUR (Page 1986), which enables investigators to determine when a juice does not fit the normal composition.

In-house Methods

The purpose of this chapter is to acquaint citrus quality control personnel with the problem of citrus juice adulteration and some of the things that have been and are being done to correct the problem. Every quality control laboratory should have some sort of screening system for all inbound products used in the formulation of its products. Simple chemical tests such as those mentioned here can constitute preliminary screenings. If these tests or information gathered from other sources suggest the possibility of adulteration, more elaborate tests should be performed. Periodic testing by professional laboratories is highly advisable, and such spot checks are best done randomly.

Some observers have suggested the establishment of in-house or company standards for the products the company purchases. The establishment of specifications is a common practice, but it rarely if ever includes standards specifi-

Table 19-4. Specifications for orange juice
according to the matrix method of determining
authenticity used commercially in the United States
(Brause et al. 1984).

Parameter	Acceptable Ranges
Fructose (%)	2–3
Glucose (%)	2–3
Sucrose (%)	3–5.5
Total sugars (%)	7.2–10.8
Glucose/fructose ratio	1.00
Total sugar of Brix (%)	61–91.5
Sucrose of total sugars (%)	30–60
Sodium (ppm)	≤ 50
Potassium (ppm)	≥ 1400
Calcium (ppm)	65–120
Magnesium (ppm)	95–170
Phosphorous (ppm)	120–310
Naringin (ppm)	<2
Sodium benzoate	0
Sorbitol	0
Isocitric acid (ppm)	≥44
UV-VIS-fluorescence (A_{443}/A_{325})	≥0.100
Fluorescence	2 peaks @ 270–306 nm
Stable isotope ratio	≥ +10.0 (at 63°Brix)
as $\delta^{18}O$ (OSIRA) (%)	≥0 (at 42°Brix)
	≥0 (fresh juice)
	≤0 (recon juice)
$\delta^{13}C$ SIRA	≤ −22.0

cally designed to prevent juice adulteration. The concept behind establishing adulteration standards is the idea of rejecting a product on the basis of violation of specifications, rather than having to prove that the product was adulterated. On the other hand, if the supplier were unaware of the exact adulteration screening that was being used, it would be more difficult to cover up any falsification. An appropriate screening program will consist of a blend of tests that can be performed easily and economically, the use of professional laboratories for more expensive and complex tests, cooperation from upper management including purchasing and marketing, thoroughness, and common sense.

QUESTIONS

1. When was the Food, Drug, and Cosmetic Act enacted, and in what section of the Act is adulteration specifically mentioned?

2. What are the four types of adulteration according to the FDA? Give an example of each.
3. What are the four charges or descriptions of adulteration according to Section 402(b) of the Act? Give an example of each.
4. What are the four main types of economic adulteration that occurs in the citrus industry?
5. What citrus products can one legally add water to?
6. What is the surest way to detect water addition to citrus juices? Then why are other methods used?
7. Why are sugar or sugar products added to citrus juices?
8. What limitations does the carbon isotope ratio determination have in detecting illicit carbohydrate addition to citrus juices?
9. What is the surest method to detect adulteration cover-up, and why is it so definitive?
10. How much grapefruit juice can one blend with concentrated orange juice for manufacturing? How much tangerine juice?
11. How much pulp wash (%) can be contained in orange juice according to Florida standards? U.S. standards?
12. What are the 3 standards used in Europe, and to which countries do they apply?
13. What is the procedure used by the U.S. government (FDA)?
14. Who does most of the policing of adulteration in the United States?
15. What is the adulteration screening system used by your company? What feasible screening process would you recommend?

PROBLEMS

1. What could you say for an orange juice sample that had the following mineral composition?
Na: 55 ppm
Ca: 126 ppm
K: 1565 ppm
Mg: 165 ppm
P: 306 ppm
2. Suppose that in three reducing sugar tests you titrated the following volumes (ml). Give the % reducing sugars and interpret the results. What would be the appropriate action in each case?

ml Titrated	% Reducing Sugars	Action
5.21		
6.03		
3.12		

3. What would $\delta^{18}O$ value of -0.15 indicate? What would a $\delta^{13}C$ value of -20.5 indicate?
4. A typical analysis of single-strength orange juice in the United States might be 0.65% w/w acid as citric acid, with a Brix/acid ratio of 18.2. Would this product pass the RSK screening? Explain why or why not, and give reasons why the RSK may be faulty in some circumstances.
5. Would the product in problem 4 meet the Dutch standards? How about the AFNOR standards?

REFERENCES

AOAC 1980. *Official Methods of Analysis, 13th Edition*, 14.154. Association of Official Analytical Chemists, Washington, D.C.

AOAC. 1984. *Official Methods of Analysis, 14th Edition*, 31.034–31.036. Association of Official Analytical Chemists, Washington, D.C.

Association of the German Fruit Juice Industry. 1987. *RSK Values, The Complete Manual*. Flüssiges Obst GmbH, Bonn, Federal Republic of Germany, 1987.

Brause, A. R., Raterman, J. M., Petrus, D. R., and Doner, L. W. 1984. Verification of authenticity of orange juice, *J. of the A.O.A.C.*, *67*, 535.

Bricout, J. and Koziet, J. 1987. Control of the authenticity of orange juice by isotopic analysis, *J. Agric. Food Chem.*, *35*, 758–760.

Fruit Juice Adulteration Workshop, August 9–10, 1989, Herndon, Va. Sponsored by General Physics Corporation, Columbia, Md.

Hofsommer, H. 1989. Analytical methodologies of detection as practiced in Europe. Presented at the Fruit Juice Adulteration Workshop, August 9–10, 1989, Herndon, Va. Sponsored by General Physics Corporation, Columbia, Md.

Low, N. 1989. Detection of beet medium invert sugar adulteration in high carbohydrate foods. Presented at the Fruit Juice Adulteration Workshop, August 9–10, 1989, Herndon, Va. Sponsored by General Physics Corporation, Columbia, Md.

Nagy, S., Attaway, J. A., and Rhodes, M. E. eds. 1988. *Adulteration of Fruit Juice Beverages*. Marcel Dekker, Inc., New York.

Page, S. W. 1986. Pattern recognition methods for the determination of food composition, *Food Tech.*, *40(11)*, 140–109.

Stutsman, M. J. 1988. Fruit juice adulteration, an overview of compliance and regulatory issues. Presented at the Fruit Juice Adulteration Workshop, July 21, 1988, Herndon, Va. Sponsored by General Physics Corporation, Columbia, Md.

Ting, S. V. and Rouseff, R. L. 1986. *Citrus Fruits and Their Products*. Marcel Dekker, Inc., New York, 188.

Union Nationale des Producteurs de Jus de Fruits. *AFNOR-UNPJF Fruit Juices Specifications*. Association Francaise de Normalisation, 1989.

Warenwet P. B. O.-voorschriften (E-17a), Verordening van 9 December 1976, Vb.Be.afl. 9, d.d. 7-3-1977 van bei Produkischap van Groenten en Fruit.

Winters, K., Scalan, R. S., and Parker, P. L. Detection of orange juice adulteration by stable isotope analyses. Presented at the Fruit Juice Adulteration Workshop, July 21, 1988, Herndon, Va. Sponsored by General Physics Corporation, Columbia, Md.

UNIT THREE

CITRUS JUICE BY-PRODUCTS

Chapter 20

Food-Grade Nonjuice Products

Besides citrus juices, other food-grade products are manufactured from citrus fruits. The most common among them are the citrus oils and essences, discussed in Chapter 6. Pulp products also are very common, and were discussed in Chapter 7. These flavoring and texturizing materials are used extensively in juice products themselves, as well as in other foods. Other food-grade by-products will be discussed in this chapter, including pectin, jellies, jams, fruit sections, juice drinks, and purees. Peel extracts are commonly used as beverage bases in many parts of the world, mainly as a result of improved debittering techniques. Citrus flavors, including encapsulated flavors, are popular, but their production involves technology beyond the scope of this book. Much of this product information is proprietary, as confidentiality is necessary in order to maintain a healthy competitive climate. However, basic principles and products common to the industry are discussed here.

PECTIN, JELLIES, AND JAMS

The manufacturing of jellies and jams using pectins has been in existence since the French chemist Braconnat used pectins to make jellies as early as 1820. However, it was not until the 1900s that large-scale pectin production emerged, with the first pectin plant being built in Corona, California by the California Fruit Growers Exchange.

Pectin Production

Pectin has been called ''nature's glue.'' It consists of long and complex chains primarily in the form of polygalacturonic acid units (a linear galacturonglycan of α-(1–4)-linked D-galactopyranosyl-uronic acid) with molecular weights of 100,000 to 200,000 (see Chapter 8). Pectin has a solubility in hot water of about 2 to 3%, yielding a pH of 2.0 to 3.5. It belongs to a group of related compounds such as protopectin, the water-soluble parent or precursor that becomes pectin upon hydrolysis. Also, related to pectin are pectinic acid, with a higher degree of methoxylation, and pectric acid, with fewer or no methoxy groups. Pectins are classified as high ester (7-8% methoxy groups) or low ester (3-5% methoxy groups). Citrus pectins are used as thickening agents in a large variety of food, pharmaceutical, and cosmetic products, but 75% of the pectin produced is used in the manufacture of jams, jellies, marmalades, and similar products. Pectin blends better than other texturizers such as gums, starches, and carbohydrate derivatives, and exceeds gelatin in firmness up to temperatures of 120°F (49°C). Most pectin is produced in the United States, Europe, and Israel.

Most of the pectin in citrus pulp is generally found in juice sac material, with large amounts occurring in the albedo of the peel. Lemon peel is the most common source of commercial pectin. Several methods for pectin extraction have been used, including the procedure discussed in the following paragraphs.

Peel used for pectin extraction usually must undergo extraction of peel oils prior to or during juice extraction. The peel then is shredded in order to enhance pectin extraction efficiency, and the shredded peel is washed with water to remove sugars, glycosides, and other water-soluble material. Some pectin may be lost in this washing, but that amount is usually insignificant and of low grade. If the peel is not to be processed right away, it should be heated at 95 to 98°C (203-208°F) for 10 minutes in order to deactivate the pectinase enzymes that cleave the $\alpha(1-4)$ linkage and degrade the pectin. The peel then is extracted with hot dilute acid (hydrochloric or sulfuric acid at a pH of about 2.0 at 40-100°C (104-212°F) for 45-60 minutes). Lower pH values and hotter temperatures shorten the time needed for extraction. Multiple extractions under milder conditions produce higher-grade pectin. Care should be taken in monitoring and removing heavy metals, which are common impurities in technical grade acids.

This acid extraction aids in the conversion of protopectins in the peel to the water-soluble pectins. The acidic solution is filtered from the peel by using shredded paper, diatomaceous earth, or the equivalent, with the aid of filter presses or similar equipment. The filtrate contains about 1% pectin and should be concentrated to about 3 to 4% pectin. Isopropyl alcohol, ethanol, methanol, or isobutanol then is added to the solution, to 50 to 70% by weight, resulting in a precipitation of the pectin in a gelatinous mass. A hydraulic press usually is used to separate the pectin from the solution. Several 50 to 70% alcohol-in-water solutions are used to wash the pectin, followed by a final 80 to 90% alcohol aqueous washing. These washings remove about 50% of the alcohol and water. Final drying involves the use of warm air, which results in a moisture content of about 6 to 10%. The solid pectin then is pulverized and packaged as a yellowish white powder. Rapid-set pectin can be made by rinsing the pectin with an acid solution prior to drying, whereas a slow-set pectin requires standing in an acidified alcoholic bath for 10 to 20 hours. Common yields of pectin

Table 20-1. *Food Chemicals Codex* specifications
for commercial pectins (*Food Chemicals Codex*
1931).

Acid-insoluble ash	≤ 1%
Arsenic (as As)	≤ 3%
Total ash	≤ 10%
Degree of esterification of high-ester pectin	≥ 50%
Degree of esterification of low-ester pectin	≤ 50%
Degree of amide substitution of low-ester pectin	≤ 40%
Heavy metals (as Pb)	≤ 0.004%
Lead	≤ 10 ppm
Loss on drying	≤ 12%
Sodium methyl sulfate	≤ 0.1%
Total anhydrogalactouronides	≥ 70%

are in the neighborhood of 3% of the peel weight. The high cost of the solvents used in pectin extraction justify the recovery and reuse of the wash solutions.

The *Food Chemicals Codex* (FCC) has established specifications for commercial pectin along with procedures for measuring the required parameters (*Food Chemicals Codex* 1931), and these specifications are illustrated in Table 20-1. As mentioned earlier, one of the quality problems of using technical grades of acids in processing is the presence of heavy metal contaminants. The FCC specifications for acid-insoluble ash, total ash, heavy metals, and lead are designed to account for these contaminants and keep them at safe levels. Arsenic contamination also must be accounted for in processing acids. The solvent content and the degree of esterification are parameters of the pectin itself, and the sodium methyl sulfate specification provides for measuring the degree of demethoxylation that has occurred in the pectin as a result of pectinase enzymes or acid hydrolysis. The FCC procedures for the determination of arsenic, lead, and sodium methyl sulfate are not given here because they involve the use of highly toxic chemicals or complex procedures not suited to routine quality control use by laboratory technicians. It is recommended that atomic absorption or ion chromatography be used instead of FCC procedures. Sending spot samples to professional laboratories also is a viable alternative if adequate in-house facilities do not exist. The following procedures are given by the FCC (*Food Chemicals Codex* 1931).

Total Ash

Equipment and Supplies

- Crucible.
- Crucible tongs.

- Heat source (550°C or 1022°F).
- Desiccating chamber.
- Balance.
- Ethanol (15 ml).
- Glass stirring rod.

Procedure

1. Weigh 3 g of pectin in a tared crucible.
2. Ignite the heater, and heat the sample at around 550°C (1022°F), not to exceed a dull redness, until evidence of carbon (lumps) has been removed by combustion.
3. If carbonaceous material still remains, wet the charred mass with hot water, filter it with ashless filter paper, and ignite the paper and charred mass. Add the filtrate to the ash, evaporate it to dryness, and heat to dull redness.
4. If carbonaceous material still persists, cool the crucible and add 15 ml of ethanol, break up the lumps with a glass stirring rod, and then burn off the alcohol. Heat to dull redness.
5. Cool the crucible in a dessicator, and weigh it.
6. The % total ash is found by using:

$$\text{Ash}_t = \frac{W_{ash}(100\%)}{W_{sample}} \qquad (20\text{-}1)$$

For example:

$$\frac{(0.301 \text{ g ash})(100\%)}{(2.977 \text{ g sample})} = 10.1\% \text{ ash}$$

Acid-Soluble Ash

Equipment and Supplies

- Same as with the previous ash test.
- Boiling flask and heat source.
- Dilute HCl (25 ml).
- Gooch crucible or ashless filter paper.
- Hot water.
- Ignitor.

Procedure

1. Add the ash remaining from the total ash determination to a boiling flask, and add 25 ml of dilute HCl. Boil the mixture for 5 minutes.
2. Collect the insoluble material on a tared ashless filter paper or in a Gooch crucible, and wash it with hot water.
3. Ignite the solid material, and after complete combustion cool it in a desiccator.
4. Weigh the remaining ash, and calculate the acid-soluble ash in the same way as done with the total ash in the previous procedure.

Degree of Esterification

Equipment and Supplies

- 250 ml beaker.
- Magnetic stirrer.
- 250 ml Erlenmeyer flask.
- 100 ml graduated cylinder.
- 500 ml distillation flask with a Kjeldahl trap and a water-cooled condenser.
- Acid titration setup shown in Fig. 3-8.
- Balance.
- Drying oven (105°C or 221°F).
- 30 to 60 ml fritted disc filter funnel of coarse porosity and vacuum source.
- About 50:50 HCl, ethanol washing solution.
- 60% isopropyl alcohol (IPA) solution in water (about 100 ml).
- Anhydrous isopropyl alcohol (20 ml).
- Ethanol (2 ml).
- Phenolphthalein indicator solution (1% in IPA).
- Methyl red indicator solution (1% in IPA).
- 0.1N NaOH solution.
- 0.5N NaOH solution.
- 2.5N NaOH solution.
- 0.1N HCl solution.
- 0.5N HCl solution.
- Concentrated HCl.

Procedure

1. Weigh 5 g of the pectin sample into a 250 ml beaker, add 5 ml of concentrated HCl and 100 ml of 60% IPA, and stir the sample for 10 minutes.

2. Vacuum-filter the solution through a fritted disc filter funnel, washing with six 15 ml portions of HCl: ethanol solution, followed by a rinse of 60% IPA solution until no further dissolution is observed (or the chloride is removed). Wash with 20 ml of anhydrous IPA, dry the pectin in an oven (105°C or 221°F) for 2.5 hours, cool it, and weigh it.

3. Transfer 500.0 mg of the dried filtrate to a 250 ml Erlenmeyer flask, and add 2 ml of ethanol followed by 100 ml of boiled (CO_2-free) water. Stopper the flask, and swirl it until the pectin is evenly dispersed or hydration is complete.

4. Add 5 drops of phenolphthalein indicator solution, titrate with $0.1N$ NaOH, and record the volume titrated as V_1 or the initial titer.

5. Add 20.0 ml of $0.5N$ NaOH to the flask, stopper it, shake it vigorously, and allow it to stand for 15 minutes.

6. Add 20.0 ml of $0.5N$ HCl and shake the flask until the pink color disappears. Add 3 drops of phenolphthalein indicator, and titrate with $0.1N$ NaOH to a faint pink color that persists on vigorous shaking. Record the titrated volume as V_2. If the pectin is thought to contain amide groups, continue with steps 7, 8, and 9. Otherwise, skip to step 10.

7. Transfer the flask contents to a boiling flask fitted to a Kjedahl trap, filled with 150 ml of boiled (CO_2-free) water and 20.0 ml of $0.1N$ HCl, and a condenser. Add 20 ml of $2.5N$ NaOH solution to the boiling flask, and heat it carefully to avoid foaming until 80 to 120 ml of distillate has been collected.

8. Add a few drops of the methyl red indicator to the receiving flask, and titrate with $0.1N$ NaOH solution. Record the volume titrated as S.

9. Repeat the titration using the same indicator in 20.0 ml of $0.1N$ HCl as a blank, and record the volume titrated as B. The amide titer ($B - S$) should be recorded as V_3.

10. Add $V_1 + V_2 + V_3$ ($V_3 = 0$ if no amide groups are determined) to get V_t. The degree of esterification can be calculated by using:

$$E° = \frac{V_2(100\%)}{V_t} \tag{20-2}$$

11. The degree of amide substitution can be calculated by using:

$$A° = \frac{V_3(100\%)}{V_t} \tag{20-3}$$

12. The % by weight of the anhydrogalacturonides can be calculated by using:

$$TAG° = 3.52V_1 + 3.80V_2 + 3.5V_3 \tag{20-4}$$

Heavy Metals

Equipment and Supplies

- Two 50 ml color comparison tubes.
- 50 ml graduated cylinder.
- 10 ml graduated cylinder.
- Filter funnel and vacuum source and trap.
- Crucible.
- Crucible tongs.
- Ignitor.
- Lead acetate standard solution. (Make a stock solution of lead nitrate [159.8 mg ACS $Pb(NO_3)_2$ in 100 ml water and 1 ml nitric acid diluted to 1000.0 ml in lead-free glass containers with water], and dilute 10 ml of the stock solution on the day of use to 100 ml with water.)
- Dilute acetic acid.
- Dilute ammonium hydroxide solution (dilution of 400 ml of ACS reagent grade ammonium hydroxide with water to 1000 ml.)
- H_2S solution. (Pass H_2S into cold water to saturate the solution, and store it in small, dark, amber bottles with little head space in a cold dark place. The solution is good only if it has a strong odor and forms a sulfur precipitate when added to an equal volume of ferric chloride (9 g ferric chloride hexahydrate / 100 ml water).)
- Concentrated H_2SO_4.
- Concentrated HNO_3.
- HCl solution (one-half concentrated HCl and one-half water).
- Freshly prepared hydrogen sulfide.
- Steam bath.
- Litmus paper or equivalent.
- Short-range pH paper (3.0–4.0 pH).
- Muffle furnace (500–600°C or 932–1112°F) (preferred).

Procedure

1. Prepare a lead solution by adding 2.0 ml of the lead acetate standard solution to a 50 ml color comparison tube along with enough water to make 25 ml. Adjust the pH with dilute acetic acid or dilute ammonium hydroxide to a pH of 3.0 to 4.0, using short-range pH paper. Dilute the solution with water to 40 ml. This is solution A.
2. Place 500 mg of pectin in a crucible with the lid loosely on it, add enough sulfuric acid to wet the sample, and heat it at a low temperature (about 550°C or 1022°F) until the pectin is thoroughly charred or carbonized.

3. Add 2 ml of nitric acid and 5 drops of sulfuric acid, and heat the crucible until dense white sulfuric acid fumes evolve.
4. Ignite the sample, preferably in a muffle furnace (500–600°C or 932–1112°F), until complete combustion of the organic material occurs, and cool it.
5. Add 4 ml of dilute hydrochloric acid, cover the sample, and heat it on a steam bath for 10 to 15 minutes.
6. Uncover the sample, and evaporate it to dryness on the steam bath.
7. Moisten the residue with 1 drop of hydrochloric acid, add 10 ml of hot water, and heat it for 2 minutes.
8. Add ammonium hydroxide dropwise until the sample is just alkaline to litmus, dilute it to 25 ml with water, and adjust the pH to 3.0 to 4.0, using short-range pH paper and dilute acetic acid.
9. Filter the sample if a solid or suspended material exists, rinse the crucible and filter with 10 ml of water, and transfer the washings to a 50 ml color comparison tube, diluting the solution to 40 ml. Mix the solution to a uniform consistency. This is solution B.
10. Add 10 ml of freshly prepared hydrogen sulfide to each tube (solutions A and B), and allow the tubes to stand for 5 minutes.
11. View the sample tube (B) and the lead standard tube (A), looking downward from the top over a white surface. If the sample solution is darker than the lead standard, then the pectin exceeds the 10 ppm lead or heavy metal specification, and vice versa.

Solvent Content

Equipment and Supplies

- Glass-stoppered weighing bottle.
- Balance.
- Drying chamber or oven (105°C or 221°F).
- Desiccator.

Procedure

1. Heat the empty dry weighing bottle and stopper in a drying chamber for 30 minutes, and weigh the bottle and stopper.
2. Place 1 to 2 g of pectin, accurately weighed, in the weighing bottle, and evenly distribute the pectin to a depth of about 10 mm.
3. Place the opened bottle and stopper in the same drying chamber (105°C or 221°F) for 2 hours.
4. After drying, stopper the bottle immediately, cool it in a desiccator, and weigh it.

5. The % solvent can be calculated from:

$$\% \text{ solvent} = \left(1 - \frac{\text{wt after drying}}{\text{wt before drying}}\right) 100\% \qquad (20\text{-}5)$$

For example:

$$\left(1 - \frac{1.102 \text{ g}}{1.224 \text{ g}}\right) 100\% = 10.0\% \text{ solvent}$$

The most common quality control parameter of pectins, however, is the jelly grade or the pounds of sugar that one pound of pectin will support. The maximum jelly grade of pectin from lemon peel is about 300 to 350; from grapefruit peel, 250 to 300; and from orange peel about 150 to 250. Most pectins used for the commercial production of jams and jellies have a jelly grade of about 150. The Institute of Food Technologists has established a standard procedure to measure the jelly grade, which has been officially accepted by the industry (IFT Committee on Pectin Standardization 1959).

Jelly Grade of Pectin

Equipment and Supplies

- Exchange Ridgelimeter with square plate glass accessories and Hazel-Atlas No. 35 glass tumblers ground down to a height of 3.125 inches with lids and spatula.
- Glass stirring rod.
- 3/4" masking tape.
- Sugar (three times as many grams as the estimated jelly grade).
- Tartaric acid solution (48–80 g in distilled water to 100 ml).
- 3 qt stainless steel saucepan (e.g., Ekcoware pan No. 7323, Flintware pan No. 7623, or Revere copper-clad pan No. 1403).
- Potato masher (e.g., Flint No. 1905).
- Heat source to boil mixture.
- Refractometer.
- Thermometer.
- 25 ± 3.0°C storage area.
- 2 ml pipette.
- 150 ml beaker and balance.

Procedure

1. Determine the grams of pectin needed by dividing 650 by the assumed jelly grade. Subtract this number from 650 to get the grams of sugar

needed. For example, for an estimated jelly grade of 150, 650/150 equals 4.33 g of pectin, and 650 − 4.33 or 646 g of sugar would be needed.

2. Add about 20 to 30 g of the needed sugar to a clean and dry 150 ml beaker, and add the pectin. Mix the two solid materials with a stirring rod.

3. Tare the saucepan and potato masher, and then add 410 ml of distilled water. Pour the contents of the beaker into the pan all at once, and gently stir the mixture for 2 minutes. Try to get the solid material under the surface of the water as quickly as possible. Place the pan on the heat source, and heat it until the solution is fully boiling, while stirring the solution the entire time.

4. Add the remaining needed sugar and stir until all of the solid material has dissolved. Boiling should continue until the net weight of the jelly batch is 1015.0 g. If the net weight is less than this, add excess distilled water, and reboil down to the proper weight. The entire heating time should be about 5 to 8 minutes.

5. After the final weighing, let the solution sit undisturbed on a flat surface for about one minute.

6. Tip the pan, skim off the floating material, and place a thermometer in the solution. When the temperature reaches 95°C, pour the solution quickly into three previously prepared Ridgelimeter glasses. These tumblers are prepared by extending the height of the glasses, which is done by encircling the rim of the glass with masking tape extending exactly one-half inch above the rim. Also, pipette 2.0 ml of the tartaric acid solution into each tumbler. (See Fig. 20-1.) Stir each solution gently with the glass stirring rod to remove bubbles. Exactly 15 minutes after pouring, cover the jars with lids that fit snugly on the masking tape rim, and store them at 25 ± 3.0°C for 20 to 24 hours.

7. Remove the lids and the masking tape around the rim of the tumblers. A clean and wetted wire furnished with the Ridgelimeter is used to separate the excess jelly above the rim of the tumbler by drawing it across the surface. Turning the tumbler during this separation ensures an even and uniform jelly surface.

8. The remaining jelly is removed from the tumbler onto a square glass plate by inverting the tumbler at about a 45° angle, as shown in Fig. 20-1, and carefully separating the jelly from the top of the rim of the tumbler with a spatula. Thereafter the jelly should slide out of the tumbler onto the glass plate on its own. One should be careful not to drop the jelly onto the glass plate or rupture the jelly. A timer is started as soon as the jelly in on the glass plate. If the jelly leans, the plate should be tilted away from the lean to position the jelly upright. Place the glass plate in the Ridgelimeter as shown in Fig. 20-1.

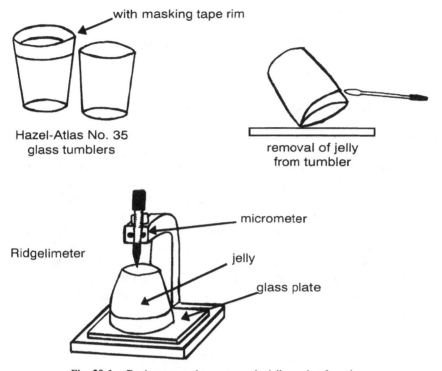

Fig. 20-1. Equipment used to measure the jelly grade of pectins.

9. Using the micrometer, lower the screw to just above the surface of the jelly. After 2 minutes adjust the screw to just contact the top of the jelly with the micrometer tip. The lowest line on the vertical scale is the percent sag, with the micrometer reading giving tenths of a percent sag. Repeat this step for each of the two remaining triplicate samples. If the percent sag differs by more than 0.6%, the jelly should be remade and the test repeated.

10. Remove a portion of the jelly from the center, place it on the prism of a refractometer, and read the Brix after about a minute. The temperature-corrected Brix (see Chapter 2) must be 65.0 $\pm$ 1.0°Brix. A 1.0°Brix error can mean an error of 3 to 4 jelly grade points.

11. The jelly grade is found by calculating a correction factor to the assumed or estimated jelly grade. The factor (F_j) is found from:

$$F_j = 2 - (\% \text{ sag})/23.5 \qquad (20\text{-}6)$$

12. This factor is used to get the true jelly grade as follows:

$$JG = F_j(\text{estimated jelly grade}) \qquad (20\text{-}7)$$

For example:

$$F_j = 2 - (24.1)/23.5 = 0.97$$

$$JG = (0.97)(150) = 146 \text{ jelly grade}$$

Production of Jams, Jellies, and Preserves

Gelation has long been used as a means of preserving various fruits; the resultant increased viscosity or decrease in water activity inhibits microbial growth and spoilage. Jams and jellies usually can be stored at room temperature until they are opened. Thereafter, the surface becomes subject to airborne molds, which literally grow on the surface of almost any food product, especially one as rich in carbohydrates and nutrients as fruit jams and jellies are.

Jellies can be defined as a clear jelled fruit product that retains its shape when cut by a knife. The federal standards of identity describe jellies as a mixture of fruit ingredients containing not less than 45 parts of fruit juice by weight to each 55 parts of saccharine ingredient (CFR Title 21 150.140). (The word "saccharine" is defined as a sweetening material.) For example, if you used 2.5 pounds of 65.0°Brix orange concentrate, the amount of saccharine ingredient that you could use while still calling the product fruit jelly would be:

$$(2.5_{\text{lb conc.}}) \left(\frac{65.0°\text{Brix}}{100\%} \right) (8.0_{\text{factor}}) \left(\frac{55 \text{ parts sach.}}{45 \text{ parts juice}} \right) = 16 \text{ lb sach.}$$

If saccharine ingredients were used, another 16 pounds of sugar would be required to make the jelly. The 8.0 factor is from the federal code for orange juice. The factor for grapefruit juice would be 11.0. There is also a federal standard of identity for artificially sweetened fruit jelly, which requires at least 55% of the final weight of the jelly to be fruit juices. For this artificially sweetened product, the allowable ingredients are more restricted, especially the sweeteners used. Only saccharine ingredients that are not sugar can be used.

Jams or preserves contain unclarified juice, usually containing thicker or larger portions of the solid material of the fruit than found in jellies. The federal code defines fruit jams or preserves as containing 47 parts of fruit ingredients by weight to 55 parts of saccharine ingredient for citrus products (CFR Title 21 150.160). Artificially sweetened fruit preserves and jams have standards of identity that are similar to those of jellies except that the term "fruit" is used instead of "fruit juice."

The process of gelation of jams and jellies depends on the hydrogen bonding balance or water activity in the solution. It is believed that the negatively charged, highly hydrated pectin chains are bound together by hydrogen bonds. Sugar molecules compete for the hydrogen bonding with water molecules, while acids are used to control the negatively charged pectin chains. By combining the right proportions of pectin, sugar, water, and acid, gelation will occur, even at temperatures as high as the boiling point of water (Joseph 1955). Also, the mineral content will affect the balance and thus the gelation. Divalent cations may act as linking agents between pectin chains, resulting in gelation. In fact, low-methoxy pectins require polyvalent cations in order to gel. For the reason, the mineral content of the water used in processing should be considered. Also, the natural pectins, sugars, acids, buffering salts, and other soluble material inherent in the juices or fruit used should be taken into consideration.

Slow-set pectins require the pH to be less than 3.30, whereas rapid-set pectins require a pH of less than 3.55. The slow-set pectins, however, are more tolerant of excess acid levels than the rapid-set pectins. Jellies generally utilize slow-set pectins so that gelation does not occur until the product is in the container. Also, jellies generally use more pectin than jams to obtain a firmer gel. For jams it is better to use a rapid-set pectin in order to avoid floating pieces of fruit that separate if gelation is too slow. Most jellies and jams are made from a standard 150 jelly grade pectin.

The general process of making jams or jellies involves premixing the pectin with about three to five times as much sugar as pectin to help disperse the pectin, and then dissolving this mixture in the fruit juice or pulp. When this mixture is dissolved in fruit juice concentrates, the amount of sugar to be premixed with the pectin should not exceed about 20 to 25% of the total solution. This mixture is cooked or evaporated until it reaches an acid- and temperature-corrected Brix of 65 to 68. The acid is added just before filling in order to postpone the gelation as long as possible. Two examples of fruit jellies and jams are presented in Tables 20-2 and 20-3.

The USDA divides the grade standards for fruit jelly into two areas. Standard fruit jelly meets the FDA standards of identity (CFR Title 21 150.140.), whereas unstandardized jelly includes fruit ingredients not allowed in the same standards

Table 20-2. Orange jelly formulation (Swisher and Swisher 1977).

Slow-set pectin (150 jelly grade)	11.0 lb
65.0°Brix concentrated orange juice	2.5 gal
Water	7.0 gal
Sugar	100.1 lb
Acid solution (7.9 lb citric acid monohydrate in 1.0 gal of water)	591 ml
Orange oil	44 ml

Procedure
 1. Mix pectin with 5 pounds of sugar, and add the mixture to warm water with stirring and bring to a boil. Boil for 30 seconds.
 2. Add the rest of the sugar, dissolve it, and add orange concentrate without further heating.
 3. Add acid solution and orange oil and fill the container. Makes about 165 pounds of jelly.

Table 20-3.   Orange marmalade formulation (Swisher and Swisher 1977).

Rapid-set pectin (150 jelly grade)	0.5 lb
65.0°Brix concentrated orange juice	0.5 lb
Commercially prepared orange peel strips	20.9 lb
Water	30.0 gal
Sugar	100.1 lb
Acid solution (7.9 lb citric acid monohydrate in 1.0 gal of water)	414 ml
Orange oil	30 ml

Procedure
1. Mix pectin with about 5 pounds of sugar, and add the mixture to warm water with stirring and bring to a boil. Boil for 30 seconds.
2. Add the rest of the sugar and peel strips, boil to dissolve, and then turn off the heat source.
3. Add the concentrated orange juice, oil, and acid solution. Yields about 152 pounds of marmalade.

of identity. Also, an unstandardized jelly can contain as much as 27 parts of black currant juice to 55 parts of sugar, whereas standardized jelly cannot contain any black currant juice.

The USDA further classifies jellies into four types: Type I jellies are standardized jellies containing fruit juice from a single variety of fruit, Type II jellies are standardized jellies that contain juices from more than one fruit variety, Type III jellies are unstandardized jellies containing juices from one variety of fruit, and Type IV jellies are unstandardized jellies that are made from more than one fruit variety.

The USDA scoring system is shown in Table 20-4. Grade A texture is described as tender to slightly firm and retains a compact shape without excessive weeping. Grade B may lack firmness but cannot be syrupy; it also can be more than slightly firm but not tough or rubbery. Grade A color is described as including a sparkling luster and is no more than slightly cloudy. Grade B color may range from slightly cloudy to slight dullness. Grade A flavor must be free

Table 20-4.   USDA scoring system for grades of fruit jelly.

	Consistency	Color	Flavor	Minimum Score
Maximum points	40	20	40	—
Grade A	36–40	18–20	36–40	90
Grade B*	32–35	16–17	32–35	80
Substandard*	0–31	0–15	0–31	—

*If the fruit jelly falls into this category in any one area, it cannot receive a higher grade regardless of the total score.

from any caramelized or other objectionable flavor, whereas Grade B may have only a slight caramel-like flavor.

FRUIT SECTIONS

The canning and the bottling of fruit sections have become a popular means of preserving citrus fruit. Grapefruit sections have found acceptance in the United States, and mandarin sections are well accepted in the Far East. Canned and bottled citrus fruit sections have the advantage of long shelf lives and retention of most of the whole-fruit nutrition, and can be conveyed to noncitrus or off-season geographical areas that otherwise might not be able to partake of the benefits of citrus.

Fruit Preparation

Prior to processing, citrus fruit should be held in storage bins for 4 to 6 days to allow it to soften. A prescaling of the fruit may shorten this holding time to 2 to 3 days. The fruit then is scalded in hot water (90–100°C or 196–210°F) for 5 to 8 minutes for grapefruit or 4 minutes for oranges, which further softens the fruit. Then the fruit is sprayed with low pressure steam that penetrates the peel, which is followed by additional bin storage until the fruit is ready for peeling. The softened fruit requires gentle handling, but the additional holding time can reduce scalding requirements by 30%. It also results in less fruit spoilage and less cooked off flavors.

Peeling

About 5% of sectionized fruit is peeled by hand, and the rest by mechanical means. Hand peeling generates greater amounts of juice in waste streams than mechanical peeling, which may be diverted to juice processing. Hand peeling also generally requires less scalding and thus gives better quality. The mechanically or hand-peeled fruit then undergoes a hot caustic spray (0.5–2.5% NaOH or Na_2CO_3) at 88 to 100°C (190–212°F) for 12 to 25 seconds. (These lye solutions can be recycled, but must be replenished daily because of the leaching of fruit-soluble solids.) The fruit then is allowed about a 15- to 25-minute reaction time, followed by water sprays to remove the basic solutions. The peeled fruit is chilled in water baths or sprays at 2 to 5°C (35–40°F), a step that facilitates sectionizing of the fruit.

Sectionizing

Manual sectionizing is done by placing the fruit against a spindle, and the separated sections are tapped or brushed against each other to remove the seeds.

Mechanical sectionizing may be done in various ways. Fruit generally is placed in a cup with the blossom end up, and aligned automatically or by hand using a light beam guide. The sectionizing blades break apart the segments, and shakers are used to remove the seeds. The separated sections usually are graded by hand prior to filling.

Filling

After the sections are placed in the container, a liquid sweetener is added, which can be a mixture of water, juices, and sugar at vaious concentrations. Allowable filling solutions include water, grapefruit juice and water, grapefruit juice, slightly sweetened sirup or slightly sweetened water (12.0-15.9°Brix), light sirup (16.0-17.9°Brix), heavy sirup ($\geq$ 18.0°Brix), slightly sweetened grapefruit juice and water (12.0-15.9°Brix), lightly sweetened grapefruit juice and water (16.0-17.9°Brix), heavily sweetened grapefruit juice and water ($\geq$ 18.0°Brix), slightly sweetened grapefruit juice (12.0-15.9°Brix), lightly sweetened grapefruit juice (16.0-17.9°Brix), and heavily sweetened grapefruit juice ($\geq$ 18.0°Brix). Usually sections from grapefruit that pack 70 to 80 to a box in fresh fruit packaging are packed in 8-ounce and 303 cans, and larger grapefruit that pack 54 or 64 to a box are packaged in 26-ounce glass containers.

There are two basic methods of filling citrus sections into containers. One is cold fill, where the container is chilled after sealing in a 35 to 40°F (2-4°C) water bath. The container then is dried by an air stream, and dry ice sometimes is sprinkled on it to maintain its temperature on the way to cold storage at 36 $\pm$ 2°F. About 0.04 to 0.05% sodium benzoate sometimes is added as a preservative, along with less than 0.035% calcium chloride or calcium lactate to enhance the firmness of the sections.

Hot-fill packaging includes heating the sealed container to 149 to 190°F (63-88°C) for 20 to 45 minutes, depending on its size. Larger containers require more heat for a longer time. The heat-treated container then is passed through a 60 to 75°F (16-24°C) water bath, which cools the contents to about 90 to 100°F (32-38°C). The residual heat dries the can.

Analytical Procedures

The Brix and acid levels of the sirup used in canning or bottling citrus fruit sections are measured as explained in Chapters 2 and 3. The integrity of the fruit sections themselves generally is determined by direct inspection. If the section maintains its form when suspended above the liquid, it can be considered firm. The main quality parameter is the drain weight, which is determined by using the following procedure.

Drain Weight

Equipment and Supplies

- No. 8 circular sieve (8" diameter).
- Collection pan at least as large as the sieve.
- Scale or balance.

Procedure

1. Pour the contents of the container through the sieve, collecting the liquid in the pan or tray under the sieve.
2. Drain for 2 minutes, and tap the sieve to remove adhering liquid.
3. Weigh the liquid and the screened fruit sections separately.
4. Calculate the percent drain weight by using:

$$\% \text{ drain wt} = (W_{\text{fruit}})(100\%)/(W_{\text{fruit}} + W_{\text{liquid}}) \qquad (20\text{-}8)$$

For example:

$$(125 \text{ g}_{\text{fruit}})(100\%)/(125 \text{ g}_{\text{fruit}} + 102 \text{ g}_{\text{liquid}}) = 55\% \text{ drain wt}$$

Standards of Identity (CFR Title 21 145.145)

The federal standards of identity require that at least 50% of the drain weight be in the form of whole sections in order for the product to be called fruit sections or segments. If less than 50% of the drain weight is whole sections, the product must be called broken sections or broken segments. Also, they must have less than 20 square centimeters (3.1 square inches) of tough membrane or albedo on the segments per 500 grams (17.6 ounces) of the product. Also, there can be no more than four developed seeds (9.0 mm or 0.35 inch or larger), and no more than 15% of the drain weight can contain injured, discolored, or otherwise blemished fruit.

USDA Grade Standards (CFR Title 21 52.1141–52.1151)

Grade A canned grapefruit sections must have at least 53% of the water capacity of the container in the form of fruit, with at least 65% of the fruit being in the form of whole sections. Grade B also requires that 53% of the water capacity of the container be in the form of fruit, with only 50% or more of the fruit being in the form of whole sections. Substandard canned grapefruit fails the above requirements. The fill of the container or the composition of the liquid medium is not considered in USDA grades. Table 20-5 illustrates the USDA scoring system for canned grapefruit.

JUICE DRINKS, BASES, AND PUREES

Citrus juice drinks can be defined as beverages that contain less than 100% fruit juices. They usually contain varying amounts of sweeteners and organic acids, and perhaps also carbonation, flavors, colors, emulsifiers, texturizers, or other components designed to enhance the beverage. A drink base can be defined as the concentrated components of a drink minus the major portion of water, sweetener, and perhaps even the acid, although many drink bases already contain the needed acid. In the United Kingdom, drink bases are called squashes; in Australia they are known as cordials and usually contain about 20 to 50% juice. The main consumer appeal of juice drinks is their low cost. Drinks and drink bases can be used for nonbeverage products as well, such as popsicles and purees. Purees are a type of fruit juice base used as a flavoring material for sherbets, candy, baking products, and many other products. Purees once were made from macerated whole fruit but now are made from juice concentrate in a manner similar to that used for drink bases. Also included in puree bases are

Table 20-5.   USDA grades for canned or bottled grapefruit sections.

	Wholeness	Color	Defects**	Character	Minimum Score
Maximum	25	25	25	25	—
Grade A	21–25	23–25	23–25	23–25	90
Grade B*	17–20	21–22	21–22	21–22	80
Broken*	0–16	21–25	21–25	21–25	70
Substandard*	—	0–20	0–20	0–20	—

*Sections falling in this classification in any one parameter cannot receive a grade higher regardless of total score.
**Defects are determined according to the scheme below.

	Extraneous Veg. Matter*	Total Seeds*	Large Seeds*	Albedo and Membrane*	Damaged Units
8 oz can$_{336\ ml}$					
Grade A	0.25(1) piece	1.6(3)	0.4(1)	$\frac{1}{2}(1)$ in.^{2}	$\frac{1}{4}$ oz.
Grade B	0.50(2) piece	4.8(6)	1.2(2)	$\frac{3}{4}(1\frac{1}{2})$ in.^{2}	$\frac{3}{4}$ oz.
No. 303 can$_{492\ ml}$					
Grade A	0.50(1) piece	3.2(4)	0.8(2)	1.0(2.0) in.^{2}	$\frac{1}{2}$ oz.
Grade B	0.75(2) piece	9.6(12)	2.4(3)	2(3) in.^{2}	$1\frac{1}{2}$ oz.
No. 3 Cyl.$_{1455\ ml}$					
Grade A	0.75(1) piece	(10)	(3)	4(5) in.^{2}	$1\frac{1}{2}$ oz.
Grade B	1(2)	(20)	(5)	$6(7\frac{1}{2})$ in.^{2}	4 oz.

*The first number is the maximum allowed average for all samples. The second number in parentheses is the maximum allowed per sample.

juice sacs, sugar, color, flavors, citrus oils, essences, and citric acid to adjust the pH to about 3.0. Sherbet bases generally are manufactured as 50-fold concentrates and are required by law to contain at least 2% juice. Sugar and color usually are added by the food processor using the puree base. Oil levels of puree bases usually occur in the range of 0.030 to 0.035%, as measured by the procedure outlined in Chapter 6. The pulp level ranges from 6 to 12.5%, as determined by procedures mentioned in Chapter 7.

The most popular beverages in the United States are carbonated soft drinks, and the increased concern in the late 1900s for healthful beverages has made juice-added soft drinks especially appealing. Also, juice drinks with high levels of juice have been developed that emphasize added nutrients such as calcium and low calorie sweeteners. The flavor and the color quality of the juices used in these drinks play a lesser role in the overall quality of the drink than they do in citrus juices; so a lower quality can be used in these drink bases, such as pulp wash juices or bitter juices as well as high-acid juices such as lemon and lime juices. Components that make up citrus drinks can be broken down into 11 groups: juice, sweeteners, acids, colorants, flavors, texturizers, clouding agents, nutrients, foaming agents, defoaming agents, and preservatives; but these components may not all be contained in an individual juice drink or base.

Juices

Juices are used for several reasons, the main one being to enhance the health appeal of the product. Depending on the amount of juice used, color and flavor enhancement are the next most important contributions the juices make to the drink. Cost often determines which juices are used in a drink, with cheaper juices such as apple, pear, or white grape juices frequently used. Some highly colored juices, such as pomegranate or cranberry juice, can add a desirable deepening of the citrus color at around the 1% level, but greater amounts impart an unsightly brownish tinge to the product. Lemon and lime juices have been used as citric acid sources for drinks, enhancing their citrusy flavor at the same time. Pulp wash juices and peel extracts are ideal juice sources for citrus drinks, which provide a use for these products that would be difficult to market otherwise.

With the advent of debittering techniques, the use of bitter juices in drinks is expected to decline. The most bitter navel juice can contain limonin levels as high as 40 ppm, well over the 6 to 7 ppm generally accepted taste threshold. A tenfold dilution in making a 10% juice drink would dilute the limonin level to 4 ppm maximum, below the taste threshold. Care must be taken in using bitter juices with nonbitter citrus juices because even nonbitter citrus juices contain some limonin. Because pulp wash juice (or the juice produced from washing the core material recovered from FMC-like juice extractors), and peel extracts are highly bitter, they may require debittering before significant quantities can be used in juice drinks.

Sweeteners

Sugar (sucrose), or its equivalent in high fructose corn syrup (HFCS) or low calorie sweeteners, is a major component in most juice drink formulations. Commercial sugars generally cost only 5 to 15% as much as the sugars or soluble solids found naturally in citrus juices, a difference that makes juice drinks a more economical product for consumers. The only known nutritional disadvantage of citrus juices is their high calorie content, which can be largely overcome by using low calorie sweeteners in juice-based drinks. The *Federal Register* allows sugar (sucrose), invert sugar, fructose, corn syrup, glucose syrup sorbitol, or any combination of these ingredients to be used as sweeteners in carbonated soft drinks (FR June 26, 1969, 9867). Because sugar is readily available and comprises the major portion of most juice drinks, it is not added until just prior to packaging of the final drink, along with the necessary water. This procedure saves a considerable amount of money in shipping, storage, and conveying.

Acids

Acids are another major component of citrus juices and juice drinks, and give the beverage a citrusy tangy taste. Because the hydronium ion is what imparts this taste, theoretically most acids can be used; but not all acids are approved for food use. Citric acid, which is the most common acid found in fruits and juices, is considered to be one of the best acids to use in foods. One reason for this is that it is easily buffered, being a triprotic weak acid. Also, citrus processors who have quality control programs based on citric acid (citric acid corrections to the Brix and lemon and lime inventories based on the weight of citric acid) are better equipped to handle citric acid than other acids. However, the *Federal Register* permits the use of acetic acid, adipic acid, fumaric acid, gluconic acid, lactic acid, malic acid, phosphoric acid, or tartaric acid in addition to citric acid in carbonated soft drinks (FR June 26, 1969, 9867). Citric acid can be buffered by adding tricalcium phosphate, sodium citrate, and citric acid in the ratio of $1:5:30$ by weight. Again, the *Federal Register* allows acetate, bicarbonate, carbonate, chloride, citrate, gluconate, lactate, orthophosphate, or sulfate salts of calcium, magnesium, potassium, or sodium to be used as buffering agents in carbonated soft drinks (FR June 26, 1969, 9867).

Colorants

Colorants play a minor role in regard to the Brix, but they have a significant influence on the drink's overall appeal and therefore can be formulated independently of the sugar and the acid. Highly pigmented juices from other varieties of fruit offer a natural and nutritional way to produce desirable colors in fruit juice drinks. As mentioned above, the addition of 1% of pomegranate or cranberry juice to citrus juice drinks can deepen the color. Also, tangerine juices are known for their deep orange color, as is juice produced in dry Mediterranean climates such as that in California, Australia, and South Africa and around the Mediterranean basin. Natural pigments can be used such as turmeric and anatto (see Chapter 19).

Other common natural pigments used in citrus drinks include the carotenoids. One of the most common carotenoids is β-carotene, which is extracted from citrus peels or other natural sources. β-Carotene imparts a light yellow to orange color to citrus juice drinks. It is not very soluble in water but is more soluble in fats and oils. Its molecular structure is as follows:

FD&C dyes, described in the Chapter 19, also can be used to enhance the color of citrus drinks. FD&C No. 5 is used in lemon drinks in the range of 10 to 15 ppm, in lime drinks at 20 to 30 ppm, and in grapefruit drinks at 5 to 10 ppm. In orange drinks FD&C No. 5 and FD&C No. 6 can be used together, each at about 0.008 to 0.2%, depending on the amount of juice used in the drink. Packaging also has an effect on the need for color. Generally, the bright colors of citrus juices are a good selling point. However, if opaque cans or containers are used, color enhancement becomes less important.

Flavors

Citrus flavors come in various forms. Natural citrus oils and essences extracted from the peel of the fruit and recovered during juice evaporation or pasteurization are the most common source of citrus flavors. Citrus oils can be concentrated or folded in order to strengthen the citrus flavor by concentrating the oxygenated compounds in the oils. Again, the addition of these flavors has little effect on the Brix measurements and can be formulated independently of the sugars and acids. The most prevalent component of citrus oils is d-limonene, which is easily measured by using the Scott method, described in Chapter 6. Ideal oil ranges are 0.015 to 0.020% by volume. Other compounds that contribute to the citrus-like taste of juice drinks include ethyl butyrate and various aldehydes, such as decanal, citral, acetaldehyde, and octanal.

Texturizers

Texture or mouth-feel is important in attempts to duplicate the sensation of authentic citrus juices. Natural juices contain pectins and proteins that act as natural emulsifiers. Pulp and other solid material also contribute to the general viscosity and mouth-feel of the juice. Juice sac material can be added to citrus drinks to give a fresh squeezed juice appearance and mouth-feel. Some texturizing agents used in the industry include glycerol, ester of wood rosin, guar gum, hydroxylated lecithin, lecithin, methyl cellulose, mono- and diglycerides of fat-forming fatty acids, polyglycerol esters of fatty acids, propylene glycol alginate, sodium alginate, sodium metaphosphate (sodium hexametaphosphate), gum acacia, gum arabic, xanthan gum, vegetable gums and brominated oils, gum tragacanth, carbo bean gum (locust bean gum), carrageenan, ester gum, sodium carboxymethylcellulose, cottonseed oil, or even pectin in a range of around 0.1 to 0.3%. The texturizer of choice depends on the desired texture, price, and availability. Even though texturizers appear to have a significant effect on the density of the juice, their contribution to the Brix measurement is generally insignificant.

Clouding Agents

Clouding agents give citrus drinks the opaque appearance of juice. This is especially important in lemonades and limeades, where the natural cloud levels in the juices must be extensively diluted in order to bring the acid levels in the juice down to the desired levels. About 30% of the natural cloud in juice is made up of proteins, which come primarily from the cytoplasm and organelles of the juice cells during extraction. About 10 to 15% of the cloud is made up of hemicelluloses from the membrane material of the fruit, and about 5% of the juice cloud is composed of pectin (Bennett 1985). Natural flavonoids, such as hesperidin and naringin, also contribute to the cloud material. Protein/pectin combinations have been shown to produce cloud similar to that found in citrus juices. Casein, a common protein, has been found to produce cloud with highly methoxylated pectin in a ratio of about 500 to 1000 μg/mg pectin (Bennett 1988). Soy flour or other suitable protein sources can be used with pectins as a starting point in most drink cloud formulations.

Nutrients

As explained in Chapter 11, natural juices contain a healthy supply of certain nutrients, especially vitamin C. Because vitamin C is associated with citrus juices, consumers usually expect to find levels of the vitamin equal to that found naturally in juices in citrus drinks. The average content of vitamin C in pure juices is about 0.05% or 50 mg/100 ml. To calculate the amount of vitamin C that must be added to these drinks to equal the levels found in pure juices, the following equation can be used:

$$\text{wt vitamin C} = 0.0005(1 - p/100)W \qquad (20\text{-}9)$$

where p is the percent of juice soluble solids in the final drink compared to the total soluble solids in the drink, and W is the total weight of the drink. The weight of the drink can be found from the gallons by multiplying by the density found from the Brix, using Equation 2-9 or 2-11. For example, if we had 1000 gallons of an 11.8°Brix juice drink that contained 10% orange juice soluble solids, the weight of drink after converting the 1000 gallons of drink to pounds would be 8717 pounds drink ($d = 8.717$ lb/gal). Equation 20-9 then would give:

$$0.0005(1 - 10\%/100)8717 \text{ lb} = 3.9 \text{ lb of vitamin C}$$

Calcium addition to juice drinks containing high levels of citrus juices has recently become a popular way to increase the nutritional value of citrus products. The USRDA for calcium is 1 gram. Assuming a serving size of 6 ounces (177 ml), a maximum of 24 grams per gallon of drink or 53 pounds per 1000 gallons of drink should constitute the maximum level of calcium needed to enhance the drink's nutrition. Because most people consume other foods that are high in calcium (such as milk), an amount that is less than this maximum level should be adequate. It must be remembered, as explained in Chapter 11, that whenever nutrients are added, nutritional labeling is mandated by law.

Preservatives

Unlike nutrients, the presence of preservatives has a negative connotation in foods. Concentrated drink bases naturally inhibit microbial growth by osmotic pressures and cold temperatures. Single-strength drinks often are packaged in aseptic bottles, cans, or paperboard or fiberboard containers using hot-fill technologies and/or hydrogen peroxide sterilization. In order to minimize the risk of contamination and the heat required for sterilization, preservatives are often added. Sulfites act not only as preservatives but as antioxidants that help inhibit

Maillard oxidative reactions, which can cause off flavors and colors to develop. Ascorbic acid or vitamin C is sometimes used as an antioxidant as well. However, the use of sulfites is becoming more restricted, and they have been banned in many areas of food processing. Some of the most common preservatives used in citrus juices and beverages are benzoates and sorbates. These preservatives are more effective against bacteria in single-strength juices and are less effective against the slower-growing yeasts, with the effectiveness of sodium benzoate against yeast depending on the pH of the juice. For pH values of 2.5 to 3.5, a sodium benzoate level of 0.05% is effective against yeasts; at pH values of up to 4.5, 0.10% sodium benzoate is required. Common sodium benzoate levels of 0.15% are found in many juice drinks, as well as levels of 0.05% potassium sorbate.

The *Federal Register* permits the following preservatives to be used in carbonated soft drinks: ascorbic acid, benzoic acid, BHA, BHT, calcium disodium EDTA, erythorbic acid, glucose-oxidase-catalase enzyme, methylparaben or propylparaben, propylgallate, potassium or sodium benzoate, potassium or sodium bisulfite, potassium or sodium metabisulfite, potassium or sodium sorbate, sorbic acid, sulfur dioxide, tocopherols, and, in the case of canned soft drinks, stannous chloride at less than 11 ppm as tin (FR June 26, 1969, 9867).

Foaming and Defoaming Agents

Some drinks have a greater appeal when associated with foam, such as some soft drinks, but others do not. Foaming and defoaming agents can be used to control this characteristic. The *Federal Register* permits the use of glycyrrhizin, gum ghatti, licorice or glycyrrhiza, yucca (Joshua tree or Mojave), and guillaia (soap bark) as foaming agents in carbonated soft drinks (FR June 26, 1969, 9867). Dimethylpolysiloxane can be used as a defoaming agent in carbonated soft drinks at levels equal to or below 10 ppm.

Juice Base Formulation

Fruit drinks usually are made in steps. Microcomponents can be preblended in the laboratory by experts to help protect proprietary combinations. Preblending also makes the addition of juice drink components easier by reducing the number of components needed during batch processing. Flavor extracts utilize water, ethanol, propylene glycol, glycerol, and/or other surfactive agents as solvents. Flavor emulsions are made by adjusting the right density combinations of citrus oils (about 0.84 g/ml) with the heavier oils (about 1.33 g/ml) to obtain about 1.02 g/ml so that the emulsion will mix with the other aqueous-phase components of the drink. Water-soluble acids, buffers, preservatives, colorants, texturizers, clouding agents, nutrients, and other components can be preblended as well. The blending of these aqueous and nonaqueous components with juice concentrates constitutes a drink base, which can be used by a bottler to make a finished drink by the addition of sugar and water. However, the amounts of sugar, acid, juice, and water significantly affect each other's concentrations; so determining the right amount of these macrocomponents in relation to one another can become confusing.

In order to determine how the four main macrocomponents of juice drinks affect each other, their relationship must be defined. The sugar or soluble solid balance in juice-like solutions can be defined as follows:

$$(V_c D_c B_c)/100 + W_s + 1.19W_a = BVD/100 \qquad (20\text{-}10)$$

where V_c, D_c, and B_c are the volume, density, and Brix of the concentrate used in making the base or drink, W_s and W_a are the weights of sugar and acid to be added to the drink, and B, V, and D are the Brix, volume, and density of the final juice drink. The 1.19 factor is used to obtain the sugar equivalent of the added acid correction to the Brix as calculated by Equation 2-2. The first and last terms in Equation 2-2 dropped because they are negligible in acid corrections to juice drinks. To understand the validity of this assumption, consider making 1000 gallons of drink at 11.8°Brix and 1% acid using 10% soluble solids as orange juice concentrate with a Brix of 65.0 and an acid level of 3.0. The error in calculating the amount of sugar needed would be about a third of a pound of sugar out of 828 pounds needed. This suggests that the first and last terms in Equation 2-2 are negligible. The percent of acid plus the correction is $1 + 0.19$, so that 1.19 times the weight of the citric acid is equal to the sugar contribution to the Brix or soluble solids.

The acid balance is:

$$(V_c D_c A_c)/100 + W_a = VDA/100 \qquad (20\text{-}11)$$

with V_c, D_c, and A_c being the volume, density, and acidity of the juice concentrate used in the drink and V, D, and A the volume, density, and acidity desired in the final drink, with W_a again being the weight of the acid needed. Unlike the Brix, which includes acid levels in its value, the percent acid does not include any sugar concentrations or sugar corrections. Therefore, the weight of the sugar is not included in Equation 20-11.

The percent juice in the drink is an important parameter, which is defined as:

$$P = (V_c D_c B_c)100\%/VDB \qquad (20\text{-}12)$$

using the same variables as above.

By using Equations 20-10, 20-11, and 20-12, other equations can be derived that enable the calculation of the amount of sugar and acid needed in a particular drink:

$$W_s = VD\big(B - 1.19A - PB(1 - 1.19A_c/B_c)/100\big)/100 \quad (20\text{-}13)$$

$$W_a = VD(A - PBA_c/100B_c)/100 \quad (20\text{-}14)$$

To find the volume of water needed, we can utilize a simple weight balance equation as follows:

$$VD = V_w D_w + V_c D_c + W_s + W_a \quad (20\text{-}15)$$

where V_w and D_w are the volume and density of water. Using Equations 20-12 through 20-15, we can obtain:

$$V_w = \frac{VD}{100 D_w}\left(100 + \frac{PB}{100}\left(1 - \frac{100 + 0.19A_c}{B_c}\right) - B + 0.19A\right) \quad (20\text{-}16)$$

The volume of concentrate needed in a base is easily derived from Equation 20-12:

$$V_c = PVDB/100 B_c D_c \quad (20\text{-}17)$$

Equations 20-13, 20-14, 20-16, and 20-17 are all independent equations based on the same drink parameters, which are usually known prior to formulation. Care must be taken to use consistent units for densities, volumes, and weights. We use mainly pounds and gallons here.

Let us consider an example to illustrate how these equations can be used. Suppose that we want to make 1000 gallons of a 10% juice drink with a final Brix of 11.8 and a final acid level of 1.0% from a 65.0°Brix concentrate with an acid level of 3.0%. How much concentrate, water, sugar, and acid are needed to make the drink? Because Equations 20-13, 20-14, 20-16, and 20-17 are independent equations, we can calculate any of the four parameters in any order. The densities can be obtained from Equation 2-9 or 2-11 from the Brix. Using Equation 20-17 to get the volume of concentrate needed, we obtain:

$$\frac{(10\%)\,(1000\text{ gal})\,(8.717\text{ lb/gal})\,(11.8°\text{Brix})}{(100)\,(65.0°\text{Brix})\,(10.977\text{ lb/gal})} = 14.4\text{ gal conc.}$$

The amount of acid needed, using Equation 20-14, is found as follows:

$$\frac{(1000\text{ gal})\,(8.717\text{ lb/gal})}{100}\left(1.0\% - \frac{10\%\,(11.8)\,(3.0\%)}{100\,(65.0°\text{Brix})}\right)$$

$$= 82.4\text{ lb citric acid}$$

The amount of sugar can be calculated from Equation 20-13:

$$\frac{(1000)(8.717)}{100}\left(11.8 - 1.19(1.0) - \frac{10\%(11.8)}{100}\left(1 - \frac{1.19(3.0)}{65.0}\right)\right)$$

$$= 827.7 \text{ lb sugar}$$

The needed water can be found from Equation 20-16:

$$\frac{1000(8.717)}{100(8.322)}\left(100 + \frac{10(11.8)}{100}\left(1 - \frac{100 + 0.19(3.0)}{65.0}\right)\right.$$

$$\left. - 11.8 + 0.19(1.0)\right) = 919.1 \text{ gal water}$$

Although it may be handy to calculate the needed parameters in any order, especially if only some of them must be calculated, it often is sufficient to calculate the parameters in sequence by using dependent equations. These dependent equations are simpler to use and easier to program into computers than the above equations. Using the above example, we must begin with Equation 20-14 to obtain the 82.4 pounds of citric acid needed and Equation 20-17 to get the 14.4 gallons of concentrate needed. We then can use the following dependent equation to get the amount of sugar needed:

$$W_s = (VDB - V_cD_cB_c)/100 - 1.19W_a \qquad (20\text{-}18)$$

For example:

$$\left(1000(8.717)(11.8) - (14.4)(10.977)(65.0)\right)/100 - 1.19(82.4)$$

$$= 827.8 \text{ lb sugar}$$

compared to the 827.7 pounds of sugar calculated by Equation 20-13. The volume of water needed can be calculated by using a form of Equation 20-15:

$$V_w = (VD - V_cD_c - W_s - W_a)/D_w \qquad (20\text{-}19)$$

In the example problem this dependent equation gives:

$$\left(1000(8.717) - (14.4)(10.977) - 827.8 - 82.4\right)/8.322 = 919.1 \text{ gal water}$$

It should be remembered that the weight of citric acid determined above is the weight of pure anhydrous citric acid. If other acids are used, their effect on the Brix must be determined and substituted into the equations accordingly.

Citric acid has a solubility in water of 54.0% at 10°C (50°F) to 84.0% at 100°C (212°F). Usually citric acid monohydrate is added to juice drink bases because it is easier to dissolve than anhydrous citric acid. The weight of citric acid calculated above can be converted to the weight of citric acid monohydrate by multiplying by 1.0938. In the above example, this would give 90.1 pounds of citric acid monohydrate instead of 82.4 pounds of citric acid. The water added in hydrate form with citric acid monohydrate amounts to less than a gallon in the example and is generally negligible in regard to the amount of water needed in the drink. Volumetric measurements in large tanks generally have an error in measurement that is larger than this.

In the manufacture of drink bases, all or part of the citric acid can be added to the base. In order to do this, it is necessary to add some water to predissolve the citric acid and perhaps other water-soluble drink base components. Too much water addition will overly dilute the drink base, and not using enough water will prevent solubilizing of the citric acid. It may be important to know what effect this acid and water addition will have on the concentration of the base. The Brix of the base after the addition of a 50% solution of citric acid can be calculated as follows:

$$B_b = (V_c D_c B_c + 119 W_a)/(V_c D_c + 1.19 W_a + V' D_w) \qquad (20\text{-}20)$$

where V' is the volume of water used to dissolve the acid. This value can be calculated by using:

$$V' = W_a(1 - X)/D_w X \qquad (20\text{-}21)$$

where X is the fraction of acid used in predissolving the acid. In the above example, if we want to dissolve 82.4 pounds of citric acid to make a 50% ($X = 0.50$) acid solution, Equation 20-21 becomes:

$$V' = 82.4(1 - 0.5)/8.322(0.5) = 9.9 \text{ gal water}$$

To calculate the fraction of citric acid in a predissolving solution, one can use the following equation:

$$X = W_a/(V_w D_w + W_a) \qquad (20\text{-}22)$$

Using Equations 20-20 and 20-21 in the above example gives:

$$\frac{(14.4)(10.977)(65.0) + 119(82.4)}{(14.4)(10.977) + 1.19(82.4) + 9.9(8.322)} = 59.3°\text{Brix for drink base}$$

More water may be needed to dissolve the citric acid. The volume of the base also needs to be known, and can be calculated by using the following equation:

$$V_b = (V_c D_c B_c + 119W_a)/B_b D_b \qquad (20\text{-}23)$$

where the density of the base, D_b, can be found from Equation 2-9 or 2-11 according to the Brix, B_b. In the above example, Equation 20-23 becomes:

$$\frac{(14.4)(10.977)(65.0) + 119(82.4)}{(59.3)(10.693)} = 31.7 \text{ gal of drink base}$$

The user of the base then would need to use 31.7 gallons of base with 919.1 − 9.9 or 909.2 gallons of water and 827.7 pounds of sugar. This formulation can be converted to the use of one 52-gallon drum of base by means of simple proportions, to give 1 drum of drink base + 1491 gallons of water + 1358 pounds of sugar to make 1640 gallons of finished drink.

As you may have surmised, these calculations can become complex and tedious if performed often. Here again, computer programming can be used to automate the calculations. Figure 20-2 shows a flow chart that can be used to write a program to automate this procedure.

One problem that emerges from time to time is that the Brix and/or % acid of the final drink does not come out right. Suppose, in our example, that for some reason the Brix came out to 11.5°Brix instead of the needed 11.8°Brix and with 0.9% acid instead of the needed 1.0% acid. If citric acid were added, the Brix would be affected; so addition of sugar and water would be required to achieve the proper balance of components. To find the amounts of acid and sugar needed to correct the situation, the following equations could be used:

$$W_a = \frac{V_i D_i\big(A - A_i + (A(B - B_i)/(100 - B)) + AD_w V_w(1 + B/(100 - B))\big)}{100 - A}$$

$$(20\text{-}24)$$

$$W_s = \frac{V_i D_i(B - B_i) + B(W_a + D_w V_w)}{100 - B} \qquad (20\text{-}25)$$

where the subscript i refers to the parameters for the initial erroneous conditions, and absence of a subscript indicates the desired final conditions. In the example above we do not want to add water to the final drink; so the last term

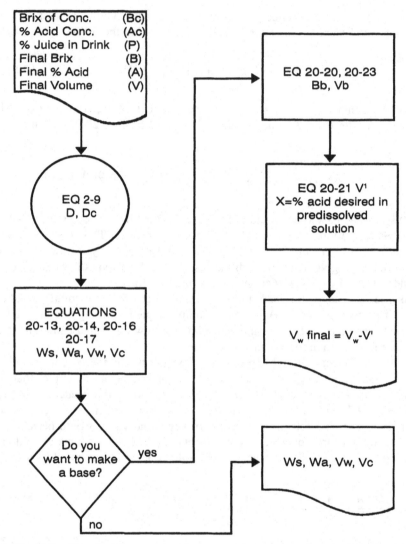

Fig. 20-2.   Flow chart that can be used to program computers to formulate drink bases.

in the numerator of the two equations above goes to zero, and we get:

$$\frac{(1000)\,(8.707)\left(1.0 - 0.8 + \left(1.0(11.8 - 11.5)/(100 - 11.8)\right)\right) + 0}{100 - 1.0}$$

$$= 17.9 \text{ lb citric acid}$$

$$\frac{(1000)(8.707)(11.8 - 11.5) + 11.8(17.9 + 0)}{100 - 11.8}$$

$$= 32.0 \text{ lb sugar}$$

The total volume of the resulting drink becomes:

$$V = (V_i D_i + W_s + W_a + V_w D_w)/D \qquad (20\text{-}26)$$

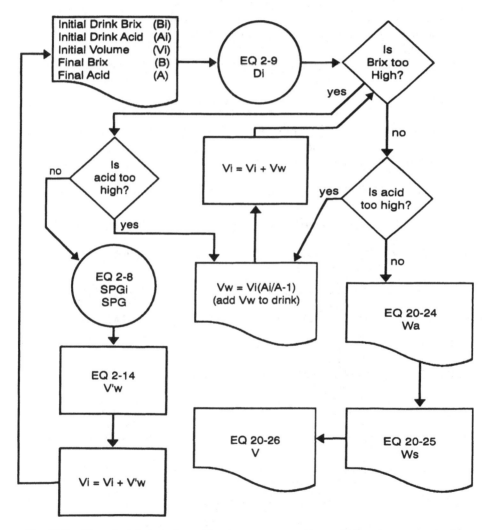

Fig. 20-3. Flow chart that can be used to program computers to calculate the sugar and acid needed to adjust errors in any type of drink.

or, in the example:

$$\frac{(1000)(8.707) + 32.0 + 17.9 + 0}{8.717} = 1005 \text{ gal of total drink}$$

Again, computer programming can simplify the use of these equations. Figure 20-3 shows a flow chart that can be used to program computers to automate the adjustment of sugar and acid in citrus or any type of drink.

Because a great amount of water is used in making juice drinks, it is important to monitor it so that contaminants do not enter drinks through the water. The alkalinity of the water is one of the major problems, as it can neutralize some of the added acids as well as affect the drink flavor. Chemicals such as

Table 20-6. Sample formulation of a 10% juice citrus drink.

Oil emulsion (296 gal)	
Citric acid	7.4 lb
Orange oil	23.7 lb
Cottonseed oil	308.9 lb
β-Carotene	23.7 lb
Gum	355.6 lb
FD&C Yellow #5	1.2 lb
FD&C Yelloc #6	0.9 lb
BHT	0.3 lb
Acid solution (736 gal)	
Citric acid	753.6 lb*
Tricalcium phosphate	28.3 lb
Sodium citrate	127.3 lb
Thiamine hydrochloride	95.6 g
Water	650.0 gal
Drink base	
Orange pulp wash concentrate**	274.0 gal
Tangerine concentrate***	7.4 gal
Lime concentrate	7.4 gal
Acid solution	736.0 gal
Oil emulsion	296.0 gal
Potassium sorbate	74.1 lb
Orange oil flavor	2.8 gal

*This value is the amount calculated in Equation 20-14 minus the citric acid added in the oil emulsion and minus the acid in the lime concentrate, assuming 400 GPL lime concentrate.
**60.0°Brix and 25 B/A ratio or 2.40% acid.
***60.0°Brix and 10.0 B/A ratio or 6.00% acid.
Note: 1000 gal of base makes 17,381 gal of drink, requiring 15,169 lb of sugar and 15,282 gal of water in addition to the 650 gal of water used to make the acid solution. The final drink should have a Brix of 11.8 and 0.56% acid.

chlorine also may influence the final product. In severe cases, water treatment may be required in citrus drink production.

Table 20-6 shows a complete sample formulation.

QUESTIONS

1. How long has pectin been used to make jellies?
2. What function does pectin serve in fruit?
3. What is the solubility of pectin in hot water?
4. In what foods are pectins mainly used?
5. What variety of fruit is the best source of pectin?
6. What are some common contaminants of pectins, and what are generally their sources?
7. What is the difference between jams and jellies?
8. What are the four main ingredients of jellies?
9. What do low-methoxy pectins require in order to gel?
10. Why is some of the sugar premixed with the pectin before water is added?
11. What is the primary quality control parameter of citrus sections?
12. What percent of the drain weight must be whole sections in order to meet a USDA Grade A standard?
13. What are puree bases primarily used for?
14. What are the 11 main components used in making juice drinks?
15. What are two advantages of making juice drinks compared to 100% juices?
16. What are some of the main advantages of using citric acid over other acids in juice drinks?
17. What is the basic composition of juice cloud?

PROBLEMS

1. In making 1000 gallons of a 50% orange juice drink, how much vitamin C do you need to add in order to have about the same level of vitamin C as found in 100% orange juice?
2. Suppose you wanted to make a 60% juice drink base that was 23.0°Brix, and that would make 10,000 gallons of final drink at 11.8°Brix and 0.53 acid, using 62.0°Brix orange concentrate with an acid level of 2.12%. How much citric acid, water, and orange juice concentrate would be needed to make the base, and how much water, base, and sugar would be needed to make the final drink? (*Hint*: You can use Equation 20-21 to find the volume of water V' needed to get the desired 23.0°Brix.)
3. Suppose that you wanted to make a 30% juice drink, using pasteurized single-strength juice (12.4°Brix and 0.95% acid) to make 20,000 gallons of 12.0°Brix, 0.67% acid drink. How much citric acid, water, sugar, and juice would you need? (*Hint*: Treat the pasteurized juice in the same way as concentrate.)

4. In making a 25,000-gallon juice drink, it was found that the Brix came out to 11.1 instead of the desired 11.9°Brix, but the acid level came out correct at 0.72%. How much sugar and acid would be needed to correct the Brix, and what would the final volume of the drink be?

5. In making a 10,000-gallon drink, it was found that the Brix came out to 12.6 instead of the desired 11.8°Brix with an acid level of 1.60% instead of the desired 1.00%. How much acid, water, and sugar will be needed to correct the situation? (*Hint*: Consider setting the W_a or W_s to zero and use Equation 20-24 and/or Equation 20-25 to solve for V_w.)

REFERENCES

Bennett, R. D. 1985. From presentation to the Citrus Products Technical Committee at the USDA Fruit and Vegetable Laboratory, Pasadena, Calif., December 16, 1985.

Bennett, R. D. 1988. From presentation to the Citrus Products Technical Committee at the USDA Fruit and Vegetable Laboratory, Pasadena, Calif., June 9, 1988.

Food Chemicals Codex, 1931. National Academy Press, Washington, D.C., 215–217.

IFT Committee on Pectin Standardization. 1959. Pectin standardization, final report of the IFT committee, *Food Tech.*, *13*, 496–500.

Joseph, G. H. 1955. Pectic substances, *Adv. Chem. Ser.*, *12*, 49–56.

Swisher, H. E. and Swisher, L. H. 1977. Specialty citrus products. *Citrus Science and Technology Vol II*, S. Nagy, P. E. Shaw, and M. K. Veldhuis, eds. The AVI Publishing Company, Inc., Westport, Conn., 321–322.

Chapter 21

Animal Feed and Fuel By-Products

The waste products of the agricultural and food industry have long been used as fertilizers and animal feeds, and to make special nonfood products such as fuels, in the form of methane gas and alcohols. Methane gas production usually is generated from waste water and will be discussed in Chapter 22. Fuel-grade alcohols usually are manufactured from peel products and are discussed in this chapter. Citrus peel and peel products have been and continue to be used as a source of high-quality animal feeds, which require careful quality control.

FEEDS FROM CITRUS PEEL

Citrus peel often is viewed as a waste product even though some citrus processors are able to make substantial profits from it. The most popular peel products are those used in animal feeds. Citrus plants frequently are located in rural areas where livestock are raised that can use the citrus wastes as feed or a feed supplement. Fertilizers from livestock or poultry wastes, in turn, are often used in citrus groves, creating a compatible symbiotic relationship. Dried peel products can be conveyed to distant markets as well, which provide the juice processor with numerous opportunities for disposing of waste products economically.

Many factors affect the quality control of feeds, such as the type and breed of animal, the desired end results or animal products, the effect of these products on humans, supplemental feeds, the complexities of animal nutrition, the economics involved, and the storage capabilities of the feed. Many of these parameters are too complex for routine quality control. Also, the processor has control over only a few of them. Most of these characteristics are naturally inherent in the feed, and once the general characteristics of the feed have been established, it is up to the quality control department to guard against excessive contamination or alteration of the basic attributes of the feed.

Most citrus processors guarantee their feeds to have at least a minimum amount of protein, crude fiber, fat, and ash. Another important parameter is the amount of nitrogen-free extract, primarily carbohydrates, which varies somewhat and usually is not guaranteed. Even so, these parameters rarely need routine monitoring, and the analyses are best performed by professional laboratories when they are required. A general estimate of the total quality of the feed can be expressed in several ways, the most common method being the total digestible nutrients (TDN) of the feed. This is determined by adding the % nitrogen-free extract, % protein, % crude fiber, and the % fat times 2.25. For example:

N-free extract	15.9%
Protein (N × 6.25)	1.7%
Crude fiber	2.7%
Fat (0.2 × 2.25)	0.5%
TDN = 20.8	

Although the TDN is a simple number to determine, the proportions of the four components that actually contribute to animal nutrition varies. Also, these proportions are based upon humans and dogs, and do not apply directly to ruminants. However, the TDN still is commonly used to estimate the general quality of citrus feeds.

Unprocessed Peel

The use of unprocessed or wet peel as a cattle feed is a widespread practice, which usually requires that the animals be located relatively close to the citrus processor. The acidic nature of the wet peel is corrosive to the metal in most trailers, inevitably resulting in messy leaks. This characteristic, combined with the wetness of the peel, limits the distance that the peel can be transported. Wet peel generally contains anywhere from 78% to 90% moisture, depending on the juice extraction procedure used. Because this product usually is sold "as is," moisture tests or other routine tests rarely are performed. An example of the nutrient guarantee of wet peel is as follows:

Protein: not less than 1.1%.
Crude fat: not less than 2.3%.
Crude fiber: not less than 2.6%.
Ash: not less than 1.3%.

Transportation costs are the main costs of processing wet peel. They usually determine the price and the need for additional quality control.

Pressed Peel

Peel that has been shredded, limed, and pressed to remove about 10% of the moisture has the advantages of lower transportation costs and a more appealing and uniform texture and composition. Such treatment provides for the production of press liquor and/or citric molasses, and allows the feed to be transported longer distances because of a reduction in trailer leaks. Typical nutrient levels of pressed peel are shown in Table 21-1. Little to no routine quality control is performed on this lightly processed peel product.

Dried Citrus Peel or Pulp

Dried citrus peel, one of the most common feeds manufactured from citrus peel, is second only to corn in nutritive value for dairy and beef cattle and sheep (Kirk and Davis 1954). The drying of citrus peel is somewhat involved, requiring more quality control monitoring than does production of unprocessed or pressed peel. Dried citrus peel is made by passing pressed peel through a rotary dryer, with the consistency and speed of the peel through the dryer determining the drying efficiency. Studies have been made of the rate of drying at various temperatures (Braddock and Miller 1978).

Citric molasses is often added to the pressed peel to aid in the drying process and to help prevent burning of the peel. Too much molasses will cause the peel to stick to the edges of the dryer, which will also cause burning. Molasses addition also darkens the color of the feed. Some investigators have even used the feed color to determine the amount of molasses that has been added (Bissett 1950). When molasses is used in the drying process, the dried pulp is referred to as sweetened dried citrus pulp. In warmer drier weather, wet or pressed peel can be spread on the ground and solar-dried. This is a common and economical practice in areas with a dry Mediterranean climate such as California. Even though most peel dryers are equipped with dust collectors, dried citrus peel usually is associated with fine dried particles. FMC juice extractors break up more juice cells than other juice extractors, causing an increase in this fine dust material in the dried feed.

The most important parameter to monitor in dried peel manufacturing is the moisture content. Moisture levels must be below 10%, or micorbiological spoilage can set in along with a buildup of heat. This heat can reach a point where

Table 21-1.   Typical nutrient levels for pressed peel.

Nutrient	California Orange	Florida Orange*	Florida Grapefruit*
Protein (N × 6.25) (%)	1.7	2.01	2.24
Crude Fat (%)	0.2**	2.65	1.18
Crude fiber (%)	2.7	4.36	4.61
Ash (%)	1.5	1.45	1.46
Moisture (%)	78	72	75
Nitrogen-free extract (%)	15.9	17.80	15.7

*Kesterson and Braddock 1976.
**ether extract

the dried peel will spontaneously combust, not only causing a fire hazard but destroying the product. Smoldering peel that has not been dried sufficiently can create a serious disposal problem. Unless the peel is spread out and air- or solar-dried immediately, it will continue to burn until it becomes ash. Some bulk dried peel warehouses have thermocouple monitors with air vents in the floor to adjust the temperature and correct moisture problems. The relative humidity in storage areas should be kept at 52% or less in order to avoid reabsorption of moisture by the peel (Braddock and Miller 1978). Oil recovery systems sometimes add moisture to the peel through peel washings, which should be accounted for in drying operations. Grapefruit peel generally contains more moisture than orange peel and thus would require more drying. There are several ways to perform moisture tests, and one of the simplest follows:

Peel Moisture Test

Equipment and Supplies

- Triple beam balance.
- Heat lamp.
- Drying pan or dish.
- Shredder (blender, food processor, or knife and cutting board).

Procedure

1. If the peel has not been shredded or chopped fine, this needs to be done with a shredder. Peel particles should be about 20 to 30 mesh or about $1/8''$ in diameter or smaller.
2. Place the empty drying pan or dish on the triple beam balance, and record the weight (W_{pan}).
3. Place the shredded peel in the drying pan so as to just cover the bottom of the pan, and record the weight (W_b).
4. Leaving the drying pan on the balance, radiate the pan with a heat lamp until no further change in weight is noticed. Record the weight (W_a). Care should be taken not to overheat or scorch the sample.
5. The % moisture can be calculated from:

$$\%M = 100\%\left(1 - (W_a - W_{pan})/(W_b - W_{pan})\right) \qquad (21\text{-}1)$$

For example:

$$100\%\left(1 - (184.54 - 179.10)/(185.10 - 179.10)\right) = 9.3\% \text{ moisture}$$

This procedure can be used with wet, pressed, or dried peel. Dried citrus peel has the general nutrient values shown in Table 21-2. Dried pulp is slightly

Table 21-2. Typical nutrient levels of dried citrus peel (Kesterson and Braddock 1976).

Nutrient	Unsweetened Orange	Sweetened Orange	Grapefruit
Protein (%)	5.9–6.2	5.3	6.4–7.0
Fat (%)	3.1–4.9	2.8	5.5–5.9
Crude fiber (%)	11.5–12.0	9.3	12.1–15.3
Ash (%)	4.9–6.9	8.0	6.5
Nitrogen-free extract (%)	62.7–64.0	66.6	56.9

higher in fiber, nitrogen-free extract, and ash than wet peel, but it is lower in fat (Kirk and Davis 1954).

Pellets

Because of the dusty nature of dried citrus pulp and its low bulk density, many processors pelletize the dried pulp. This compacting of the dried pulp volume cuts shipping and storage costs about in half and facilitates a cleaner and more efficiently conveyed product. The higher-density pellets are more easily consumed by ruminants as well. Pellets made from dried pulp fines are generally about $\frac{1}{4}$ inch in diameter and about 1 to $1\frac{1}{4}$ inches long. Whole dried citrus pulp is made into pellets about $\frac{3}{8}$ inch in diameter and about $\frac{1}{2}$ to $\frac{3}{4}$ inch long in order to retain the pulp fiber and seeds (Kesterson and Braddock 1976).

Several factors affect the mechanical durability of the pellets, including the thickness of the die, holding time, pellet size, energy used in pelletizing, and bonding agents used. Citrus molasses has been found to be an excellent bonding agent when used in the proportions of about 5 to 15% of the total weight. However, molasses usually is added before the peel is dried instead of during pelletizing in order to aid in the drying process, as mentioned previously. The extrusion rate and the bulk density of the pellets vary inversely with the die length, and the energy needed during the extrusion is inversely proportional to the die thickness.

Some additional advantages of pelletizing dried citrus peel include the reduction of microbial spoilage. It has been found that pellets with twice the moisture content of dried pulp still are resistant to microbial deterioration (Dean 1966). However, pellets wetter than this lose their mechanical stability. Because pelletizing involves friction that can heat the product, and thus dry the pellets, dried citrus pulp destined to become pellets must be dried to only 10 to 12% moisture levels. After pelletizing, the moisture levels will drop to below the needed 10% level. Like dried citrus pulp, pellets should be kept in areas with 52% or less humidity to avoid reabsorption of moisture into the peel. A relative humidity of 90% for 30 days has been shown to induce mold growth (Braddock and Miller 1978).

Insect Infestation

Another problem in the storage of dried citrus pulp or pellets is the occurrence of insect infestation. The almond moth, *Cadra cautella* (Walker) (Hagstrum and Sharp 1975) and the saw-toothed grain beetle, *Oryzaephilus surinameusis* (L.), have been associated with dried citrus pulp, with the latter found in coarse pulp (Laudens and Davis 1956). Pellets and fine meals have been found to contain the cigarette beetle, *Lasioderina serricorne* (F.) (Laudens and Davis 1956). Other insects have also been found in dried citrus material and justify thorough cleaning of warehouses as often as possible. Residual insecticides, traps, and barriers should be used, as well as professional services if needed. (See Chapter 16.)

Press Liquor

Press liquor is the solution expressed from shredded and limed peel. Unlimed peel is very slimy and retains moisture. To remove moisture, about 0.15 to 0.25% lime is added to shredded peel, and is allowed to react for at least 15 minutes. The exact mechanism by which the lime helps to release water and remove the slimy texture of the peel is not known, but probably the basicity of the lime demethoxylates the pectins in the peel in a manner similar to the action of pectinase enzymes:

$$R-\overset{\overset{\textstyle O}{\|}}{C}OCH_3 \xrightarrow{OH^-} R-\overset{\overset{\textstyle O}{\|}}{C}O^- + CH_3OH \qquad (21\text{-}2)$$

which is similar to Equation 8-1. The divalent cations, Ca^{+2}, from the lime probably combine with the demethoxylated pectin in a manner similar to that illustrated in Fig. 8-2. The end result is synersis, or the natural expression of fluid, which is not unexpected considering that 80 to 85% of the peel is water. The slimy nature of the unlimed peel is probably due to hydrogen bonding of the ester groups of the pectin with water. Lime demethoxylation increases the ionic strength of these bonds, producing a less slimy texture. About 10% of the peel moisture can be removed in primary presses. Secondary presses generally can remove only about another 2% of the moisture from the peel.

Press liquor usually contains about 9 to 15% soluble solids expressed as °Brix, with about 60 to 70% of these soluble solids being in the form of sugars. Peel oil levels range from 0.2% to 0.5% in Florida press liquor and up to 0.8% in California press liquor made from orange peel. The drier climate in California produces peel with a higher oil content. Typical press liquor constituents are shown in Table 21-3. Press liquor is occasionally sold as is for animal feed production, or fermented to produce fuel-grade alcohols, or concentrated to citric molasses, the last being its most common use.

Table 21-3.   Typical composition of citrus press liquor (Nolte, von Loesedke, and Pulley 1942).

Constituent	Range	Average
pH	5.4–6.4	5.7
Brix (17.5°C)	6.1–12.6	10.1
Sucrose (%)	1.20–3.09	2.40
Reducing sugars (%)	2.82–5.81	4.23
Protein (N × 6.25)	0.40–0.59	0.47
Pectin (alcohol ppt, %)	0.27–0.88	0.66
Ash (%)	0.43–0.94	0.72
Acid as citric acid (%)	0.15–0.30	0.21
Alcohol (% by volume)	0.00–0.39	0.22
Peel oil (% by volume)	0.12–0.58	0.23

Citric Molasses

The manufacture of citric molasses offers several advantages over the manufacture of press liquor alone. Storage and shipping costs are greatly reduced because molasses takes up less space; concentrated sugar solutions are more resistant to microbial spoilage; and citric molasses can be mixed with pressed peel to aid in peel drying, as mentioned previously, in addition to acting as a bonding agent when pellets are made from dried pulp. The use of evaporators that utilize waste heat from peel dryers can greatly reduce the cost of molasses production. An example of the composition of citric molasses is given in Table 21-4.

Citric molasses generally has about the same characteristics as Blackstrap molasses except that it contains half as much ash and twice as much protein. Florida requires that at least 45% of the total sugars be invert sugars (glucose

Table 21-4.   Typical analysis of Florida citric molasses (Kesterson and Braddock 1976).

Constituent	%	Constituent	%
Brix	72.0	Calcium	0.8
Sucrose	20.5	Sodium	0.3
Reducing sugars	23.5	Magnesium	0.1
Protein (N × 6.25)	4.1	Iron	0.04
N-free extract	62.0	Phosphorus	0.06
Fat	0.2	Manganese	0.002
Fiber	0	Copper	0.003
Ash	4.7	Niacin	35 ppm
Pectin	1.0	Riboflavin	11 ppm
Potassium	1.1	Pantothenic acid	10 ppm
pH	5.0	Viscosity (25°C)	2000 cps

and fructose) with a Brix of at least 35.5 after mixing with an equal weight of water (Florida State Dept. of Agriculture 1950).

Two common problems in the manufacture of citric molasses are foaming and scale formation. Commercial antifoaming agents can be used to minimize foaming problems; and in addition to foaming agents, removal of the suspended solids through centrifuging or other suitable means can reduce foam formation. Foaming can be due to microbial spoilage, alkaline processing, or the mixing of old and new molasses (Hendrickson and Kesterson 1971). Scale formation is due to calcium from the lime used to treat pressed peel, which often precipitates on evaporator surfaces as calcium citrate. Using preheaters will provide a site for the precipitation of this scale in an area that is easier to clean, and the use of two such preheaters allows the use of one while the other is being cleaned. The scale can be removed by using a 5 to 15% hot caustic solution containing 2 to 3% chelating agents such as alkaline ethylene diaminetetraacetic acid (Kesterson and Braddock 1976).

Most citric molasses has been sold as a feed supplement for animal feeds. Storage of the molasses usually will decrease the pH by about 0.4 unit per season (Hendrickson and Kesterson 1950), and viscosity will increase with storage. Vigorous agitation generally will restore fluidity, but prolonged storage can result in solidification of the molasses. Molasses that is increasing its viscosity at a rate of less than 500 cps per day can be stored for at least 3 months (Hendrickson and Kesterson 1952).

The microbial flora of citric molasses would be expected to resemble that of juice concentrates. Various species of *Lactobacillus* and *Leuconostoc* have been isolated from citric molasses including *L. fermentum*, *Le. mesenteroides* subsp. *mesenteroides*, and *Le. paramesenteroides* (Parish and Higgins 1988).

A final footnote on citric molasses concerns the suspended matter. Citric molasses manufactured in California generally contains a large amount of settleable sludge material, which has been linked to the commercial waxes applied in the fresh fruit packing houses. In Florida, where most citrus fruit is processed directly from the field, this waxy sludge is less common. Also, the waxes used in Florida and California are not the same. In California, juice fruit usually is heavily coated with these waxes, which find their way into the press liquor and then into the molasses. Commercial centrifuging to remove this sediment has proved ill advised because sand impurities from the lime end up in the press liquor and produce excessive wear in centrifuges. This sediment adds to the viscosity of the molasses and may cause a conveyance problem if allowed to settle out in storage tanks.

FUEL-GRADE ALCOHOL

With the increase in price of fossil fuels over the past several decades, alternative sources for fuels have been sought; and some of the more promising sources are found in renewable agricultural products and waste materials. The food industry has always tried to achieve economical recovery of waste materials. Now anhydrous ethanol is being used in premium unleaded gasolines, as ethanol is an efficient fuel that can be used to enhance octane ratings.

Citrus press liquor and reconstituted citric molasses have been used as a feedstock for commercial alcoholic fermentations. A major problem encountered in such fermentations is the presence of citrus peel oil, which is toxic to most microorganisms. As mentioned previously, press liquor contains at least 0.2% oil (based on 11.8°Brix). Citric molasses usually contains much less oil, as most of it has been stripped out during evaporation. Oil levels must be 0.08% or less for a successful alcoholic fermentation; so the oil in citrus press liquor must be removed before fermentation can take place. A common way to do this is to use direct steam injection into press liquor (see Chapter 6 and section on stripper oil). The viscosity of citric molasses also can inhibit fermentation, including the contribution of waxy sludge material found in California citric molasses, as mentioned. Citric molasses usually is diluted to about 20°Brix or less prior to fermentation.

Like most microorganisms, yeasts undergo four main stages of growth, as shown in Fig. 21-1. In the initial phase, they begin adjusting to conditions and start to grow. Growth is rapid during the log phase, where reproduction occurs exponentially. As the feedstock is depleted and the yeast waste material (alcohol here) is concentrated, growth slows until fermentation reaches a steady state. It is in the lag phase that fermentation is the most efficient. If fresh feedstock is

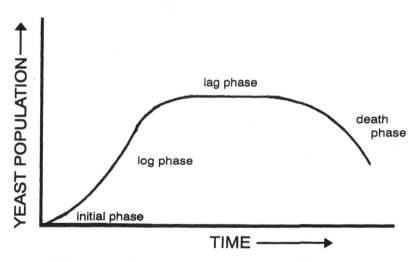

Fig. 21-1. The four stages of microbial growth. Alcoholic fermentation is most effective when the yeasts are in the lag phase.

not introduced, the growth will eventually begin to decrease, resulting in the death phase.

There are basically two types of fermentation, continuous and batch. Continuous fermentation requires that yeast be kept in the lag phase by constantly replenishing the feedstock while bleeding off the alcoholic solution without changing the yeast population or activity. A popular and efficient continuous fermentor is the Wick fermentor, shown in Fig. 21-2 (Wick 1980). The gas generated in the fermentation causes a circular flow in the solution; and this circular current not only mixes the solution to make it uniform, but it keeps the yeast within it, providing a "dead zone" that contains little yeast in the corners where the alcoholic solution can be bled off without significantly changing the yeast population. Commercial use of the Wick fermentor has proved successful (Eastman 1981). Continuous fermentation also is easier to monitor as far as quality control is concerned because its conditions do not change as much as those of batch operations. Batch fermentation requires greater tank space to produce alcohol as rapidly as the continuous methods do. Also, the constant starting and stopping of batches is very time-consuming and requires close attention to ensure a successful fermentation.

Other factors also affect the rate of fermentation, including the type of yeast used. Distiller's yeast has been found to grow faster in reconstituted citric molasses or steam-treated press liquor in California than do the natural yeasts or Montrachet yeast. Excessive bacterial growth can destroy a fermentation, pro-

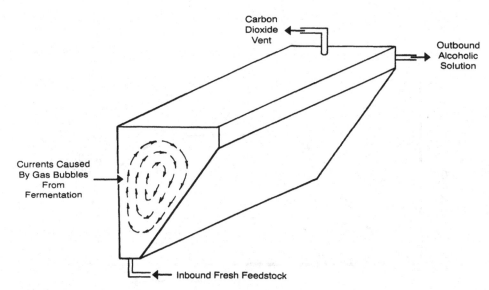

Fig. 21-2. Wick continuous fermentor used commercially to produce fuel-grade alcohol (Wick 1980).

ducing acetic acid rather than alcohol. Because bacteria grow much faster than yeast, bacterial growth can quickly take over a fermentation. However, bacteria are more sensitive than yeast to the osmotic effects in concentrated molasses, as well as to benzoates and penicillin. Steam treatment of press liquor and pasteurization of molasses during evaporation are generally sufficient to prevent bacterial contamination. Small amounts of benzoate or veterinary grade penicillin can be used to inhibit bacterial growth as well. A good cleanup and sanitation program is the best way to prevent bacterial spoilage.

In order to start up a fermentation process, using either a batch or a continuous method, the feedstock must be diluted to 12 to 30°Brix. Also, fermentation is more rapid with warm solutions. The peel oil level should be checked to ensure that it is below the 0.08% v/v limit (based on 11.8°Brix). As the container is filling with feedstock, one should add about 120 ml of veterinary grade penicillin, 3 gallons of FMC defoamer or the equivalent, and 1 gallon of pectinase if needed to reduce the viscosity of the solution, assuming a 10,000-gallon fermentation volume. One should mix about 20 pounds of distiller's yeast in a 5-gallon bucket of lukewarm water to paint consistency, and when gas evolves, add the mixture to a fermentation vessel containing about half of the feedstock (5000 gallons here). It is important to allow a vent for evolving carbon dioxide gas. The 20.0°Brix feedstock should decrease in Brix down to about 13.0°Brix after fermentation in batch processing. Fermentation of 10,000 gallons takes up to 24 hours or so to really get under way and about 36 to 48 hours for completion, depending on feedstock, yeasts, and temperature.

To determine when fermentation is complete, a reducing sugar test can be used. The Brix alone is insufficient because alcohol and other components of the solution will contribute to it. Although the reducing sugars account for only about half of the sugars present, their disappearance can be used as an indication that all the sugars have been consumed in the fermentation. Reducing sugar levels of 1 to 2% usually indicate that the fermentation is complete. A method that can be used to measure the reducing sugars is given in Chapter 19.

After fermentation, the beer is distilled in commercial stills. Aqueous alcohol forms an azeotropic mixture upon distillation that prevents more than 95.6% ethanol from being distilled at atmospheric pressure. Further heating cannot result in any enrichment of the ethanol solution. Commercial plate column stills can produce 90% ethanol solutions but generally achieve around 80% ethanol. An 80% ethanol solution is sufficiently enriched to be used directly as a fuel in automobiles with enlarged air intake jets, but 100% or anhydrous ethanol is required for blending with gasoline. Alcohol burns hotter than gasoline and thus requires a higher air mixture in the carburetor. Pure ethanol can be obtained by further drying of the 80% solution using absorbents, such as cornstarch or dried citrus peel, or by the use of a vacuum (95 torr, 0.125 atm, or 26 inches of Hg) during distillation that will shift the azeotropic composition up to as high as

99.5% ethanol. When an absorbent is used to dehydrate the alcohol solution, dry carbon dioxide from the fermentation process can be used to redry the absorbent so it can be reused. A typical continuous plate alcohol still is shown in Fig. 21-3. Some of the alcohol solution is refluxed back into the top of the second column in order to maintain the proper temperature.

The quality control of distillation involves the measurement of alcohol levels at various points. The distillage can be concentrated and added back to the peel like citric molasses, as a high-protein supplement to the feed. This distillage should not contain more than 5% alcohol; otherwise, too much alcohol is being lost. The temperature of the first column should be about 245°F (118°C) in order to remove 95% of the alcohol. If water builds up in the bottom of the second column, it can be pumped back into the first column. The temperature in the second column should stabilize at about 173°F (78.2°C). In order to avoid federal liquor license requirements, the alcohol produced needs to be denatured or poisoned by adding methanol or a similar toxic chemical that will

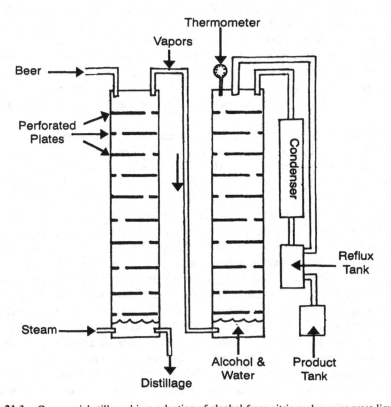

Fig. 21-3. Commercial still used in production of alcohol from citric molasses or press liquor.

not affect its performance as a fuel. Adequate signs and labeling must be used, warning that the alcohol is toxic and is not intended for human consumption.

In order to measure the alcohol level in the distillage and the final product, several methods can be used. One method involves measurement of the chemical oxygen demand (COD) and is lengthy compared to other methods (AOAC 1984); another involves measurement of the specific gravity (AOAC 1984); a third method involves the direct fractional distillation of the alcohol from the sample. The latter two methods have proved sufficient for routine quality control. The specific gravity method is best for measuring the ethanol in the distillate or final product, and the distillation method works best for measuring the ethanol in the distillage.

Ethanol by Specific Gravity (AOAC 1984)

Equipment and Supplies

- Balance.
- 25 ml pipette.
- 25 ml beaker or flask.
- 25 ml of distilled water.
- Thermometer.
- AOAC table 52.003 (optional).

Procedure

1. Pipette 25 ml of distilled water at 20°C into a clean and dry beaker or flask, and weigh it. If the same pipette and container are used each time, the same weight of the container and water can be used for each test.
2. Using the same pipette and container, pipette 25 ml of the alcoholic distillate (70% or more in ethanol) at 20°C into the container, and weigh it.
3. Divide the weight of the sample by the weight of the water to get the specific gravity.
4. Use AOAC table 52.003 to find the % alcohol according to specific gravity, or use the following regression equation, which can be easily programmed into hand-held programmable calculators or computers:

$$\%A = -1176 + 1895/S_g - 701/S_g^2 \qquad (21\text{-}3)$$

where S_g is the specific gravity, and A is alcohol. For example:

$$-1176 + 1895/0.839 - 701/(0.839)^2 = 87\% \text{ alcohol}$$

Commercial ethanol from citrus products contains a slight off coloration. In citrus alcohol produced in California, small amounts of isopropyl alcohol and ethyl esters have been found (Bennett 1981). These ethyl esters probably are coupled with inorganic salts such as phosphates or sulfates.

Ethanol by Distillation

Equipment and Supplies

- Setup shown in Fig. 21-4.
- 100 ml graduated cylinder.

Procedure

1. Add 100 ml of the sample to the boiling flask in Fig. 21-4, and apply heat. The temperature at the top of the column should reach 78.2°C (173°F) when the fractionating ethanol reaches the thermometer. This temperature should remain steady until toward the end of the distillation.
2. When no more ethanol can be seen coming over into the condenser, or when water beads are noticed in the column or condenser, record the milliliters distilled as the % v/v of alcohol. For further accuracy, multiply the milliliters of distillate by 0.95 to get the approximate true % alcohol in the sample, or use the specific gravity method on the distillate.

It should be readily apparent that the distillate cannot contain more than the 95.6% ethanol azeotropic limit. This is the reason for using the 0.95 factor in step 2 above and for the suggestion that the specific gravity method be used on the distillate. However, the % v/v of ethanol measured by using this method is adequate for most routine quality control.

QUESTIONS

1. What are the main nutritional parameters of citrus peel animal feeds?
2. What is the most common feed produced from citrus peel?
3. Why is citric molasses added back to pressed peel before peel drying?
4. What is the most important quality control parameter in the manufacture of dried citrus peel, and why?
5. What are the advantages of pelletizing dried citrus pulp?
6. Why is lime added to press peel?
7. What is the major use of citric molasses?

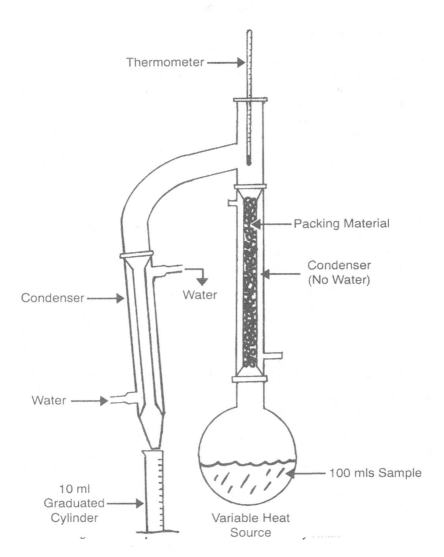

Thermometer

Packing Material

Condenser
(No Water)

Water

Condenser

Water

10 ml
Graduated
Cylinder

100 mls Sample

Variable Heat
Source

8. What basic citrus peel component inhibits alcoholic fermentation of peel extracts?
9. Why is bacterial growth a problem in alcoholic fermentations?
10. What two solutions require quality control monitoring during distillation of fermented citrus beers?

PROBLEMS

1. What is the TDN for wet peel that just meets a typical nutrient guarantee as mentioned in the chapter, assuming a nitrogen-free extract of 10.3%?
2. What would be the TDN of pressed California orange peel, sweetened dried citrus pulp, and citric molasses, according to the data in the chapter?
3. Is dried citrus pulp dry enough if a 6.832 g sample in a 165.132 g pan weighed 171.178 g (pan and sample) after drying under a heat lamp?
4. What would be the % alcohol in a distillate with a specific gravity of 0.832?
5. What would be the exact % alcohol in the distillage if 6.7 ml were distilled off in the alcohol determination described in the chapter, yielding a distillate with a specific gravity of 0.812? Would this be considered an acceptable distillation efficiency?

REFERENCES

AOAC. 1984. *Official Methods of Analysis, 13th Edition*, 9.012, 9.014, 10.024, 11.005, 11.007, 11.008. Association of Official Analytical Chemists, Washington, D.C.

Bennett, R. D. 1981. Private communication.

Bissett, O. W. 1950. A method for estimating soluble solids in dried citrus pulp, *Proc. Fla. State Hort. Soc.*, *63*, 174–179.

Braddock, R. J. and Miller, W. M. 1978. Some moisture properties of dried citrus peel, *Proc. Fla. State Hort. Soc.*, *91*, 106–109.

Dean, W. W. Jr. *Pelleting Wet Citrus Pulp with the Aid of a Bonding Agent.* M.S. Thesis, University of Florida, August 1965, University of Florida Agriculture and Engineering Mimeo Report EG66-1, February 1966.

Eastman, R. 1981. Private communication.

Florida State Dept. of Agriculture. Lett. of December 8, 1950, *Supplementary Feed Bull.*, *97*.

Hagstrum, D. W. and Sharp, J. E. 1975. Population studies of *Cadra cautella* in citrus pulp warehouse with particular reference to Diapause, *J. Econ. Entomol.*, *68(1)*, 11–14.

Hendrickson, R. and Kesterson, J. W. 1950. Storage changes in citrus molasses, *Proc. Fla. State Hort. Soc.*, *63*, 154–162.

Hendrickson, R. and Kesterson, J. W. 1952. Viscosity of citrus molasses, *Proc. Fla. State Hort. Soc.*, *65*, 226–228.

Hendrickson, R. and Kesterson, J. W. 1971. Citrus molasses, *Fla. Agr. Exp. Sta. Bull.*, *677*, 3–27.

Kesterson, J. W. and Braddock, R. J. 1976. *Byproducts and Specialty Products of Florida Citrus.* University of Florida, Gainsville, Fla., 10, 22, 31.

Kirk, W. G. and Davis, G. K. 1954. Citrus products for beef cattle, *Fla. Agr. Exp. Sta. Bull.*, *538*, 5–16.

Laudens, H. and Davis, D. F. 1956. Dried citrus pulp insect problem and its possible solution with insecticides—coated paper bags, *Proc. Fla. State Hort. Soc.*, *69*, 191–195.

Nolte, A. J., von Loesedke, H. W. and Pulley, G. N. 1942. Feed yeast and industrial alcohol from citrus wastes press juice, *Ind. Eng. Chem.*, *34*, 670–673.

Parish, M. and Higgins, D. 1988. Isolation and identification or lactic acid bacteria from samples of citrus molasses and unpasteurized orange juice, *J. Food Sci.*, *53(2)*, 645–646.

Wick, E. 1980. New vessel design for rapid, continuous, fermentation. USDA Science and Education Administration, *Advances in Agricultural Technology*, AAT-W-10/February.

Chapter 22

Wastes from Citrus Plants

Citrus processing plants, like most other plants, generate materials of little or no commercial value that generally need to be removed from the processing site. It usually is the responsibility of the quality control department to monitor and manage waste accumulation and treatment, especially in small processing plants. These materials, which can be classified as solid, liquid, gaseous, or hazardous wastes, can be conveyed from the plant through waste streams, by air flow into the atmosphere, or by vehicle. Because these materials affect persons in the vicinity of the plant, and because federal, state, and local regulations apply to the disposal of these wastes, the quality control of wastes is an important function.

The degree of pollution regulation depends largely upon the type and amount of pollution as well as the surrounding environment. Citrus processing plants and bottlers exist in a variety of locations, from rural to metropolitan, with a wide range of existing pollution. Many plants are located next to large bodies of water, which often act as disposal sites for wastes. Rural sites include large land areas, which also may serve as disposal sites. Metropolitan plants often take advantage of municpal disposal facilities, whereas plants located in suburban areas must treat their own wastes or convey them long distances. The federal government has passed several laws to control the pollution of the environment, including the Clean Air Act of 1977, the Clean Water Act, the Safe Drinking Water Act, the Resource Conservation and Recovery Act, and the Toxic Substance Control Act. State and local governments are likely to have even stricter requirements and to act as watchdogs for the standards of the federal Environmental Protection Agency (EPA). It is estimated that except for preconstruction environmental impact reports, a properly running citrus processing plant that processes under 500,000 tons of fruit a year probably will not need to deal with EPA on a continual basis. With good waste management, even larger plants can escape close regulatory scrutiny.

AIR POLLUTION

Citrus processors and bottlers are not considered nationally to be major polluters of the atmosphere. However, as with all manufacturers, their pollutant emissions are significant enough to require some form of monitoring. Some of the main sources of air pollution include boilers, evaporators, and furnaces, as well as minor sources such as storage tanks, cooling equipment, and dried peel conveyance equipment. Rotting refuse and waste water are another source of off odors that can contribute to atmospheric pollution.

The EPA has issued standards for ambient air quality, as summarized in Table 22-1. Primary emissions are those that immediately escape the processing equipment. Secondary emissions are those formed from primary emissions in the air, which are of less importance to citrus processors than primary emissions. Pollution problems often arise from errors in the feedmill where peel is burned or volatile organic compounds (VOCs) are not sufficiently scrubbed. A bluish gray smoke is indicative of d-limonene emission. Many local regulations limit the VOCs that can be emitted, including the VOC content of paints and thinners. For example, in Tulare County, California, only 420 grams of VOC/liter of paint (excluding water) is allowed during drying below 194°F (90°C). The stringency of local regulations can be seen by comparing this value to the federal maximum of 1470 grams VOC/liter of paint used in painting automobiles (45 FR 85414, December 24, 1980).

Sulfur compounds are considered one of the major pollutants in the combus-

Table 22-1. Primary ambient air quality standards set by EPA as of 1983 (Perry and Green 1984).

Pollutant	Maximum Primary Emission	Per Time Period
SO_x	0.03 ppm	Annual mean**
	0.14 ppm*	24 hours
CO	8.7 ppm*	8 hours
	35 ppm*	1 hour
O_3 (corrected for NO_2 and SO_2)	0.12 ppm*	1 hour
NO_x	0.05 ppm	Annual mean**
Lead	0.08 ppb*	1 hour
Hydrocarbons (corrected for CH_3)	0.24 ppm*	3 hours
Particulate matter	0.04 ppm	Annual geometric mean***
	0.13 ppm*	24 hours
Photochemical oxidants (expressed as O_3)	0.08 ppm*	1 hour

*Not to be exceeded more than once per year.
**$(\Sigma_{i=1}^{n})/n$
***nth root of $(\Pi_{i=1}^{n})$.

tion of fossil fuels. In citrus processing, natural gas and oils, which contain some sulfur, are the most common sources of energy. In Tulare County, California, a rural area, the sulfur content must be below 0.7% in fuels. In Metropolitan Los Angeles, the sulfur content must be 0.5% or less in fuels. These maximum values change from time to time, depending on existing conditions. Local authorities generally require notification of changes from one fuel to another or any other changes that may affect the pollution of the area. Good communication and consideration for one's neighbors and the environment generally will prevent pollution regulation from becoming a serious issue.

SOLID WASTES

Even though citrus peel generally is sold as an animal feed, it sometimes is referred to as a major solid waste from citrus processing plants. Discarded peel, rot bins, and even pooled waste streams can cause a stench and should be removed. Piles of discarded peel are an unsightly, reeking mess. Other forms of solid wastes may include broken glass that can be shipped back to the glass factory; shipping cartons, crates, fiber containers, and so on which can be burned or sent to recycling stations or municipal disposal sites; and discarded equipment, drums, metal cans, and so forth, which usually can be sold as scrap metal. Collections of these solid wastes provide harborages for rodents and insects that may spread to processing areas.

LIQUID WASTES

The primary waste from citrus processing plants and bottlers is in the form of liquid or aqueous wastes. Aqueous wastes assume a variety of forms, from relatively clean refrigeration water to highly organic discharges from oil centrifuges. Regardless of the source, citrus waste streams basically contain four types of contaminants: suspended and settleable solids (such as juice sac material, pulp, and waxes), soluble organics (primarily sugars and acids), soluble inorganics (caustic sodas), and volatile organics (d-limonene from peel oils, etc.). Various methods and method combinations can be used to treat this type of waste, depending on the need and the resources available.

Filtration

All waste treatment methods list filtration as the initial step. Self-cleaning gravity-fed screens or shaker screens remove particulate matter, which can clog equipment or lines as well as prove difficult to treat. Solid material thus removed can be added to citrus pulp and processed in the feedmill, or collected and transported to a remote disposal site.

Reuse Water

Waste effluents can be greatly reduced by reusing some of the waste streams. For example, the relatively clean refrigeration water can easily be recycled within the refrigeration system itself. Condensate water from the first effect in evaporators can be reused in boilers as makeup water. Other condensate streams from cooling towers can be used in belt sprayers to facilitate fruit conveyance, as well as in some clean up hoses. This water may not be suitable for other drinking or food processing purposes and should be so labeled, but potable water that has been completely treated can be used anywhere in the plant. Because juice evaporators produces water, the waste water effluents will always exceed water usage.

Dissolved Air Flotation (DAF)

The introduction of fine air bubbles into static waste water is used extensively in the food industry to separate grease, fats, and waxy materials. Under such conditions, these materials have a tendency to float so that they can be skimmed off and removed. In citrus processing plants DAF is used to help separate suspended pulp and waxes from screened waste water and to aid in sludge settling after aerobic treatment. As mentioned in Chapter 21, waxes washed from fresh market culls produce a waxy sediment in waste streams, which can be separated by DAF. The skimmed residue then can be added to citrus pulp entering the feedmill. It is estimated that 10 to 20% of the effluent waste loading can be removed by using DAF.

Flocculation

Another method used to reduce waste effluent loading is flocculation of heavier dissolved or suspended material in the waste streams. A flocculating agent is introduced into the water, and it complexes, aggregates, or precipitates with heavier components of the water, which then proceed to settle out. The settled sludge can be removed by centrifugation and added to feedmill material. Laboratory tests have shown that lime addition followed by centrifugation can remove up to 21% of the effluent loading, and the use of lime and alum can reduce the loading by as much as 43% (Kimball 1981). In another study, lime and polyelectrolyte floc used with DAF reduced the loading by 10 to 20% (Ratcliff 1974). Lime addition (4.4 lb/10,000 gal of water) to clarifiers after aerobic treatment can aid in sludge settlement as well.

Irrigation

In many areas where citrus is grown, irrigation is used extensively in agriculture as a means of watering various crops. Citrus waste streams are being used as a source of irrigation water in many such areas, especially in California. The San Joaquin Valley's alkaline soils are ideally suited for the generally acidic citrus waste streams. Even in the sandy soils of Florida, citrus effluents have been shown to be effective water sources for citrus groves (Wood 1973). The sandy Florida soils act like sieves that filter particulate matter from the water. It has been reported that Florida soils can receive up to 500 lb BOD/acre day (Wheeler 1977).

Basic effluents have a tendency to "tie up" or bind the soil, forming a mat that decreases soil permeability; so caustic cleanup solutions should be neutralized or mixed with other acidic effluents before land deposition. Unaerated ponding or pooling of citrus waste streams causes anaerobic off odors to develop, which may be undesirable in less rural areas. Care should be taken in depositing onto cropland those effluents from special processing that contain high sodium levels from alkaline solutions. Additional sodium in the soil can leech into the water table and can increase the salt level in water extracted from wells. If necessary, pH control systems can be used to control the acidity of the water used in irrigation.

Generally, little quality control is needed for such water, which is an advantage of the method although local regulatory agencies may require monitoring of sodium, nitrate, and BOD loading of water deposited on soils. Another advantage is that the water is being used for something useful rather than just going down the drain. A common practice is to mix citrus effluents with well water in order to dilute acidity and other contaminants before deposition on the soil. This is usually done by the user of the irrigation water rather than the processing plant.

Municipal Disposal

Processing plants that exist in less-rural areas often use municipal sewage treatment plants as a means of waste water disposal. This usually requires some quality control monitoring, as disposal fees usually are based on the loading of the effluent and the volume. Outside laboratory analyses often are used in order to ensure the impartiality of the analysis results. In larger communities, citrus effluents make minor contributions to overall waste streams; but in smaller communities, citrus effluents run the risk of becoming political footballs, whether or not that is warranted. Careful records should be kept regardless of actual requirements.

Sample collection for such monitoring can be done in several ways. The easiest and least accurate method is to take periodic grab samples. Waste water composition varies dramatically throughout processing, and such grab samples are not likely to be representative of the total stream. Periodic autosamplers are costly and, again, may not accurately represent the waste streams. The most accurate sampling system is a bleed line attached to the main waste stream, which empties into a composite tank. This continuous sampling gives the closest representation of the total stream. When sampling is to be done, the tank contents are thoroughly mixed, a sample is taken, and the tank is drained so that it can collect another composition of the waste stream.

Daily sampling usually is sufficient. Daily samples can be blended into weekly composites, which in turn can be blended into monthly composites according to the volumes of the waste water per time period. For example, if the daily volumes were 50,142 gallons, 64,321 gallons, 36,478 gallons, 84,168 gallons, and 47,171 gallons, you could make a weekly composite by adding 50.1 ml of sample one, 64.3 ml of sample two, 36.5 ml of sample three, 84.2 ml of sample four, and 47.2 ml of sample five, to get 282.3 ml of total weekly composite. Samples should be well mixed before blending in order to contribute a representative amount of settleable solids to the composite. Also, all waste water samples should be refrigerated to avoid changes in composition before analysis. The common tests performed on such samples can be found in the following sections.

Aerobic Treatment

There are many types of microorganisms, each growing under a certain range of conditions. In waste water treatment, aerobic microbes—those that require oxygen to grow—are known to consume most organic wastes without the production of off odors. Those organic wastes that can be thus consumed are called biodegradable. Anaerobic microbes—those that do not require oxygen to grow or require that oxygen be absent—produce waste products associated with rotten odors, which are environmentally undesirable. For this reason, *aerated lagoons* are used to induce aerobic growth in treating most organic wastes.

Microbial growth is very sensitive to certain conditions, one of them being the acidity or pH. Aerobes grow best at neutral pH levels, but citrus effluents generally have pH values of around 3 to 4, which means that *pH control systems* may be required prior to aerobic treatment. Most municipal effluents are basic and are partially neutralized by citrus waste streams. Shock loadings of low pH or high pH values from caustic cleanups can seriously upset aerobic activity. Also, waste streams that are left standing without aeration for long periods of time, especially during warm weather, may undergo anaerobic growth that can significantly reduce the pH. After the pH has been adjusted, the waste stream should be treated promptly; otherwise, the acid neutralization can be totally reversed in a matter of hours.

Another important parameter to monitor during aerobic treatment is the *dissolved oxygen* (DO). In order for aerobes to function, a DO level of 1 to 2 ppm must exist in the solution. The DO is usually measured by using an oxygen meter equipped with an oxygen electrode. Oxygen meters can be calibrated in ambient air according to the air pressure or altitude and the maximum solubility in moisture-saturated air. First, the electrode is wrapped in a damp cloth or inserted into a BOD bottle with a few drops of water. Care should be taken not to let anything touch the membrane of the electrode, and about 10 minutes should be allowed for temperature equilibration. The reading to which the meter should be adjusted can be calculated by using either the atmospheric pressure or the altitude, in the following equations:

$$O_2(ppm) = (0.001316)P/(0.0666 + 0.00216T) \qquad (22\text{-}1)$$

or:

$$O_2(ppm) = (0.9916 - 3.049 \times 10^{-5}A)/(0.066 + 0.00216T) \qquad (22\text{-}2)$$

where T is the temperature in degrees centigrade, A is the altitude in feet, and P is the atmospheric pressure in mm Hg. These equations are sufficiently accurate for most needs. For a little more accuracy, the following more complex equations can be used:

$$O_2(ppm) = (0.001316)P/(6.02 \times 10^{-6}(T + 156.8) - 0.07955) \qquad (22\text{-}3)$$

or:

$$O_2(ppm) = \frac{1.136 \times 10^{14}e^{-(A + 1764110)^2/9.616 \times 10^{10}}}{6.025 \times 10^{-6}(T + 156.8)^2 - 0.07955} \qquad (22\text{-}4)$$

These equations are good from around 0°C to 45°C, 502 mm Hg to 775 mm Hg, and from −540 feet to 11,273 feet above sea level. Once the oxygen meter is calibrated, it can be used for several tests before recalibration is needed, depending on the care of the electrode. The manufacturer's instructions should be followed in the operation and care of the oxygen meter and the electrode. When the DO levels in the aeration lagoon are too low, the aeration equipment should be adjusted accordingly. The higher the loading, the higher the rate of oxygen consumption will be. If insufficient aeration is performed, the water will generate anaerobic growth, off odors, and low pH levels, and aerobic activity will cease.

It also is important to monitor the *d-limonene* level in aerobic treatment. This substance, obtained primarily from oil processing wastes, is toxic to most microbes and may affect aerobic activity (Murdock and Allen 1960; McNary, Wolford, and Patton 1951). Water entering the aeration lagoon should have an oil level of 0.005% v/v or less as measured by the Scott method (see Chapter 6). Oil strippers can be used on high-oil streams, or these oil-laden streams can be isolated and treated separately. If the water content of high-oil streams is not too high, they can be added to the press liquor in the feedmill and processed into molasses, and then added back to the dried peel.

Determination of the *biological oxygen demand* (BOD) is the most common way to find the loading or the necessary aerobic treatment for waste streams. This is done by growing certain microbes in waste water samples and measuring the rate of decrease of the DO, which usually takes 5 days. The seed bacteria used in the test can be obtained from the aeration lagoon itself. The seed water should be stored at 20°C for 24 to 36 hours, and 3.0 ml of the clear undisturbed top layer should be used to seed the samples. The water used in the dilution of the sample should be distilled from alkaline permanganate solution in order to eliminate all organic residues. The use of water deionized by ion exchange is not recommended because interfering organic residues may be contained therein. The dilution water needs to be saturated with oxygen. This can be done by filling one-gallon jugs three-quarters full with the distilled water and incubating them at 20°C for at least 24 hours while sealed. The air in the head space above the water will saturate the water. BOD nutrient buffer pillows can be purchased, which can be added to a certain volume of dilution water to ensure proper pH values and nutrients for microbial growth. The following procedure can be used.

Five-Day BOD

Equipment and Supplies

- Calibrated oxygen meter and electrode.
- Five 300 ml BOD bottles with screw caps and room to insert the oxygen electrode. (The bottles should be made of glass.)
- 1, 2, 3, 4, and 5 ml pipettes with pipette bulb or the equivalent.
- Incubator (20 ± 1°C).
- 15 ml seed water from aeration lagoon.
- Dilution water free from extraneous organics, buffered and supplied with nutrients using commercial BOD nutrient buffer pillows, and saturated with oxygen (see text above).
- Stoppers for each BOD bottle.

Procedure

1. Using the O_2 meter, measure the DO of the sample, and record it.
2. Using the pipette bulb, pipette 1, 2, 3, 4, and 5 ml of stirred sample each into corresponding 300 ml BOD bottles.

3. Add 3 ml of seed water to each bottle, and fill the bottles to just below the lip with the dilution water. When pouring water, decant it along the sides of the bottle to avoid the formation of air bubbles. Stopper each bottle, being careful not to entrap air bubbles, and invert the bottles several times to mix the contents.

4. Remove the stoppers, and fill the bottles to the top so that no head space will remain when they are capped. Incubate the samples ($20 \pm 1°C$) for 5 days in the dark.

5. After 5 days, remove the cap of each bottle, and insert the oxygen electrode and measure the DO of each. The oxygen meter should have been previously calibrated, as described above in the text. When DO readings are made, some stirring may be required, but excess stirring will cause atmospheric oxygen to dissolve in the solution and distort the results.

6. Plot the DO readings versus milliliters of sample taken. At least three points should be on a straight line. Discard any extraneous points. Determine the DO consumed per milliliter of sample (the slope of the plot) and the dissolved oxygen level at "0" sample, or where the line crosses the DO axis. This should be between 8 and 9 ppm DO, depending on the elevation of the laboratory. The BOD can be calculated from:

$$BOD = \left(\frac{DO_{larger} - DO_{smaller}}{V_{smaller} - V_{larger}} \right) 300 - DO_i + DO_s \quad (22\text{-}5)$$

where DO_i is the DO where the line intersects the DO axis, and DO_s is the DO measured in the pure sample. The 300 refers to the BOD bottle volume. The oxygen consumed is found by selecting two points used in the plot in Fig. 22-1 and subtracting the DO of the larger DO (DO_{larger}) from the DO of the smaller DO ($DO_{smaller}$) and dividing by the volume of the smaller DO ($V_{smaller}$) minus the volume of the larger DO (V_{larger}). For example, if we obtained DO readings as shown in Fig. 22-1 and measured a DO value of 1 ppm in the original sample, we would obtain the following by using Equation 22-5:

$$BOD = \left(\frac{7.5 \text{ ppm} - 2.3 \text{ ppm}}{4 \text{ ml} - 1 \text{ ml}} \right) 300 - 9.2 \text{ ppm} + 1 \text{ ppm} = 512 \text{ ppm}$$

Some problems in BOD analysis can result from insufficient seed bacteria and/or excessive loading. The volumes of the samples used can be increased for lighter waste loads and decreased for heavier loading. Also, the original sample can be diluted and a factor applied to the final result. The pounds BOD can be calculated by using:

$$P_{BOD} = (BOD_{ppm}) \left(\frac{453.6 \text{ g}}{lb} \right) \left(\frac{8.322 \text{ lb}}{\text{gal water}} \right) \left(\frac{1 \text{ lb}}{453.6 \times 10^6 \text{ }\mu g} \right) (\text{gal})$$

$$(22\text{-}6)$$

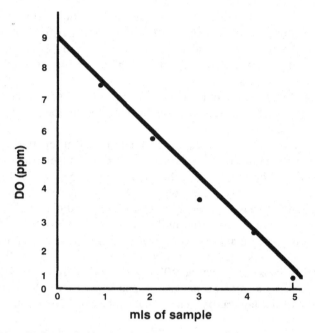

Fig. 22-1. Plot of DO (ppm) of incubated BOD wastewater samples versus milliliters of sample used in the sample BOD problem.

or:

$$P_{BOD} = (BOD_{ppm})(8.322 \times 10^{-6})(gal) \qquad (22\text{-}7)$$

For example:

$$(512_{ppm\,BOD})(8.322 \times 10^{-6})(50,000\ gal) = 213\ lb\ BOD$$

Even though BOD values give the amount of oxygen needed to aerobically treat waste water, the analysis is time-consuming, requiring 5 days for completion. A more rapid method that can be used to estimate the BOD is the *chemical oxygen demand* (COD). The COD can be measured in about 3 hours and provides for a good estimate of the BOD. Generally, BOD values range from 60 to 65% of the COD values. In the COD determination, the organic material is oxidized by the acid-catalyzed oxidizing agent potassium dichromate. After incubation, the unreacted dichromate is titrated with a solution of ferrous ammonium sulfate, which produces a greenish color. The endpoint is determined by using a ferroin indicator, which reacts with excess ferrous ion to form a red complex. The appearance of this red complex marks the end of the titration.

The chemical reactions involved in the titration are as follows:

$$Cr_2O_7^{2-} + 14H^+ + 6Fe^{2+} \rightarrow 6Fe^{3+} + 2Cr^{3+} + 7H_2O \quad (22\text{-}8)$$

dichromate   sulfuric   ferrous   chromic
(orange)   acid   ammonium   ion
                    sulfate   (green)

$$Fe^{2+} + (Ph)_3Fe^{3+} \rightarrow (Ph)_3Fe^{2+} + Fe^{3+} \quad (22\text{-}9)$$

ferrous   ferroin   (red)
ammonium   (colorless)
sulfate

COD kits are commercially available and are recommended. These kits involve less chemical handling and are faster and safer to use than the method below. In the absence of such kits, the following procedure can be used.

COD Test

Equipment and Supplies

- Three 100 ml volumetric flasks.
- 5 ml pipette.
- 10 ml pipette.
- 20 ml pipette.
- Three 200 ml+ boiling flasks with water-cooled condensers attached for refluxing each flask.
- Heat source to boil samples.
- Distilled water.
- 90 ml of 97.2% sulfuric acid.
- 50 ml buret and magnetic stirrer.
- $K_2Cr_2O_7$ solution. (Heat crystalline $K_2Cr_2O_7$ at 108°C for 2 hours to remove hydrated water. Cool it in a desiccator, and dissolve 12.259 g in distilled water in a 1-liter volumetric flask filled to the mark with distilled water.)
- Ferrous ammonium sulfate solution (Dissolve about 39 g of $Fe(NH_4)_2SO_4 \cdot 6H_2O$ in about 100 ml of distilled water. Add 20 ml of the concentrated sulfuric acid and cool. Dilute the solution to 1 liter. The solution is standardized in the procedure.)
- Ferroin indicator. (Dissolve 1.591 g of 1,10-phenanthroline monohydrate together with 0.695 g of $FeSO_4 \cdot 7H_2O$ in distilled water, and dilute the solution to 100 ml.)
- One flask for the first blank.
- 50 ml graduated cylinder.

Procedure

1. Pipette 5 ml of the stirred waste water sample into a 100 ml volumetric flask, and fill it to the mark with distilled water. More than one sample can be run at one time by using the same two blanks (see below) in the procedure.
2. Pipette 5 ml from each volumetric flask containing a diluted sample into a boiling flask fitted with a condenser for refluxing. Add about 15 ml of distilled water to each boiling flask. Make a blank solution by adding 20 ml of distilled water to a separate boiling flask to constitute the first blank.
3. Pipette 10 ml of the dichromate solution into each flask as well as into an empty container to serve as a second blank. Fill the second blank up to about 100 ml. This blank will be used to standardize the FAS solution.
4. Add about 30 ml of the sulfuric acid to each flask, including the blanks, using a graduated cylinder. Be careful not to get *any* of the acid on your skin or clothing. If you do, wash the area with water for 15 minutes.
5. Heat the flasks to boiling, and reflux them for 2 hours.
6. Cool the flasks to room temperature, and titrate each solution with the ferrous ammonium sulfate solution. The solution will begin to turn green with the production of the chromic ion, and then it will turn blue. At this point 3 drops of the ferroin indicator should be added. Just before the endpoint the solution will turn gray, and then the final development of a red color will indicate the endpoint.
7. The normality of the ferrous ammonium sulfate solution is determined by titrating the second blank, and is calculated as follows:

$$N_{FAS} = (10 \text{ ml dichrm})(0.25N \text{ dichrm})/(\text{ml FAS titrd}) \quad (22\text{-}10)$$

8. The COD is calculated as follows:

$$\text{COD} = \frac{(N_{FAS})(8000)(\text{ml FAS}_{1\text{st blank}} - \text{ml FAS}_{sample})}{(0.5 \text{ ml of sample})} \quad (22\text{-}11)$$

For example:

$$N_{FAS} = (10 \text{ ml})(0.25N)/(25.00 \text{ ml FAS}_{2\text{nd blank}}) = 0.100N$$

$$\text{COD} = \frac{(0.100N \text{ FAS})(8000)(28.30 \text{ ml}_{1\text{st blank}} - 24.32 \text{ ml}_{sample})}{(0.5 \text{ ml of sample})}$$

$$= 6368 \text{ ppm}$$

9. Using the COD to estimate the BOD, you can multiply by 65% or 0.65 to get 4139 ppm BOD in the example.

The amount of *settleable solids* is another important parameter in aerobic treatment. Once the aerobic microbes have incorporated soluble material into their bodies, they become more dense than the solution and can be settled out and be removed from it. The level of settleable solids in the water stream prior to aerobic treatment can affect the efficiency of the microbial digestion, as well as the settleable solids loading in the clarifying ponds or tanks. Clarifiers should be designed for about 2 hours of retention time with 5 to 10% settleable solids in the sludge slurry being removed, depending on the type of pump used. This sludge can be dewatered by using centrifuges or filters, and added to the feed-mill; it is higher in nitrogen and vitamin B than citrus pulp and thus adds to the nutritional value of the feed. DAF and/or lime addition sometimes is used to aid in the separation of the sludge in the clarifier. Some of the sludge may be recirculated back into the aeration lagoon or chamber to stabilize the microbial activity.

Aerobic flora and fauna have the ability to adjust to a wide variety of conditions as long as changes are slow enough to permit adaptation. If citrus effluents are the major wastes being treated, it may be possible to generate microbes that are pH- and oil-tolerant, which could reduce the need for some of the pretreatment. Gradually reducing pretreatment operations may make it possible eventually to eliminate them all together.

Anaerobic Treatment

Primary anaerobic treatment of citrus processing effluents is not a common practice, one reason for this being the toxic effect of the peel oils on anaerobic bacteria. The big advantage of anaerobic primary treatment is the recovery of methane gas, a microbial by-product, which can be used as a fuel for boilers and peel dryers. Assuming that the waste streams can be sufficiently stripped of *d*-limonene, an anaerobic digester can reduce the BOD by 90%, producing gas that is 60 to 70% methane, and can generate about 6000 Btu's per pound of COD removed. Less sludge is produced in anaerobic treatment compared to aerobic treatment, with about 0.02 to 0.10 pounds sludge produced per pound of BOD removed.

Activated charcoal or sand often is used in the reactor as a site for bacterial growth in fluidized-bed sealed reactors. As in aerobic treatment, some of the sludge can be recycled back into the reaction chamber in order to maintain reactivity. Precautions must be taken in anaerobic primary treatment to keep the gases and odors that are generated from escaping into the atmosphere, as they can cause environmental problems. Also, some of the anaerobes produce volatile acids that can lower the pH and thus inhibit anaerobic activity. Buffering agents can be used to modify the pH (such as bicarbonate). As with aerobic methods, gradual changes in effluents can allow anaerobic flora and fauna to adapt to low-pH and high-peel-oil conditions that can reduce the need for pretreatment.

Anaerobic treatment also is used as a secondary treatment in aerobic methods to remove those contaminates that aerobic treatment could not remove, especially volatile organics. Carbon granules sometimes are added to help absorb volatile organics as well as to provide a site for anaerobic growth. Sludge produced in this secondary treatment generally needs to be disposed of in a remote site in order to prevent off odors from lowering the quality of feed coming from the feedmill. However, as mentioned above, less sludge is produced with anaerobic treatment than with aerobic treatment. After secondary anaerobic treatment, a tertiary chemical oxidation may be employed, using chlorine gas to remove remaining organics and nitrogenous compounds. A final carbon filter, used to remove residual chlorine gas and volatile organics, generally will produce potable water that usually is of a higher quality than that used by the plant from other sources.

Reverse Osmosis

The process of high pressure filtration (about 800 psi) through semipermeable membranes has been investigated in the treatment of citrus effluents (Kimball 1982). This method, using both cellulose acetate membranes and non–cellulose acetate membranes, can reduce COD levels by about 75%. The filtered permeate can undergo further processing or may be recycled as belt sprays or in other suitable applications. About 10% of the original volume of the untreated waste water will emerge as a concentrate with the Brix increased about fourfold (~0.5°Brix untreated to ~2.0°Brix treated), which can be added to press liquor and can be concentrated to molasses and added back to the feed in the feedmill. About half of the peel oil can be removed from the waste stream as well, by using reverse osmosis. This method requires frequent cleaning and durable membranes.

HAZARDOUS WASTES

Hazardous wastes may be defined as substances that can ignite, corrode, react in a dangerous manner, or be toxic. Citrus juice plants utilize materials such as cleaning agents, lubricants, fuels, refrigerants, boiler treatment chemicals, processing chemicals, compressed gases, and laboratory chemicals that fall into this category. Whenever these chemicals are spilled, contaminated, or decompose, they become waste material that must be properly disposed of by using licensed disposal services. Empty containers that contained hazardous material also must be disposed of in proper fashion, according to local laws. Empty caustic containers can be triple-rinsed to render them safe in most areas. Empty closed-head oil or solvent drums that contained flammable material should never be opened with a torch. Most chemical suppliers will pick up used drums or partially used drums that contained hazardous materials for recycle. Empty containers (generally over 5 gallons) should never by simply discarded, with no regard for proper and legal disposal requirements. Care should be taken in the disposal of hazardous chemicals from laboratories. Direct disposal of such chemicals down the drain, so that they end up in municipal treatment plants, may cause a hazardous condition. Used compressed gas cylinders should be promptly returned to the vendor and not allowed to accumulate. When food containers are used for nonfood purposes, they should be labeled as such and not used to contain food material again.

QUESTIONS

1. What are some of the sources of air pollution in citrus processing plants?
2. What is the "total sulfur oxides" maximum level permitted by EPA in ambient air during a 24-hour period?
3. Why should inert solid wastes be removed from plant sites?
4. How can DAF aid in the treatment of aqueous citrus plant effluents?
5. How many pounds of BOD/acre day can Florida's sandy soils receive, according to text?
6. What is the advantage of aerobic treatment?
7. What are the quality control parameters of aerobic treatment?
8. What DO levels must be maintained in aerobic treatment?
9. What maximum peel oil content is permissible for successful aerobic treatment?
10. What is the advantage of COD measurements over BOD measurements?

PROBLEMS

1. Suppose that daily samples of waste water were collected for 5 days, as listed below, corresponding to the given daily meter readings for gallons of effluent. How would you make 200 ml of a weekly composite that would represent the week's effluent?
   Monday: 61,452 gallons
   Tuesday: 85,967 gallons
   Wednesday: 31,002 gallons
   Thursday: 46,972 gallons
   Friday: 55,012 gallons
2. In air calibration of an oxygen meter, what would you set the meter to if you were at an altitude of about 1600 feet at a temperature of 28°C?

3. Suppose you obtained the following data in a BOD analysis. What would be the BOD (ppm)?

	DO
Initial sample	1.6
After incubation using:	
1 ml of sample	6.5
2 ml of sample	4.0
3 ml of sample	1.6
4 ml of sample	0.0
5 ml of sample	0.0

4. Suppose we obtained the following data in COD a analysis. What would be the COD (ppm) and the estimated BOD (ppm)?
   29.27 ml FAS titrated in 2nd blank
   25.60 ml FAS titrated in 1st blank
   24.78 ml FAS titrated in sample
5. How many estimated pounds of BOD are there in 156,472 gallons of waste water with a COD value of 4367 ppm?

REFERENCES

Kimball, D. A. 1981. Unpublished data. California Citrus Producers, Inc., Lindsay, Calif.

Kimball, D. A. 1982. Unpublished data. California Citrus Producers, Inc., Lindsay, Calif.

McNary, R. R., Wolford, R. W., and Patton, V. D. 1951. Experimental treatment of citrus waste water, *Food Tech.*, *8*, 319–323.

Murdock, D. I. and Allen, W. E. 1960. Germicidal effect of orange peel oil and *d*-limonene in water and orange juice, *Food Tech.*, *14*, 441–445.

Perry, R. H. and Green, D. 1984. *Perry's Chemical Engineer's Handbook*, 6th edition. McGraw-Hill Book Company, New York, 26.11.

Ratcliff, M. W. 1974. Unpublished research. Citrus World, Lake Wales, Fla.

Wheeler, R. W. 1977. Unpublished communication. FMC, Riverside, Calif.

Wood, C. 1973. Recycling citrus waste water, *Sunshine State Agr.*, September–October, 6–7.

UNIT FOUR

CITRUS JUICE MANAGEMENT

Chapter 23

Quality Control Statistics

Quality control of any product inherently involves the use of statistical methods. Statistical mathematics is used to evaluate parameters that cannot be determined absolutely. For example, it would be impossible to determine the precise diacetyl level in chilled juice being processed and packaged continuously from fresh extracted fruit, as the juice composition is constantly undergoing change. However, statistical or spot checking could give a good indication of what the quality would be. Although the field of statistics is broad, only a few techniques that may be of use to quality control personnel in citrus processing plants are discussed here.

SIGNIFICANT FIGURES

Errors in regard to significant figures often occur in the citrus industry. Often labels are encountered with Brix readings to two decimal places, or acid levels are reported to one decimal place. The subject of significant figures is one of the first studied in science classes, as the number of significant figures represents a way to show the accuracy of a measurement. For example, if we report the milliliters titrated as 23, do we mean exactly 23 ml or 23 ± 0.5 ml? According to the rules of significant figures, the latter is true. If exactly 23 ml is intended, we should write 23.0 or 23.00, depending on the accuracy of the measurement. The reading of a standard buret generally is ±0.02 ml, which means that all buret readings should be expressed to the hundredths decimal place. Using the proper number of significant figures should not replace the need to express error ranges. If we were to report the milliliters titrated as 23.234, we would be implying that we could make the measurement accurate to the thousandths decimal place, which would be false. On the other hand, if we were to report the milliliters titrated as 23.4, we would not be reporting the accuracy of the measurement as close by as we could. A 23.4 ml reading really means 23.4 ± 0.05 ml, whereas 23.40 means exactly 23.40 ml or 23.40 ± 0.005.

The number of significant figures in a measurement or in raw data is determined by the assigned accuracy of the measurement. When the data are processed, however, there are nine rules that can be used to determine the number of significant figures in an answer after calculations have been made. Significant figures are basically those figures that add accuracy or quantify a number; figures that are not significant are generally place holders and do not define the quantity or accuracy of the number directly. The nine rules of significant figures are as follows:

1. All nonzero figures are significant. For example, 123.45 = five significant figures.
2. All zeros between nonzero figures are significant. For example, 3,076,002 = eight significant figures.
3. All zeros to the left of the first nonzero figure are *not* significant. For example, 0.013 = two significant figures.
4. All zeros to the right of the last nonzero figure and to the right of the decimal point *are* significant. For example, 3.2300 = five significant figures.
5. All zeros to the right of the last nonzero figure and to the left of the decimal point *may* or *may not* be significant. If the decimal point appears (expressed decimal), then the zeros are significant. For example, 3700 = two, three, or four significant figures, but 3700. = four significant figures.
6. Use scientific notation, when in doubt, to clarify the proper number of significant figures in rule 5. For example, 3700 can be expressed as 3.700×10^3, 3.70×10^3, or 3.7×10^3 to indicate the exact number of intended significant figures.
7. Some numbers have unlimited significant figures. For example, 5 people means exactly 5 people because it generally is impossible to have a partial person. In other words 5 people is the same as 5.0000000000000000000 people. Also, 60 seconds in one hour means the same as 60.00000000 seconds in one hour because exactly 60 seconds are in one hour. Absolute or exact numbers have an unlimited number of significant figures.
8. When one is adding or subtracting numbers, the answer can be expressed only to the number of significant figures that are completely known. To begin with, one can insert any necessary zeros needed in the calculation and then must round off to the nearest fully known figure. For example, if we add or subtract the following:

$$
\begin{array}{r}
25.3? \\
+3.45 \\
\hline
28.75
\end{array}
\qquad
\begin{array}{r}
3455.346 \\
-321.1?? \\
\hline
3134.246
\end{array}
$$

$$
\textit{or} \quad 28.8? \qquad\qquad 3134.2??
$$

QUALITY CONTROL STATISTICS 373

The last figure(s) in the answer are not known for certain. Rounding off to the nearest known figures would give 28.8 and 3134.2 for the answers expressed to the correct number of significant figures. The least accurate data determine the number of significant figures in the answer.

9. When one is multiplying or dividing, the total number of significant figures in the answer can be no more than the least number of significant figures in the numbers multiplied or divided. For example, if we multiplied 34.56789 by 0.023, we would get an answer of 0.80. This is so because even though the first number has seven significant figures, the number with the least significant figures has only two. Thus, the answer can have only two significant figures. In dividing 6.234 by 3.45633567, we would get 1.804 because the least number of significant figures in the data is four.

Most data reported in routine quality control measurements should be expressed to the proper number of significant figures. The most common parameter in citrus processing is the Brix measurement, whose standard error range is $\pm 0.1\,°$Brix. Thus Brix readings should be reported rounded off to the nearest tenth of a °Brix; but because acid and temperature corrections often are determined to the hundredths decimal place, many processors report their Brix values to the hundredths decimal place as well. This is incorrect according to the rules of significant figures; such reporting implies deceptive accuracy in measurement. Generally reports of oil levels give significant figures to the thousandths of a percent, which is generally correct because all the data used have at least two significant figures, and oil levels usually are expressed in two significant figures (0.011%, 0.025%, etc.). However, oil levels below 0.010% usually are expressed as one significant figure (0.009%, 0.003%, etc.), a practice that technically violates the rules of significant figures. However, it is convenient in the industry to use a standard decimal place and this practice has overridden the rules of significant figures. The same condition applies to pulp levels. The rules of significant figures dictate that a centrifuge tube can be read to ± 0.2 ml in measurements ranging from 0.2 ml to 9.0 ml; so the number of significant figures can be one or two. The number of milliliters of pulp in the centrifuge tube determines the number of significant figures in the final pulp level. Most pulp levels are expressed as two significant figures (9.4%, 8.1%, etc.), but pulp levels of 10% or more generally are expressed with three significant figures (10.4%, 11.6%, etc.), a convention that is technically incorrect. Again, the need of the industry to have a consistent number of decimal places has overridden the rules of significant figures. Computers and calculators always disregard rules of significant figures, and their use in the industry has led to the need to use standard decimal places in reporting data.

ERROR ANALYSIS

The expression of error is an inherent part of any precise science. All measurements contain error, regardless of the effort or care exercised to avoid or minimize such. Error can be reported in many ways, including the following four ways in which the error range can be expressed.

Actual Range

The simplest expression is the actual error range (AR). For example, if we have three acid determinations of 3.41%, 3.48%, and 3.49%, we can say the range is between 3.41 and 3.49. This type of error analysis does not take into consideration the fact that two of the three data points are close together with the third somewhat distant from the others. It is useful for bracketing data within groups of absolute occurrence, however. Mathematically it can be expressed as:

$$AR = X_{max} - X_{min}$$

Average Deviation

The average deviation, (AD) is calculated by averaging the deviation from either the average value or the median value. (The median value is the value exactly midway between the two extremes.) In the acid example above, the median value would be 3.45% and the average 3.46%. In calculating the average deviation, the average difference from the median is the preferred method, rather than the difference from the average value. In our example this would be:

$$(|3.45 - 3.41| + |3.45 - 3.48| + |3.45 - 3.49|)/3 = 0.04\%$$

Mathematically this could be expressed as follows:

$$AD = \left(\sum_i |(X_i - X_m)|\right)/n \tag{23-1}$$

with X_m representing the median and X_i the ith component of n components. The average deviation is easily used for small samples. However, the absolute values are hard to use in mathematical analysis and can be seriously affected by extreme variations.

Root Mean Square Deviation

The square root of the average of the squared deviations from the mean is called the root mean square deviation $(RMSD)$. It can be calculated by using the following expression:

$$RMSD = \sqrt{(\Sigma_i \, (X_i - X_m)^2 / n}$$ (23-2)

This is similar to the average deviation except that one does not need to deal with absolute values. However, the RMSD has an inherent bias and so is not commonly used.

Standard Deviation

The standard deviation (SD) is the most commonly used means of determining the error range of data. It is similar to the RMSD except that it eliminates the inherent bias by multiplying the RMSD by the factor $\sqrt{n/(n-1)}$. The SD can be expressed as:

$$SD = \sqrt{(\Sigma_i \, (X_i - X_m)^2)/(n-1)}$$ (23-3)

For example, in our acid example, the SD would be:

$$\sqrt{((3.41 - 3.46)^2 + (3.48 - 3.46)^2 + (3.49 - 3.46)^2)/(3-1)} = 0.04$$

The above equation requires calculation of the data mean (X_m). A more convenient form of the equation, which does not require the calculation of the mean, is:

$$SD = \sqrt{[n(\Sigma \, (X^2) - (\Sigma \, X)^2]/n(n-1)}$$

In the % acid example, this would become:

$$\sqrt{[3(35.92) - (10.38)^2]/3(2)} = 0.05$$

You will notice that the above answer is given to only one significant figure. This is done because the % acid data is reported to the hundredths place; the error range is reported likewise. Otherwise, if the error range were reported to the proper number of significant figures (four here), it would appear to be more accurate than the original data, which is not the case. Also, more accurate results are obtained if one does not round off the summations or any intermediate answer until the final error range. If this practice had been followed, we would have obtained a standard deviation of 0.04 instead of 0.05, using 35.9186 for the first summation instead of 35.92. The former (0.04) is the more accurate standard deviation, as seen in the use of Equation 23-3 with the sample data. The calculation of standard deviations is a built-in feature of many calculators and computers.

CONTROL CHARTS

Control charts have been used to monitor on-line quality throughout the food industry. A control chart consists of upper and lower allowable limits, depicted on graph paper as shown in Fig. 23-1. The actual levels of the target parameter are plotted in relation to these limits to determine not only if the product is within specifications, but if undesirable trends are developing.

The median value and limits of control charts are determined mathematically by statisticians. Often, however, in the food industry, the upper and/or lower limits are determined by specification or quality thresholds and limits. For example, the maximum level of diacetyl acceptable in citrus juices is determined by the taste threshold and not by the probability of finding a diacetyl level more than 95% away from the mean value. This absolute nature of quality in food processing requires changes in the interpretation of quality control charts.

Runs

A run on a control chart is a consecutive series of plots that are on the same side of the median value. Runs found above the median value are called "runs up" and those below the median value "runs down." Long runs that consist of many consecutive points are indicative of a trend, and such trends may result in the product drifting out of specifications if not corrected. The use of control charts enables quality control personnel to detect such trends early enough to

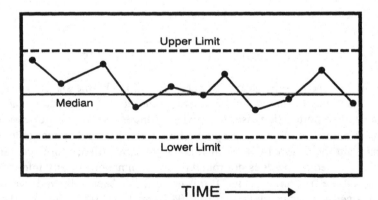

Fig. 23-1. A typical control chart, including the median and upper and lower control limits.

correct them before the product is seriously damaged or contaminated. The statistical determination of excessive runs or runs that cannot be described to random occurrence is very simple. The following table shows the total number of data points and the corresponding maximum number of points in a consecutive run (up or down) that can be considered random or acceptable statistically at the 0.05 probability level:

Total data points	Maximum points in a run
10	5
20	7
30	8
40	9
50	10

For example, if we made 30 measurements of % pulp from line or freshly extracted juice, we could have up to 8 pulp levels above (or below) the generally accepted median of 10.0% pulp consecutively before we could attribute the run to a nonrandom effect in processing. In this case, the main adjustable processing parameter in pulp levels is the finisher pressure. If more than 8 pulp levels are above 10% pulp, then the finisher pressure is too high and should be decreased to avoid a drift of pulp levels perhaps out of specification. If more than 8 pulp levels are below the 10% median, then the finisher pressure should be increased to avoid loss of juice yield with expelled unwashed pulp. Some quality control engineers use 7 as a standard maximum length of runs before corrective action is taken, regardless of the number of total data points.

Applications

Quality control charts have been used in the monitoring of the diacetyl levels in single-strength juice products and pulp levels during juice extraction, and some attempt has been made to apply the use of control charts to fruit grading. In the production of single-strength juices, if the diacetyl values constitute a run up, spoilage is setting in, and the temperatures and/or retention times in the pasteurizers need to be increased. If the diacetyl values form a run down, too much heat may have been applied, and perhaps the pasteurizer temperature and/or retention time should be reduced. The example of % pulp monitoring given above is another illustration of the application of control charts to citrus processing. One can monitor fruit graders by taking counts of broken and rotten fruit leaving the grading table and plotting the results on a control chart. The upper limits of fruit grading have been established by the USDA—10% for broken fruit and 2% for rotten fruit leaving the grading table. The establishment of a median value will permit the detection of runs up, which would require the use of more graders or grading activity, or runs down, which would allow the use of fewer graders or less grading activity. Control charts can be used for packaging equipment, sample monitoring, and many other applications in continuous processing.

ANALYSIS OF VARIANCE

It may be of interest to the quality control manager or technician to know whether or not something is significantly different from something else; and many techniques can be used to determine whether or not a statistically significant change has occured. One method, where a large number of samples (>30) exists, is to convert the data to a standard normal frequency distribution and apply statistical methods to determine whether the data are significantly different from each other or a set parameter. Such large samplings of a single lot are not common in industrial practice, however, and other restrictions on the method generally prohibit use of the technique on a commercial basis. A Student's t-test is used in a similar manner for small samples (<30) and may be of more value than the former method to commercial citrus plants. A chi-square test determines the significant variance in standard deviations, and an F-test determines the significant difference between two variances. However, an analysis of variance (ANOVA) is ideally suited for most industrial quality control determinations of the significance of the difference among data and can be used as a general method.

ANOVA uses an F-distribution where a significant variance is said to occur if the test statistic (F value) falls beyond the 95% critical value. In other words, if the data exist beyond 95% of the frequency distribution, a significant difference can be assumed. If such is the case, it is said that the variance is significant at the 0.05 level. For example, suppose that three technicians are performing Brix measurements, and you notice that there seems to be a difference in their readings but are not sure if there is a difference and why that difference is occurring. In order to find out, you can ask all three technicians to analyze each of 10 samples of orange juice concentrate. The answers that you are given probably will show some differences. Are these differences due to the lab technician or perhaps the samples of concentrate? Are these differences significant? You can find out by using an analysis of variance. Such an analysis tests what is called the null hypothesis, or the hypothesis that no difference exists. The limits of the null hypothesis are expressed in terms of probability limits or levels. If there is a certain probability that the differences can lie outside a certain range, then the difference is statistically significant. As mentioned above, a common probability of 5% or 0.05 is used as the critical value to determine the acceptance or rejection of the null hypothesis.

We can illustrate the use of an analysis of variance through an example.

Suppose that three technicians measure the Brix on three samples of concentrate as shown in the table:

	Tech A	Tech B	Tech C
Sample 1	61.2	61.4	60.8
Sample 2	61.3	61.0	60.9
Sample 3	60.9	61.2	60.9

Using a one-way analysis of variance, we will try to determine if there is a significant difference between the results from each technician, or if one technician significantly varies from another. First, we can simplify our calculations; by subtracting 60.0 from each of the numbers, we get:

	Tech A	Tech B	Tech C
Sample 1	1.2	1.4	0.8
Sample 2	1.3	1.0	0.9
Sample 3	0.9	1.2	0.9

Then the following procedure can be used:

1. Sum each column, square the sum, add the squares, and divide by the number of items in each column. For example:

$$\frac{(1.2 + 1.3 + 0.9)^2 + (1.4 + 1.0 + 1.2)^2 + (0.8 + 0.9 + 0.9)^2}{3}$$

$$= 10.43$$

2. Square each number, and add the results. For example:

$$(1.2)^2 + (1.3)^2 + (0.9)^2 + (1.4)^2 + (1.0)^2 + (1.2)^2$$
$$+ (0.8)^2 + (0.9)^2 + (0.9) = 10.60$$

3. Sum all the items, square the sum, and divide by the total number of items. For example:

$$\frac{(1.2 + 1.3 + 0.9 + 1.4 + 1.0 + 1.2 + 0.8 + 0.9 + 0.9)^2}{9} = 10.24$$

4. Subtract the results in (3) from the results in (1). For example:

$$(1) - (3) = 10.43 - 10.24 = 0.19$$

5. Subract the results in (3) from (2). For example:

$$(2) - (3) = 10.60 - 10.24 = 0.36$$

6. Subtract the results in (4) from those in (5). For example:

$$(5) - (4) = 0.36 - 0.19 = 0.17$$

With this information we can construct an analysis of variance table. The sum of squares for the columns is found from step 4 above. The sum of squares for the columns is found from step 4 above. The sum of squares for the residual error is found from step 6 above. The degrees of freedom (df) for the columns is found by taking the number of columns (c) and subtracting 1, or $c - 1$. The degrees of freedom for the residual error is found from $c(r - 1)$, where r is the number of rows or Brix tests per technician. The mean square is found by dividing the sum of the squares by the degrees of freedom. The variance ratio then can be found by dividing the mean square of the columns by the mean square of the residual error. For example:

Source	Sum of squares	df	Mean square	Variance ratio
columns	0.19	2	0.10	3.33
residual	0.17	6	0.03	

The variance ratio can be compared to Table 23-1, where v_1 equals the degrees of freedom for the columns, and v_2 equals the degrees of freedom of the residual. At the 0.05 probability level, the variance ratio must exceed 5.14 in order for a significant difference to occur between the lab technicians. Because

Table 23-1. F-distribution that can be used in the analysis of variance to test the null hypothesis at the $p = 0.05$ level (Duncan 1974).

v_2	$v_1 = 1$	2	3	4	5	6	7	8	9	10
1	161	200	216	225	230	234	237	239	241	242
2	18.5	19.0	19.2	19.2	19.3	19.3	19.4	19.4	19.4	19.4
3	10.1	9.55	9.28	9.12	9.01	8.94	8.89	8.85	8.81	8.79
4	7.71	6.94	6.59	6.39	6.26	6.16	6.09	6.04	6.00	5.96
5	6.61	5.79	5.41	5.19	5.05	4.95	4.88	4.82	4.77	4.47
6	5.99	5.14	4.76	4.53	4.39	4.28	4.21	4.15	4.10	4.06
7	5.59	4.47	4.35	4.12	3.97	3.87	3.79	3.73	3.68	3.64
8	5.32	4.46	4.07	3.84	3.69	3.58	3.50	3.44	3.39	3.35
9	5.12	4.26	3.86	3.63	3.48	3.37	3.29	3.23	3.18	3.14
10	4.96	4.10	3.71	3.48	3.33	3.22	3.14	3.07	3.02	2.98

3.33 is less than 5.14, we fail to reject the null hypothesis, which means there is no significant difference between the lab technicians' from a statistical point of view. That is not to say that the lab technicians are performing the Brix measurements with sufficient accuracy; it only means that statistically they are all performing equally at the 5% level.

We can expand the analysis of variance to include the determination of whether or not there is a significant difference between the concentrate samples as well as between the lab technicians. This is called a two-way ANOVA. We use a similar example to illustrate how this is done. Suppose that four lab technicians each analyze three samples of orange concentrate for Brix, as shown below:

	Tech A	Tech B	Tech C	Tech D
Sample 1	60.0	60.1	59.9	59.9
Sample 2	60.1	60.1	60.1	60.0
Sample 3	60.1	60.2	60.1	60.0

First, subtract 60.0 from each of the measurements in order to simplify the calculations:

	Tech A	Tech B	Tech C	Tech D
Sample 1	0	0.1	−0.1	−0.1
Sample 2	0.1	0.1	0.1	0
Sample 3	0.1	0.2	0.1	0

The following procedure can be used:

1. Sum each column, square the sums, add the results, and divide by the number of cases in each column. For example:

$$((0.2)^2 + (0.4)^2 + (0.1)^2 + (-0.1)^2)/3 = 0.073$$

2. Sum each row, square the sums, add the results, and divide by the number of cases in each row. For example:

$$((-0.1)^2 + (0.3)^2 + (0.4)^2)/4 = 0.065$$

3. Square each number, and add the squares. For example:

$$0 + (0.1)^2 + (-0.1)^2 + (-0.1)^2 + (0.1)^2 + (0.1)^2 + (0.1)^2 + 0$$
$$+ (0.1)^2 + (0.2)^2 + (0.1)^2 + 0 = 0.120$$

4. Sum all the cases, square the sum, and divide by the number of cases. For example:

$$(0 + 0.1 - 0.1 - 0.1 + 0.1 + 0.1 + 0.1 + 0 + 0.1 + 0.2 + 0.1 + 0)^2/12 = 0.030$$

5. Subtract the results in (4) from those in (1). For example:

$$(1) - (4) = 0.073 - 0.030 = 0.043$$

6. Subtract the results in (4) from those in (2). For example:

$$(2) - (4) = 0.065 - 0.030 = 0.035$$

7. Subtract the results in (4) from those in (3). For example:

$$(3) - (4) = 0.120 - 0.030 = 0.090$$

8. Subtract the results in (5) and (6) from (7). For example:

$$(7) - (5) - (6) = 0.090 - 0.043 - 0.035 = 0.012$$

With this information we can construct an analysis of variance table. The sum of squares for the columns is found from step 5 above. The sum of squares for the rows is found from step 6 above. The sum of squares for the residual error is found from step 8 above. The degrees of freedom for the columns is found from $c - 1$. The degrees of freedom for the rows is found from $r - 1$. The degrees of freedom for the residual error is found from $(r - 1)(c - 1)$. The mean square is found by dividing the sum of the squares by the degrees of freedom. The variance ratio then can be found for both the rows and the columns by dividing the mean square of the columns and rows by the mean square of the residual error. For example:

Source	Sum of squares	df	Mean square	Variance ratio
columns	0.043	3	0.014	7.00
rows	0.035	2	0.018	9.00
residual	0.012	6	0.002	

The variance ratios can be compared to Table 23-1, where v_1 equals the degrees of freedom for the columns or rows, and v_2 equals the degree of freedom of the residual. At the 0.05 probability level the variance ratio must exceed 4.76 for the columns and 5.14 for the rows in order for a significant difference to occur between the lab technicians or the samples of concentrate. As both ratios are more than 4.76 or 5.14, there is a significant difference between the lab technicians and the samples of concentrate from a statistical point of view. As you may see, this difference is not obvious by inspection.

REGRESSION

Regression analysis involves the fitting of data to mathematical relationships, which then can be used for a variety of quality control functions, including the prediction of results. If, for example, a relationship can be found between the pulp levels before and after freshly extracted juice has passed through a seven-effect TASTE evaporator, then this relationship can be used to predict the pulp level change under similar conditions. Another common use of regression analysis is the conversion of data into a form that can easily be placed into a programmable device such as a calculator, computer, bar code system, or automated processing controls. This was done with sucrose Brix/density tables and Brix corrections that have been used extensively in this text.

Regression analysis simply means fitting a line or a curve to data points plotted on a graph. This can be done by simply drawing a line or a curve by eye or by seeking a somewhat simple mathematical relationship that may be made somewhat obvious by the data; however, a statistical procedure exists that enables the calculation of such relationships as well as an objective means of determining the validity of such relationships. This method is called the method of least squares.

Least Squares Analysis

Least squares analysis involves the use of differential calculus in determining the equation that best fits a set of data. However, once the relationships have been established, computers can be easily programmed to handle the bulk of the calculations. The simplest example is determining the best equation for a straight line that fits the data, of $y = a + bx$. Here b is the slope of the line, and a is the y-intercept. In order to find the best values of a and b that will fit the data, we must sum the squares of the differences between the actual y values and the predicted or y_p values:

$$y_p = a + bx \qquad (23\text{-}4)$$

$$\Sigma(y - y_p)^2 = \Sigma(y - a - bx)^2 = D \qquad (23\text{-}5)$$

The best a and b values are those obtained when this sum of the squares is at a minimum. In order to find the minimum point on this equation, we can try to find the point where the slope is zero. This we can do by partially differentiating

Equation 23-5 with respect to each coefficient, and setting the result equal to zero:

$$\partial D / \partial a = \Sigma\, 2(y - a - bx)(-1) = 0 \qquad (23\text{-}6)$$

$$\partial D / \partial b = \Sigma\, 2(y - a - bx)(-x) = 0 \qquad (23\text{-}7)$$

Rearranging the above equations gives:

$$an + b\,\Sigma\, x = \Sigma\, y \qquad (23\text{-}8)$$

$$a\,\Sigma\, x + b\,\Sigma\, x^2 = \Sigma xy \qquad (23\text{-}9)$$

This gives two equations that can be used to find the two unknowns a and b, where n is the number of (x, y) data pairs used in the analysis. Using Equations 23-8 and 23-9 to solve for a and b, we obtain the following:

$$a = \frac{\Sigma\, x^2 \,\Sigma\, y - \Sigma\, x \,\Sigma\, xy}{n\,\Sigma\, x^2 - \left(\Sigma\, x\right)^2} \qquad (23\text{-}10)$$

$$b = \frac{n\,\Sigma\, xy - \left(\Sigma\, x \,\Sigma\, y\right)}{n\,\Sigma\, x^2 - \left(\Sigma\, x\right)^2} \qquad (23\text{-}11)$$

each of which is an independent equation. The equation for a can be simplified by first calculating b and then using the following dependent equation:

$$a = y_{avg} - bx_{avg} \qquad (23\text{-}12)$$

Thus, using these two equations, we can calculate the best values for the constants a and b by performing the above summations, using the sample data points x and y and the number of data pairs n.

For example, suppose we were interested in finding a linear equation that would best describe the change of pulp level in freshly extracted orange juice before and after evaporation to 65°Brix. We first obtain the following data:

Sample	% Pulp before evap. (x)	% Pulp after evap. (y)
1	16.5	10.6
2	15.9	8.8
3	16.9	10.5
4	17.4	11.4
5	18.6	11.7

Using Equations 23-10 and 23-11, we obtain the following parameters and the resulting best a and b values:

$$\Sigma x = 85.3 \qquad a = -6.21$$

$$\Sigma y = 53.0$$

$$\Sigma x^2 = 1459.4 \qquad b = 0.985$$

$$\Sigma xy = 908.3$$

We can make a statistical prediction of pulp levels before and after evaporation using the resulting equation:

$$\% \ \text{pulp}_{\text{after}} = 0.985 \, (\% \ \text{pulp}_{\text{before}}) - 6.21 \qquad (23\text{-}13)$$

Using this equation we can calculate predicted values and compare them to the actual values:

Measured % pulp after evap.	Calculated % pulp after evap.
10.6	10.0
8.8	9.5
10.5	10.4
11.4	10.9
11.7	12.1

This is a simple way to check the accuracy of the linear least squares best fit of the data; however, the "goodness" of fit can be calculated mathematically with more objective results by using the correlation coefficient, R^2. The closer the correlation coefficient is to 1.00, the better the data fit the regression equation. If the R^2 equals exactly 1.00, the curve fits the data exactly, and exact predictions can be made. For a linear equation, the R^2 value can be calculated from:

$$R^2 = \frac{n\left(\Sigma xy\right) - \left(\Sigma x\right)\left(\Sigma y\right)}{\sqrt{n\left(\Sigma x^2\right) - \left(\Sigma x\right)^2} \sqrt{n\left(\Sigma y^2\right) - \left(\Sigma y\right)^2}} \qquad (23\text{-}14)$$

In the above example, the R^2 value becomes 0.892 or 89.2% of the y values (% pulp after evaporation here) that can be accounted for using this least squares equation.

It is important to keep in mind that "noisy" variations will give a lower correlation coefficient than smooth or stretched deviations from the regression line. Actual and calculated values should be checked to verify the correlation. Also, the regression equation usually is valid only within the range of the data.

The linear least squares equation determined above may not fit the data as closely as one would like. Because the above results represent the best linear fit to the data, the only way to obtain a better fit is to use another equation—a nonlinear equation. By using the principles illustrated above and substitution, the method of least squares can be applied to a wide variety of equations. Also, the least squares analysis can be programmed into computers for easy use. The reader is referred to Kolb (1984), who provided not only the equations needed to fit data to up to 19 different equations but also computer programs in a variety of program languages, which can be used to determine regression relationships automatically.

QUESTIONS

1. Give five areas where statistical methods can be applied to citrus quality control in routine processing.
2. What is the purpose of using significant figures?
3. What is the difference between significant figures and the determination of error ranges?
4. How are the control limits (upper and lower) usually determined in the food industry?
5. How many points can constitute a run down in 32 measurements before the process can be considered statistically influenced by some factor?
6. In what situation can an analysis of variance be useful in citrus quality control?
7. Explain the null hypothesis.
8. Explain regression analysis.
9. What is the difference between regression analysis and least squares analysis?
10. How can the use of statistical analyses be greatly simplified in routine use in citrus quality control?

PROBLEMS

1. How many significant figures does each of the following numbers have?
   2,345.34
   3000.1
   2500
   0.00230
   6 dogs

2. Express the answer to the following mathematical problems to the correct number of significant figures:

$$
\begin{array}{llll}
23.460 & 1,234.5 & 0.002340 & 0.004 \div 25,500.0 = \\
+\ 6.0 & -\ 523.000201 & \times\ 66.345 &
\end{array}
$$

3. What is the variance ratio (VR) for a one-way ANOVA analysis between lab technicians using the data below, and does it prove the null hypothesis?

Measurement	Tech A	Tech B	Tech C
1	61.2	61.4	60.5
2	61.3	62.0	60.6
3	60.9	61.4	60.4

4. What is the variance ratio for both rows (refractometers) and columns (lab technicians) for the following data, and is there a significant variance between lab technicians and/or refractometers?

Refractometer	Tech A	Tech B	Tech C	Tech D
1	59.8	60.0	60.1	59.9
2	60.1	60.2	60.2	59.9
3	60.4	60.3	60.3	60.3

5. It may be useful to have an equation that predicts the acid correction to the Brix base on the % acid. Determine the linear least squares constants and correlation coefficient using the following data. Compare the calculated Brix corrections with the actual Brix calculations.

% Acid	Brix correction
0.50	0.10
1.00	0.20
1.50	0.30
2.05	0.40
2.55	0.50

REFERENCES

Duncan, A. J. 1974. *Quality Control and Industrial Statistics, 4th edition*. Richard D. Irwin, Inc., Homewood, Ill., 659–662.

Kolb, W. M. 1984. *Curve Fitting for Programmable Calculators. 3rd edition*, Syntec Inc., Bowie, Md.

Chapter 24

Quality Control Management

Quality control involves more than just laboratories and test tubes; a good quality control department requires administration and management. Managerial policies and procedures vary widely from plant to plant, but the success of any system is dependent on basic principles. Like any other employee, quality control personnel must understand the basic policies and objectives of the company. Also, quality control departments are run by people who report to non–quality control adminstrators.

Like all employees, quality control personnel should be shown courtesy and respect. They should be given concrete goals and responsibilities as well as the resources and latitude needed to achieve those goals. Quality control personnel must be leaders, with the courage to speak up when something is going wrong. They must be competent, and their competency must be recognized. They need sufficient resources to perform their jobs, such as adequate laboratory space and equipment.

The quality control laboratory is a major focus of outside visitors, who often judge a company by the professional appearance of the quality control laboratory. The plant's "IQ" is often evalulated by the laboratory's veneer.

Good employees should have a vision of where they are going and why they are there. Just as they are required to support management in decisions regarding the operation of the plant, management should stand behind quality control personnel in their quest to protect the quality of the company's products. Moreover, quality control procedures should be part of some type of plan that ultimately will benefit the company. Such a plan must be managed; it should be efficient, personally rewarding, and profitable.

OTHER DEPARTMENTS

As complex and intricate as a quality control department may seem, it comprises only a portion of the overall manufacture of citrus products; and books could be written about the other areas as well. The quality control department must interact with all of these other areas and act as a focal point for many of the plant's decisions. It also acts as a helm that steers the course of many plant activities.

One of the first things that should be done in setting up a quality control department is to establish a realistic set of specifications, for both inbound and outbound products. Quality control personnel can and should help to define these specifications. The specifications should at least contain legal requirements as defined by the federal standards of identity, good manufacturing practice, and, if USDA grades are used, the requirements of the desired USDA grade. Quality control then should be given the authority to reject products that do not meet these specifications or sanitary conditions. A specification that is not enforced is a specification that really does not exist.

It is important to keep the quality control department separate from the production department because production often will have a tendency to cut corners with respect to quality in order to get the product made and out on time. Quality control, on the other hand, should be considerate of production's schedules, and should not unduly delay or hamper their activities any longer than is necessary to ensure quality products and sanitation.

Although company organizations may vary, the relationships between their departments may have similarities. A general example of the communication flow between departments is shown in Fig. 24-1. One should remember that communication is a two-way street, and much follow-up communication is not shown in the figure. The figure does show that quality control acts as a traffic light that determines whether a product may proceed, or whether it must stop and be rerouted.

In addition to routing quality control activities, the quality control department may be called upon to become involved in other, non–quality control activities, involving other departments. One such activity is the analysis of fruit samples for maturity and juice yields, as mentioned in Chapter 5. The office generally receives this information so that the owners of the fruit can be credited proportionally. The sales department may ask quality control personnel to assist in preparing and sending product samples to customers or potential customers. Similarly, purchasing may ask quality control to analyze inbound product samples received from suppliers or potential suppliers. Quality control may be called upon to work with research and development regarding new products, specifications, or complex problems. Quality control may and should be asked to assist customers and suppliers in regard to quality problems and/or specifications.

PERSONNEL

The hiring of quality control personnel is generally left to the discretion of the quality control manager. Job requirements and criteria usually are a function of

OUTBOUND   **INBOUND**

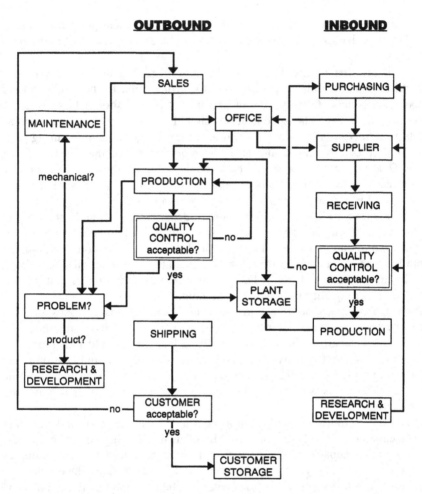

Fig. 24-1. Example of communication flow and interdepartmental relationships involving the quality control department.

supply and demand in the local job market, but job descriptions for every position in the quality control department should be written and communicated to personnel already on board as well as to potential employees. A new employee should receive a complete orientation, including a written and an oral explanation of the company and departmental policies, rules, and procedures. The new employee should be given a comprehensive plant tour and be assigned an on-the-job trainer. Every quality control department should have a manual outlining procedures and other information related to the quality control activities that are specific to the particular plant, and a copy of the manual should be made available to both the new employee and older employees for reference.

The orientation should include instruction in the use of safety gear and safety procedures (see below in this chapter). Also it should include a description of the chain of command and the relationship that should exist between other supervisors and members of management. As with any position, the one-boss rule should be invoked so that communication up and down the chain of command can go through one person in order to avoid confusion and conflicting directives. If more than one person is allowed to give quality control personnel directives, conflicts and confusion may arise.

Although training could begin right away, and even though the employee may have existing experience and skills for the job, serious training should not begin for several days in order to give the employee time to adapt to the new environment. A new employee's success is determined to a significant degree by how well he or she is allowed to adapt in the first week of employment. Once the employee can perform the tasks contained in the job description at a level that is satisfactory to the quality control manager, the initial training can be considered complete. In addition to this initial training, all laboratory technicians should undergo continuous training, which may coincide with performance evaluations. Cross training among employees of different job descriptions protects quality control activities during times of sickness, layoff, leaves of absence, vacations, and job changes.

Each employee should be evaluated by his or her supervisor at least once a year, and should be rated on job performance as well as on how well he or she has performed generally. A good employee may have poor technical ability, and a good technician may find it difficult to arrive at work on time. Annual evaluations provide a good opportunity to solicit employee input about the general activities of the department and the company. This can be done during the evaluation interview orally or by means of a written questionnaire. Many problems will not emerge without some degree of probing, and unless they do emerge, they cannot be solved. Written exams on laboratory procedures and tests as well as departmental and company policies stimulate thought and provide an excellent forum for instruction and review. Performance tests, where all technicians analyze the same sample, can demonstrate the consistency of analyses between lab technicians and ensure standardization in laboratory results. If possible, a program of incentives should be instigated to encourage employees to improve and to excel, including such rewards as days off, longer vacations, or cash awards. Morale is always high and work well done when an employee feels that he or she is growing professionally and working toward a goal. Evaluations also provide an important opportunity for the employee to account for his or her performance. It is important for every employee to receive appreciation and recognition for work well done.

One of the biggest problems in any corporate setting is communication. Both written and oral communication should be used as much as possible in quality control operations. Instructions and information should be written down, illustrated, demonstrated, and confirmed. Log books can be used for communication from one shift to another. Each employee, including the quality control manager, should be responsible for checking the log book at the beginning of each shift. Employees should all feel free to communicate frankly with the quality control manager at all times. An open door policy is important in any managerial setting.

From time to time it may be necessary to discipline errant employees. This job is difficult when rules and guidelines are obscure, or disciplinary procedures are not clearly defined. These procedures should be spelled out clearly in the company policy manual and followed precisely in order to avoid favoritism. Generally an oral warning is followed by a written warning, and records of all warnings are kept in the employee's confidental file, including the date, time, and nature of the warning, and the action taken. If undesirable conduct continues after a written warning, termination of the employee should be considered, depending on the company policy. If multiple warnings are given without definitive actions, the legal implication is that the company accepts the undesirable behavior, and that can lead to legal entanglements. Failure of management to follow company policies weakens the validity of the company policies and suggests their nonexistence even though they are written down. Mistakes made in the distant past should be generally forgiven and not brought up during disciplinary action. Some behavior may justify an employee's immediate dismissal, such as insubordination, intentional product contamination, or threatening another employee or another's safety. Any such problems that quality control personnel observe should be reported immediately for the good of the company. Problems originating in or pertaining to other departments should be reported through the proper chain of command.

When it becomes necessary to lay off or dimiss an employee, the employee should be so notified at least two weeks in advance. (An exception to this rule is if an employee is immediately dismissed for serious misconduct.) Most people need a two-week period to arrange their affairs in order to minimize the impact of the layoff. The reason for the layoff should be discussed honestly and clearly, as should such matters as how the final check will be handled, including vacation pay and so forth, as well as the turning in of safety gear, keys, or other equipment checked out to the employee. If the person has performed well, or if there are plans to rehire him or her this information should be communicated, as well as an estimated time or date of rehire. If the separation is due to misconduct or poor workmanship, this too should be frankly explained.

People sometimes become involved with problems that they are unable to solve by themselves, such as alcoholism or drug use; and it is the employer who often wields the greatest influence in changing the lives of employees with such problems. A frank termination may be the last chance for such a person to be jolted into getting the needed help. Employers must be sure that such terminations are for the violation of company policies and not just the result of someone's personal opinion about what ought to be done. It is advisable for the employer and the employee to get professional advice in questionable situations.

EQUIPMENT MANAGEMENT

Citrus quality control departments utilize equipment that is specialized in comparison to equipment used elsewhere in the plant; so plant maintenance personnel are generally unacquainted with laboratory instruments. Minor repairs, cleaning, and calibrations usually can be done by lab personnel according to instructions from the manufacturer. Improper use can damage laboratory instruments, necessitating expensive repair; therefore, only trained technicians should use laboratory equipment. Another source of equipment damage is insects and rodents, as electronic equipment provides a warm comfortable habitat for such pests. Pests may eat wire insulation, short-circuit the electronics, or wedge themselves into areas that affect the mechanical operation of the instrument. All owner's manuals should be kept on file for easy access, as well as serve as a record of maintenance so that trends can be detected. Users should be careful not to spill liquids on electronic instruments.

Like most other laboratories, citrus quality control labs utilize a wide selection of glassware, including volumetric glassware. Delivered or measured volumes are indicated by the markings of the glassware and are accurate for most purposes. Volumetric markings on beakers and Erlenmeyer flasks are generally accurate to about 5%; volumetric flasks, burets, and pipettes are generally accurate to within about 0.5%, depending on the volume measured. If greater accuracy is desired, volumetric glassware can be calibrated. One performs a calibration by delivering or measuring a precise volume of water using the volumetric glassware and weighing the water at a given temperature. The weights obtained with analytical balances are more accurate than the measured volumes obtained by using volumetric glassware, so the weight of water is used as a standard and is converted to volume using standard density tables for water at various temperatures (Weast 1968). This converted volume can be recorded as a precise volumetric measurement for the particular piece of glassware used. For example, if we delivered 24.941 grams of distilled water at 20.2°C using a 25 ml pipette, we would divide the weight by the density (0.998162 g/ml) (Weast 1968) to get 24.987 ml. The pipette then can be given a number or other identifier so that when it is used 24.987 ml can be used as the volume instead of the approximate 25 ml. One should be careful to read the meniscus properly when using volumetric glassware, as shown in Fig. 3-7.

SAFETY

Safety training and accident avoidance should be integral parts of any industrial operation. Mechanical, electrical, and fire safety should be emphasized in areas where quality control personnel work. Shelf chemicals should be restrained in case of an earthquake. Guards for belts and chains on equipment used by quality control personnel in performing fruit maturity tests and yield tests should be in place, and machinery should be locked out during cleaning. Employees should not place their hands in machinery unless absolutely necessary, and even then should do so only when the machine is locked out. Screw conveyers should always be covered, to guard against the accidental insertion of a leg or an arm.

Employees should wear goggles and gloves when cleaning with caustic solutions, and bump hats, hair nets, beard guards, and so on, should be used, not only to avoid product contamination but to prevent hair from getting caught in machinery. Likewise, loose clothing should not be worn around moving machinery. Proper hand rails and climbing equipment must be installed for sanitation inspections, the taking of samples, or the performance of other necessary tasks. Quality control personnel should not climb onto tankers, rail cars, or other moving vehicles until the vehicle has come to a complete stop, and the driver is out of the cab. Finally, quality control personnel should avoid using tools or equipment not included in their job description. Even if the lab technician is proficient in the use of such items, it could lead to unsafe and even unsanitary conditions.

HAZARDOUS MATERIAL HANDLING

The Occupational Safety and Health Administration of the U.S. government (OSHA) issued a Hazard Communication Regulation (part 1910.1200) effective May 25, 1986 that has become known as the "Employee Right to Know Law." This law was designed to require employers to inform employees of the various hazards of their work place. Most of the law is directed to manufacturers and importers of hazardous chemicals, and food processors are largely exempt from those provisions. Under this law, "hazardous chemical" is defined as "any chemical which is a physical hazard or health hazard." Health hazards are said to include "carcinogens, toxic or highly toxic agents, reproductive toxins, irritants, corrosives, sensitizers, hepatotoxins, nephrotoxins, neurotoxins, agents which act on the hematopoietic system, and agents which damage the lungs, skin, eyes, or mucous membranes." The term "chemical" is defined as "any element, chemical compound or mixture of elements and/or compounds." To a chemist, this is simply everything. "Physical hazard" is defined as a "chemical for which there is scientifically valid evidence that it is a combustible liquid, a compressed gas, explosive, flammable, an organic peroxide, an oxidizer, pyrophoric, unstable (reactive) or water-reactive.'' According to the strict definitions of these terms, everything can be considered a hazard including water and air (drowning, slippery ice, steam burns, cold air in freezers, etc.) In at least partial recognition of the ambiguity of the definitions, appendix B of the law states that the ''hazard evaluation is a process which relies heavily on the professional judgment of the evaluator.'' In other words, this hazard communication regulation must be defined by those to whom it applies, based on professional interpretation, with the ultimate authority being the OSHA inspector and/or the judicial branch of government.

Citrus juice plants utilize various cleaning agents, lubricants, fuels, refrigerants, boiler treatment chemicals, compressed gases, and laboratory chemicals that professional judgment would dictate fall under this regulation. Food products are exempt from this law. Because it is concerned with ''chemicals,'' a large portion of the responsibility for dealing with this law falls on the quality control department. Those persons trained in the use and care of chemicals, such as laboratory technicians, are best equipped to monitor the use of chemicals in the plant and to assist in emergencies. In order to comply with the law, a written hazard communication program must exist, which includes training, a hazardous material list including the location of each material, an outline of the program including who is in charge of each portion, and material safety data sheets on the hazardous materials used in the plant.

There should be a written explanation of how employees and outside contractors will be notified regarding hazardous materials in each work area. This information must be available to all employees, their designated representatives, and OSHA. Even though chemical manufacturers are responsible for proper warning labels, the employer must ensure that all such hazards are properly identified by using signs, process sheets, batch tickets, or other written materials.

Chemical manufacturers and suppliers are required to supply their customers with material safety data sheets (MSDSs) on the chemicals they sell. These MSDS's are to be made available to all employees who use or work near the hazardous material. An MSDS outlines the hazards and the emergency actions necessary in handling a material. Even though there is no fixed format for the MSDS, the following information must be included:

1. Identity of the material—chemical and common names.
2. Identity of components present at 1% or greater concentrations (or, for carcinogens, 0.1% or greater).
3. Physical and chemical hazards (vapor pressure, flash point, etc.).
4. Health hazards.
5. Exposure limits where available.
6. Whether the material is a carcinogen.

7. Known safety precautions.
8. Emergency and first aid procedures.
9. The date of preparation of the MSDS or the last change.
10. Name, address, and phone number of the supplier or the manufacturer.

The written hazard communication program must include a list of the hazardous materials in the plant and their location with an MSDS for each material. Training should include methods to detect the release of the hazardous substance, information on physical and health hazards, and ways that employees can protect themselves, especially in an emergency. It is a good idea to include an outline of the training session in the written hazard communication program, along with the date held and a list of employees who attended. For more details on the Hazardous Communication Regulation, the reader is refered to the law itself or to an appropriate consultant.

QUESTIONS

1. What is one of the first things that should be done in setting up a quality control department?
2. Should production oversee quality control, or should quality control oversee production?
3. What are some of the non–quality control activities that quality control personnel may become involved with regarding other departments within the company?
4. What is one of the biggest problems in any company?
5. Is it always true that an employee's personal problems are not the business of the company? Why or Why not?

REFERENCES

Weast, R. C. 1968. *Handbook of Chemistry and Physics*, The Chemical Rubber Company, Cleveland, Ohio, F4.

Chapter 25

Inventory Management

The management of product inventories is primarily the function of the office staff. However, some aspects of the inventory clearly cross over into the realm of the quality control department, including inbound and outbound product sampling, fill weights, blending, label preparation, tank volume measurements, and the management of laboratory equipment and supplies. The management of rejected or defective products in inventory is often under the quality control department's direction.

SAMPLING

The sampling of bulk lots (drums, tankers, etc.) has been discussed in previous chapters and is fairly straightforward. However, in continuous operations in the manufacture of retail or consumer products, various sampling techniques must be used because the product usually is not standardized or mixed to complete uniformity. Questions arise about how many samples to take and analyze in order to ensure representative sampling and how many defective samples can be allowed. Numerous mathematical and statistical methods can be used to justify a wide variety of sampling plans. However, the simplest and easiest methods to consider can be copied from those sampling plans used by the Food and Drug Administration (FDA) (CFR Title 21 145.3) and the U.S. Department of Agriculture (USDA) (CFR Title 7 52.1–52.83), which are illustrated in Tables 25-1 and 25-2. These tables give the number of samples needed per total number of containers per lot and the number of permissible rejects. The USDA sampling plan is based on container type or volume, and the FDA sampling plan is based on the net weight. Because the FDA is looking for contamination that would violate the standards of identity and/or health violations, its sampling plan is more extensive than the USDA plan, which is looking for general grade or quality characteristics. The continuous sampling performed by the USDA is an involved procedure in which fewer samples are taken than the number listed in Tables 25-1 and 25-2. Another sampling plan in the federal code is the one

Table 25-1. FDA sampling plan for canned fruits (CFR Title 21 145.3).

Lot Size (containers)	Needed Samples	Acceptable Rejects
nt. wt. ≤ 1 kg		
0–4,800	13	2
4,801–24,000	21	3
24,001–48,000	29	4
48,001–84,000	48	6
84,001–144,000	84	9
144,001–240,000	126	13
over 240,000	200	19
1 kg < nt. wt. < 4.5 kg		
0–2,400	13	2
2,401–15,000	21	3
15,001–24,000	29	4
24,001–42,000	48	6
42,001–72,000	84	9
72,001–120,000	126	13
over 120,000	200	19
over 4.5 kg nt. wt.		
0–600	13	2
601–2,000	21	3
2,001–7,200	29	4
7,201–15,000	48	6
15,001–24,000	84	9
24,001–42,000	126	13
over 42,000	200	19

used to determine compliance with labeling regulations (CFR Title 21 101.9(e)(2)). It consists of taking 12 samples each from 12 different random cases that represent one lot and making a composite from the 12 samples. This composite then is checked for nutrient levels and compared to what is declared on the label, as discussed in Chapter 11.

In addition to samples taken for analysis, it is a prudent industrial practice to take extra samples (three minimum) as so-called retain samples. These retain samples should be kept in a sealed container in cold storage and can be valuable when complaints or problems arise. If a customer claims that a particular lot is defective, the retain samples can be used to check the customer's claims. If the retain samples do not show the same defects, it is likely that the defects arose during shipment or customer storage. Retain samples also are handy when seasonal products are needed in a special research project, or when a customer or potential customer requests samples of a particular type of product. Retain samples may have a shelf life of a few weeks to a year, depending on the type of product. Edible discarded samples can be given to local charities or employees.

Table 25-2. USDA sampling plan in determining grade standards (CFR Title 7 52.1–52.83).

Lot Size (containers)	Needed Samples	Acceptable Rejects
0–3,000	3	0
3,001–12,000	6	1
12,001–39,000	13	2
39,001–84,000	21	3
84,001–145,000	29	4
145,001–228,000	38	5
228,001–336,000	48	6
336,001–480,000	60	7

Container Size (canned fruits)	Portion of Lot Size Above*
303 × 406 (16 oz) and less	Use lot sizes given above.
> 303 × 406 to No. 3 cylinder (46 oz)	Use same # of samples with $\frac{1}{2}$ lot size above.
> No. 3 cylinder to #12 can (1 gal)	Use same # of samples with $\frac{1}{4}$ lot size above.
> #12 can	Use $\frac{1}{4}$ above lot size for 6 lb equivalent.

For Fluid Comminuted or Homogeneous Fruit Products	Portion of Lot Size Above*
1 lb or less fruit juice	Use 1.5 times the lot sizes in the chart above.
1 lb to 3.75 lb (60 oz)	Use 2.0 times the lot sizes in the chart above.
3.75 lb to 10 lb	Use 2.0 times the lot sizes in the chart above.
over 10 lb	Convert to 6-lb equivalents and use chart as is.

*For example, for 100,000 303 cans take 29 samples and accept 4 rejects. For 100,000 No. 3 cylinder cans take 38 samples (or the same as 200,000 in the chart above) and accept 5 rejects. For 100,000 #12 cans take 60 samples (or the same as 400,000 in the chart above). For 10,000 60-lb pails, take 29 samples and accept 4 rejects (60 × 10,000/6 lb = 100,000 equivalent containers for 100,000 container equivalents in chart above).

FILL WEIGHT

Fill weights are important for some citrus products, and general methods for their determination are regulated by the federal government. The "general method for fill of containers" can be found in the *Code of Federal Regulations* Title 21 130.12(b), which describes how the percent fill of a container is to be determined. The lid is cut out and the distance from the double seam to the level of food (or juice) is measured. The food is removed, and the container is cleaned, dried, and weighed. The container then is filled with water to $\frac{3}{16}$ inch from the double seam and weighed, and the net weight of the water is recorded. Then the water, at the same temperature, is removed down to the previously measured level of the food (or juice), and the net weight of remaining water is determined. The latter water weight is divided by the former water weight and multiplied by 100% to give the percent fill of the container. For grapefruit juice and lemon juice, the standards of identity require a 90% fill of the container. Even though there are no other "fill of container" requirements in the standards of identity for other citrus juice products, generally a 90% fill is a standard procedure in the food industry.

LABELS

Labels come in a wide variety of shapes, sizes, and colors. The nutritional aspect of labeling has already been discussed, in Chapter 11. Retail or consumer commodity labels generally are fixed or quite inflexible from lot to lot and must comply with the Fair Packaging and Labeling Act (Public Law 89-755 89th Congress, S. 985 November 3, 1966) as well as the regulations for the enforcement of the Federal Food, Drug, and Cosmetic Act and the Fair Packaging and Labeling Act (CFR Title 21 101.1–101.105). A new addition to the law is the country of origin regulation, which states that 75% of the imported juice (citrus and/or noncitrus) must be declared by country of origin on all labels for both retail and nonretail products. The standards of identity also have labeling requirements, as do the USDA grades for retail products. In addition to governmental requirements, industrial organizations have established guidelines for labeling, such as those of the National Food Processors Association (1977). Lawyers and graphic artists are asked to design labels, which then are used to manufacture large inventories of labels and/or labeled containers. The quality control department should monitor the general quality of labels and containers as a part of sample analysis. Lot or day codes imprinted on retail containers also should be inspected, as well as the fill levels and the container seal quality.

Nonretail or nonconsumer commodities (generally bulk) sold to other commercial entities still fall under the labeling requirements of the standards of identity. Each lot of bulk juice (drums, pails, tankers, etc.) should be clearly marked as the product described in the standards of identity. Often a product will meet the criteria for several standards of identity, and the seller can choose the one he or she prefers to use to identify the product. (See Chapter 19 for brief descriptions of the standards of identity.) Included with the product name should be the name and address of the manufacturer, the lot number, the date of manufacture, the net weight, and the Brix. Perhaps the Brix/acid ratio, tare, and gross weights would be appropriate as well. If any other ingredients have been added, such as preservatives or sugar, or any material that may affect the standards of identity, those ingredients also should be declared on the label. Defective product should be marked with a highly visible tag, such as a red tag, indicating the nature of the defect and the lot number. This facilitates proper handling of the defective product. Tanker or barge lots generally have the above-listed information on the bill of lading.

Drum labels generally are preprinted or preprogrammed into a computer with a fixed or unchanging label format with spaces for variable information. Later the variable information, such as lot number, weights, Brix, and so on, is written in by hand as the product is packaged. This information often is determined by the quality control personnel because they are the ones who determine some of the parameters, such as the Brix and net weights. Also, the quality control department usually is more aware of other labeling requirements and can be more effective than production in label monitoring.

AUTOMATIC IDENTIFICATION METHODS

Automatic identification methods are the state-of-the-art methods used in commercial and retail outlets across the country to identify products by machine. The most common method employs bar codes. Other related methods include optical character recognition, voice data entry, machine vision, radio frequency data communication magnetic strips, and smart cards (Knill 1988), which are tailor-made for different applications. Bar codes are used with most retail products and have found their way into the bulk citrus industry as well. On April 3, 1973 the Uniform Product Code Council, Inc. established the Uniform Product Code (UPC), a bar code system designed for use with retail products (UPC Council, Inc.). International interest in the UPC system led to the adoption of the European Article Numbering system (EAN) in December of 1976, which expanded the UPC capabilities. Now, nearly all products in supermarkets have a UPC identifying bar code. The code consists of 10 digits divided into one or two numbers. Slot scanners, which consist of two orthogonal scanners, are used so that any orientation of the bar code will be scanned by one or the other scanner as the product is moved across the slots. The scanners read the 5-or 10-digit code and automatically associate it with a preprogrammed price and product type in the cash register.

For in-house industrial use, the UPC code can be used to count empty containers, labels, or individual product containers. However, unlike the case of the retail outlet, it is easier to count bulk quantities industrially, such as cases, pallets, or drums, rather than individual cans or bottles. The bar codes used on these bulk containers need not utilize the UPC system. Also, information such as lot number, Brix, Brix/acid ratio, and net weights is of greater concern to members of industry than to the consumer, who assumes these parameters to be constant; and it might be helpful if this information were encoded into bar code symbologies for industrial purposes.

Citrus bulk processors are beginning to realize the great benefits of automatic identification techniques. One such system involves the use of a preprogrammed thermal printer that prints labels impervious to the wet environment of industrial processing and storage. Such printers allow for computerized entry of variable data on a lot-by-lot basis, which can be converted instantly to bar code symbologies. Labels can be customized on-site to fit the needs of any manufacturer. Portable scanners can be used to scan the bar codes of bulk inventories and can be programmed to process the data into any form or format. The portable scanner then can be attached to a printer or a computer, and the finished manifest, blend sheet, or inventory listing can be printed in the desired format with no human error in counting or writing. This information also can go directly into a central processing unit to automatically adjust product inventories, again without human involvement. Examples of UPC/EAN symbology as well as an automatic identification drum label are shown in Fig. 25-1. Figure 25-2 shows a

Fig. 25-1. Example of UPC/EAN bar code symbology and an industrial bar code label used commercially for 52-gallon drums.

**CALIFORNIA CITRUS
PRODUCERS, INC.**
P.O. Box C
525 E. Lindmore Ave.
Lindsay, CA 93247
(209) 562-5169

Customer_____ Sunny Orange Company_____ ORDER NO. 6603 ROW NO. 8.N

LOT NUMBER	NUMBER OF DRUMS	BRIX	B/A RATIO	DRUM NET WEIGHT
VB1758	4	59.8	12.5	560
VA1758	18	59.9	12.6	559
VB1698	1	59.5	12.6	562
VB1728	20	59.9	12.9	559
VC1728	20	60.0	12.9	558
VB1738	21	59.8	12.5	560

TOTAL DRUMS= 84

Fig. 25-2. Example of shipping manifest printed from data transmitted directly from a bar code label scanner. Counting, sorting, and manifest formatting can be programmed into the portable bar code scanning unit.

drum manifest printed from a scanning of drum bar code labels that was processed and transmitted directly to a small inexpensive printer.

TANK MEASUREMENT

Another important area that quality control personnel often get involved with is the determination of the volume of product in bulk tanks. Fluid levels are measured from a fixed standard point, or they are measured using various types of level meters. Level meters designed for water measurements according to density usually are not adequate for juices or concentrates because their densities are quite different from that of water. Once the level is determined, equations can be used to calculate the volume of fluid in the tank. Net weights then can be determined too, if needed. Most processing tanks have the shape illustrated in Fig. 25-3, with a dome top, cylindrical center, and slant bottom. As the product fills from the bottom, each of the sections is filled, with a varying relationship existing between the product level and the volume of the tank.

The slant bottom portion is divided into two sections, the first half and the second half, as shown in Fig. 25-4, where the shaded area represents the product or fluid in the tank. As the product fills the first half of the slant bottom, the top surface area of the product may be represented as a segment of a circle,

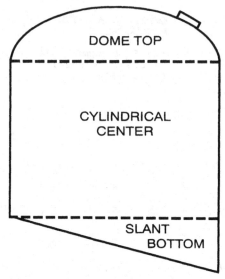

Fig. 25-3. Typical shape of bulk tanks used in citrus processing, with three main sections.

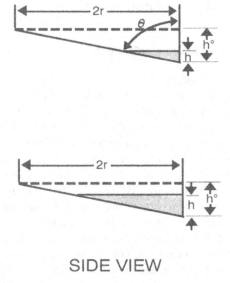

SIDE VIEW

Fig. 25-4. Isometric views of the slant bottom portion of bulk tanks and dimensions used to derive volume equations.

described mathematically by:

$$A_{\text{seg}} = \pi r^2/2 - \left(y \sqrt{r^2 - y^2} + r^2 \sin^{-1}(y/r) \right) \qquad (25\text{-}1)$$

with the variables as depicted in Fig. 25-4. Looking at the side view and using a trigonometric relationship we get:

$$(r - y)/h = \tan \theta = 2r/h_0 \qquad (25\text{-}2)$$

which leads to:

$$y = r(1 - 2h/h_0) \qquad (25\text{-}3)$$

With inches used as the unit of measurement, the volume of the slant bottom during filling of the first half becomes:

$$V_b = A_{\text{seg}} h (231 \text{ in}^3/\text{gal})/2 \qquad (25\text{-}4)$$

Substituting Equation 25-1 and 25-3 into Equation 25-4 gives the volume of the first half of the slant bottom as a function of only one variable, the measured vertical height h, as follows:

$$V_b = \frac{hr^2}{462} \left(\frac{\pi}{2} + 2 \left(\frac{2h}{h_0} - 1 \right) \sqrt{(1 - h/h_0)h/h_0} - \sin^{-1} \left(1 - \frac{2h}{h_0} \right) \right) \quad (25\text{-}5)$$

The volume is found in gallons.

Once the first half has been filled, the second half of the slant bottom will have a surface area expressed as follows:

$$A'_{\text{seg}} = \pi r^2/2 + y \sqrt{r^2 - y^2} + r^2 \sin^{-1}(y/r) \qquad (25\text{-}6)$$

which is similar to Equation 25-1 except for a negative sign. Using the same treatment as before, we obtain the following relationship for the volume of product in the slant bottom as the product fills the second half of the slant bottom:

$$V_b = \frac{hr^2}{462} \left(\frac{\pi}{2} + 2 \left(\frac{2h}{h_0} - 1 \right) \sqrt{(1 - h/h_0)h/h_0} + \sin^{-1} \left(\frac{2h}{h_0} - 1 \right) \right) \quad (25\text{-}7)$$

which is similar to Equation 25-5 except for the signs of the last terms.

Once the total bottom section is filled, the total volume can be calculated simply by using the following:

$$V_b'' = \pi r^2 h_0 / 462 \tag{25-8}$$

The cylindrical section uses a volume equation similar to Equation 25-8:

$$V_c = \pi r^2 (h - h_0) / 231 \text{ in}^3/\text{gal} \tag{25-9}$$

By adding the filled volume of the slant bottom section to V_c, the relationship between the product level and the volume while the level is within the cylindrical section becomes:

$$V_c + V_b'' = \pi r^2 (h - h_0/2) / 231 \tag{25-10}$$

When the cylindrical and slant bottom portions are full, the dome top begins to fill, as shown in Fig. 25-5. The volume of a segment of a sphere can be expressed as:

$$V_{\text{seg sphere}} = \pi v_0 (3R - v_0) / 693 \tag{25-11}$$

using the dimensions in Fig. 25-5. The radius of the sphere must be determined by measuring the curve of the dome, transposing it to paper, and graphically measuring the radius. Equation 25-11 represents the total volume of the dome portion of the tank. The product volume in the dome is determined by subtract-

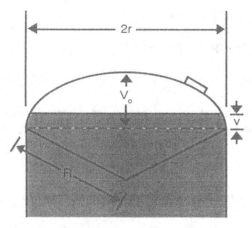

Fig. 25-5. Dimensions useful in determining the volume of the dome top in citrus bulk tanks. The radius R must be determined graphically or by measurement.

ing the volume of the spherical segment of space remaining above the product from Equation 25-11. After simplification we obtain the following:

$$V_d = \pi\left(v_0^2(R - v_0/3) - (v_0 - v)^2(R - (v_0 - v)/3)\right)/231 \quad (25\text{-}12)$$

With the cylindrical portion filled, the term h in Equation 25-10 can be expressed as the fixed constant d for the length of the cylindrical portion of the tank, and the total volume of the tank with product in the dome top can be expressed as follows:

$$V_t = \frac{\pi r^2}{231}\left(d + \frac{h_0}{2}\right) + \frac{\pi}{231}\left(v_0^2\left(R - \frac{v_0}{3}\right) - (v_0 - v)^2\left(R - \frac{v_0 - v}{3}\right)\right)$$

$$(25\text{-}13)$$

The total height from the bottom of the tank to the product level in the dome top is:

$$H_t = h_0 + d + v \quad (25\text{-}14)$$

To illustrate the use of these equations, we will take an imaginary tank and calculate the volume of fluid at different levels. Suppose the tank has a diameter of 12 feet ($r = 72$ inches), a slant bottom of 12 inches vertically, a cylindrical length of 15 feet (180 inches), a dome top with height of 3 feet, and a spherical radius of $6\frac{1}{2}$ feet (78 inches). We measure the total height H from the very bottom of the tank and determine the volume of product in the tank by using the various equations shown above. Table 25-3 summarizes the data and the results. With the hatch set at 6 inches above the cylindrical portion of the tank, the maximum v value is 6 inches. Paddle or propeller agitators in product tanks

Table 25-3. Calculated volumes of product from
fluid level measurements.

Measurement #	H (total vertical level)	Equation	# Gallons
1	3″	25-5	21
2	8″	25-7	200
3	112″	25-10	7,473
4	195″	25-13	13,284
5	198″	25-13	13,444

Tank dimensions—a radius of 72″, a vertical slant bottom of 12″, a cylindrical length of 180″, a dome top height of 36″, and a dome spherical radius of 78″. The hatch is 6″ above the cylindrical section and sets the maximum filling level of the tank.

displace a portion of the product, and their volume can be subtracted from the product volume. However, this displaced volume generally is not significant.

If a tank has a spherical bottom, the volume with the juice level in this spherical segment again can be described as in Equation 25-11, but with v_0 now representing the actual fluid level. This form of Equation 25-11 then can be used in place of the equations used for the slant bottom portion in the above procedures.

Using the above-derived equations every time a volume determination is needed is awkward and time-consuming. The best way to use such equations is to generate a table consisting of a level measurement and the corresponding volume. Such charts or tables can be used quickly and effectively, not only by quality control but by production personnel as well. Programmable calculators or computers are ideally suited for such table generation. A flow chart that can be used to program computers to generate a volume table is illustrated in Fig. 25-6.

A problem commonly found with some computer languages, such as RPG, is the inability to perform the $\sin^{-1}$ function. In such cases, a Taylor series expansion (shown below) out to six terms gives sufficient accuracy in most cases:

$$\sin^{-1} x = x + \frac{x^3}{3!} + \frac{9x^5}{5!} + \frac{180x^7}{7!} + \frac{10080x^9}{9!} + \frac{833490x^{11}}{11!} + \cdots \quad (25\text{-}15)$$

Tankers

The amount of juice in stationary tanks is most easily measured by volume. However, mobile tanks, such as tankers, usually measure the amount of product by weight, using industrial truck scales. This net weight of juice can be converted to juice volume easily by using the density conversion from Equation 2-9 or 2-11. For example, if we had a tanker that weighed 46,765 pounds and contained orange concentrate at 60.2°Brix, we would get:

$$(46,765_{\text{lb conc}}) / (10.737_{\text{lb/gal @ 60.2 Brix from Eq. 2-9 or 2-11}}) = 4355 \text{ gal}$$

LAB EQUIPMENT AND SUPPLIES

The Internal Revenue Service and general company accounting procedures generally require an annual inventory of products and supplies. Even if no inventory is required, a knowledge of equipment and supplies on hand will help the quality control manager to utilize resources efficiently and will facilitate the timely restocking of supplies. Inventories can be classified according to such

VOLUMES IN SLANT—
BOTTOM TANKS

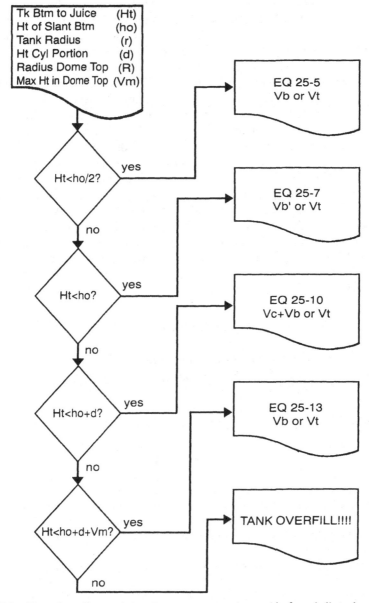

Fig. 25-6. Flow chart that can be used to generate a volume table for a bulk tank using the equations in the chapter.

categories as beakers, pipettes, burets, hardware, chemicals, and so on. Individual items on the inventory can be priced by using identification numbers found in supplier catalogues. Commercial inventories usually are performed in pairs, with one person counting and another recording the number of items and the item identification. Personnel should be careful to isolate items being counted so that they are not used or moved before the counting can be completed. A copy of the previous year's inventory can be used as a guide.

Laboratory technicians must monitor the current inventory of supplies at all times to avoid running out of needed items. The establishment of minimum supply levels can help them to determine when an item needs to be reordered. Delivery time should be a factor in determining the minimum allowable inventory. Most items can be obtained in a week or less, but some items may take much longer. Some equipment and supplies are indispensable in the operation of a quality control laboratory, and backups should always be kept on hand for them. This includes such items as refractometers, balances, and titrating equipment. If a piece of equipment must be sent away for repair, the backup can be used. Because ordered supplies may take some time to arrive, ordering records should be kept; and these records can be used as an aid in annual inventories as well as to ensure timely shipments. Upon the receipt of supplies, packing slips should be promptly and thoroughly examined to ensure that the proper item was received, as well as to ensure proper storage or distribution to those persons who need the item(s).

QUESTIONS

1. Why is a sampling plan for consumer products important?
2. Why are the USDA and FDA sampling plans different?
3. Why are retain samples important?
4. How can bar codes on labels be of use?
5. Why would a plant purchase two refractometers?

PROBLEMS

1. If the FDA sampling plan was used to determine the number of samples needed for 50,000 half-gallon containers of orange juice from concentrate, how many samples should be taken, and how many rejects would be allowed? If the half-gallon containers were produced at the rate of 100/min., at what time intervals should samples be taken using this plan?
2. Consider a bulk tank with a radius of 150 inches, a vertical slant bottom of 36 inches, a dome top height of 24 inches, a cylindrical section 250 inches high, and a dome spherical radius of 300 inches. The hatch is 8 inches above the top of the cylindrical section. What is the maximum capacity of the tank?

3. If the tank in problem 2 were filled to a height of 200 inches from the very bottom of the slant section, what would be the product volume?
4. If the tank in problem 2 were filled to a height of 30 inches from the bottom of the slant section, what would be the product volume?
5. Suppose a tank had a spherical bottom with a depth of 18 inches and a spherical radius of 120 inches and a product level of 13 inches. What volume of juice is in the tank?

REFERENCES

Knill, B. 1988. New directions in automatic identification, *Industry Week*, *237(4)*, A1–A36.
National Food Processors Association Labeling Manual, August 1977.
UPC Council, Inc. 7051 Corporate Way, Suite 201, Dayton, Ohio 45459.

Appendix A

Acid Corrections to the BRIX

% Acid	Correction	% Acid	Correction	% Acid	Correction
0.10	0.02	1.75	0.35	3.40	0.66
0.15	0.03	1.80	0.36	3.45	0.67
0.20	0.04	1.85	0.37	3.50	0.68
0.25	0.05	1.90	0.38	3.55	0.69
0.30	0.06	1.95	0.39	3.60	0.70
0.35	0.07	2.00	0.39	3.65	0.71
0.40	0.08	2.05	0.40	3.70	0.72
0.45	0.09	2.10	0.41	3.75	0.73
0.50	0.10	2.15	0.42	3.80	0.74
0.55	0.11	2.20	0.43	3.85	0.75
0.60	0.12	2.25	0.44	3.90	0.76
0.65	0.13	2.30	0.45	3.95	0.77
0.70	0.14	2.35	0.46	4.00	0.78
0.75	0.15	2.40	0.47	4.05	0.79
0.80	0.16	2.45	0.48	4.10	0.80
0.85	0.17	2.50	0.49	4.15	0.81
0.90	0.18	2.55	0.50	4.20	0.81
0.95	0.19	2.60	0.51	4.25	0.82
1.00	0.20	2.65	0.52	4.30	0.83
1.05	0.21	2.70	0.53	4.35	0.84
1.10	0.22	2.75	0.54	4.40	0.85
1.15	0.23	2.80	0.54	4.45	0.86
1.20	0.24	2.85	0.55	4.50	0.87
1.25	0.25	2.90	0.56	4.55	0.88
1.30	0.26	2.95	0.57	4.60	0.89
1.35	0.27	3.00	0.58	4.65	0.90
1.40	0.28	3.05	0.59	4.70	0.91
1.45	0.29	3.10	0.60	4.75	0.92
1.50	0.30	3.15	0.61	4.80	0.93
1.55	0.31	3.20	0.62	4.85	0.94
1.60	0.32	3.25	0.63	4.90	0.95
1.65	0.33	3.30	0.64	4.95	0.96
1.70	0.34	3.35	0.65	5.00	0.97

% Acid	Correction	% Acid	Correction	% Acid	Correction
5.05	0.98	7.50	1.44	9.95	1.90
5.10	0.99	7.55	1.45	10.00	1.91
5.15	1.00	7.60	1.46	10.05	1.92
5.20	1.01	7.65	1.47	10.10	1.93
5.25	1.02	7.70	1.48	10.15	1.94
5.30	1.03	7.75	1.49	10.20	1.95
5.35	1.03	7.80	1.50	10.25	1.96
5.40	1.04	7.85	1.51	10.30	1.97
5.45	1.04	7.90	1.52	10.35	1.98
5.50	1.05	7.95	1.53	10.40	1.99
5.55	1.06	8.00	1.54	10.45	2.00
5.60	1.07	8.05	1.55	10.50	2.01
5.65	1.08	8.10	1.56	10.55	2.02
5.70	1.09	8.15	1.57	10.60	2.03
5.75	1.10	8.20	1.58	10.65	2.04
5.80	1.11	8.25	1.59	10.70	2.05
5.85	1.12	8.30	1.60	10.75	2.06
5.90	1.13	8.35	1.61	10.80	2.06
5.95	1.14	8.40	1.62	10.85	2.07
6.00	1.15	8.45	1.63	10.90	2.08
6.05	1.16	8.50	1.64	10.95	2.09
5.75	1.10	8.55	1.65	11.00	2.10
6.10	1.17	8.60	1.66	11.05	2.11
6.15	1.18	8.65	1.67	11.10	2.12
6.20	1.19	8.70	1.68	11.15	2.13
6.25	1.20	8.75	1.69	11.20	2.14
6.30	1.21	8.80	1.69	11.25	2.15
6.35	1.22	8.85	1.70	11.30	2.16
6.40	1.23	8.90	1.71	11.35	2.17
6.45	1.24	8.95	1.72	11.40	2.18
6.50	1.25	9.00	1.72	11.45	2.19
6.55	1.26	9.05	1.73	11.50	2.20
6.60	1.27	9.10	1.74	11.55	2.21
6.65	1.28	9.15	1.75	11.60	2.21
6.70	1.29	9.20	1.76	11.65	2.22
6.75	1.30	9.25	1.77	11.70	2.23
6.85	1.31	9.30	1.78	11.75	2.24
6.90	1.32	9.35	1.79	11.80	2.24
6.95	1.33	9.40	1.80	11.85	2.25
7.00	1.34	9.45	1.81	11.90	2.26
7.05	1.35	9.50	1.82	11.95	2.27
7.10	1.36	9.55	1.83	12.00	2.27
7.15	1.37	9.60	1.83	12.05	2.28
7.20	1.38	9.65	1.84	12.10	2.29
7.25	1.39	9.70	1.85	12.15	2.30
7.30	1.40	9.75	1.86	12.20	2.31
7.35	1.41	9.80	1.87	12.25	2.32
7.40	1.42	9.85	1.88	12.30	2.33
7.45	1.43	9.90	1.89	12.35	2.34

% Acid	Correction	% Acid	Correction	% Acid	Correction
12.40	2.35	14.85	2.79	17.30	3.23
12.45	2.36	14.90	2.80	17.35	3.24
12.50	2.37	14.95	2.80	17.40	3.25
12.55	2.38	15.00	2.81	17.45	3.26
12.60	2.39	15.05	2.82	17.50	3.26
12.65	2.40	15.10	2.83	17.55	3.27
12.70	2.41	15.15	2.84	17.60	3.28
12.75	2.42	15.20	2.85	17.70	3.29
12.80	2.42	15.25	2.86	17.75	3.30
12.85	2.43	15.30	2.87	17.80	3.31
12.90	2.44	15.35	2.88	17.85	3.32
12.95	2.45	15.40	2.89	17.90	3.33
13.00	2.46	15.45	2.90	17.95	3.34
13.05	2.47	15.50	2.91	18.00	3.35
13.10	2.48	15.55	2.92	18.05	3.36
13.15	2.49	15.60	2.93	18.10	3.37
13.20	2.50	15.65	2.94	18.15	3.37
13.25	2.51	15.70	2.95	18.20	3.38
13.30	2.52	15.75	2.96	18.25	3.39
13.35	2.53	15.80	2.97	18.30	3.40
13.40	2.54	15.85	2.98	18.35	3.41
13.45	2.55	15.90	2.99	18.40	3.42
13.50	2.56	15.95	2.99	18.45	3.43
13.55	2.57	16.00	3.00	18.50	3.44
13.60	2.57	16.05	3.01	18.55	3.45
13.65	2.58	16.10	3.02	18.60	3.46
13.70	2.59	16.15	3.02	18.65	3.47
13.75	2.60	16.20	3.03	18.70	3.48
13.80	2.61	16.25	3.04	18.75	3.48
13.85	2.62	16.30	3.05	18.80	3.49
13.90	2.63	16.35	3.05	18.85	3.50
13.95	2.64	16.40	3.06	18.90	3.51
14.00	2.64	16.45	3.07	18.95	3.52
14.05	2.65	16.50	3.08	19.00	3.53
14.10	2.66	16.55	3.08	19.05	3.54
14.15	2.67	16.60	3.09	19.10	3.55
14.20	2.68	16.65	3.10	19.15	3.55
14.25	2.69	16.70	3.11	19.20	3.56
14.30	2.70	16.75	3.12	19.25	3.57
14.35	2.71	16.80	3.13	19.30	3.58
14.40	2.72	16.85	3.14	19.35	3.59
14.45	2.73	16.90	3.15	19.40	3.59
14.50	2.74	16.95	3.16	19.45	3.60
14.55	2.74	17.00	3.17	19.50	3.61
14.60	2.75	17.05	3.18	19.55	3.62
14.65	2.76	17.10	3.19	19.60	3.63
14.70	2.77	17.15	3.20	19.65	3.64
14.75	2.77	17.20	3.21	19.70	3.65
14.80	2.78	17.25	3.22	19.75	3.66

% Acid	Correction	% Acid	Correction	% Acid	Correction
19.80	3.67	22.25	4.10	24.70	4.53
19.85	3.68	22.30	4.11	24.75	4.53
19.90	3.69	22.35	4.12	24.80	4.54
19.95	3.69	22.40	4.13	24.85	4.55
20.00	3.70	22.45	4.14	24.90	4.56
20.05	3.71	22.50	4.15	24.95	4.57
20.10	3.72	22.55	4.16	25.00	4.58
20.15	3.72	22.60	4.17	25.05	4.59
20.20	3.73	22.65	4.18	25.10	4.60
20.25	3.74	22.70	4.19	25.15	4.61
20.30	3.75	22.75	4.19	25.20	4.62
20.35	3.76	22.80	4.20	25.25	4.63
20.40	3.77	22.85	4.21	25.30	4.64
20.45	3.78	22.90	4.22	25.35	4.65
20.50	3.79	22.95	4.23	25.40	4.66
20.55	3.79	23.00	4.24	25.45	4.67
20.60	3.80	23.05	4.25	25.50	4.68
20.65	3.81	23.10	4.26	25.55	4.68
20.70	3.82	23.15	4.26	25.60	4.69
20.75	3.83	23.20	4.27	25.65	4.70
20.80	3.84	23.25	4.28	25.70	4.71
20.85	3.85	23.30	4.29	25.75	4.72
20.90	3.86	23.35	4.29	25.80	4.73
20.95	3.87	23.40	4.30	25.85	4.74
21.00	3.88	23.45	4.31	25.90	4.75
21.05	3.89	23.50	4.32	25.95	4.75
21.10	3.90	23.55	4.33	26.00	4.76
21.15	3.90	23.60	4.34	26.05	4.77
21.20	3.91	23.65	4.35	26.10	4.78
21.25	3.92	23.70	4.36	26.15	4.78
21.30	3.93	23.75	4.37	26.20	4.79
21.35	3.94	23.80	4.38	26.25	4.80
21.40	3.95	23.85	4.39	26.30	4.81
21.45	3.96	23.90	4.40	26.35	4.82
21.50	3.97	23.95	4.40	26.40	4.83
21.55	3.98	24.00	4.41	26.45	4.84
21.60	3.99	24.05	4.42	26.50	4.85
21.65	4.00	24.10	4.43	26.55	4.85
21.70	4.01	24.15	4.43	26.60	4.86
21.75	4.01	24.20	4.44	26.65	4.87
21.80	4.02	24.25	4.45	26.70	4.88
21.85	4.03	24.30	4.46	26.75	4.89
21.90	4.04	24.35	4.47	26.80	4.90
21.95	4.04	24.40	4.48	26.85	4.91
22.00	4.05	24.45	4.49	26.90	4.92
22.05	4.06	24.50	4.50	26.95	4.93
22.10	4.07	24.55	4.50	27.00	4.94
22.15	4.08	24.60	4.51	27.05	4.95
22.20	4.09	24.65	4.52	27.10	4.96

% Acid	Correction	% Acid	Correction	% Acid	Correction
27.15	4.96	29.50	5.37	31.85	5.78
27.20	4.97	29.55	5.38	31.90	5.79
27.25	4.98	29.60	5.39	31.95	5.79
27.30	4.99	29.65	5.40	32.00	5.80
27.35	4.99	29.70	5.41	32.05	5.81
27.40	5.00	29.75	5.41	32.10	5.82
27.45	5.01	29.80	5.42	32.15	5.83
27.50	5.02	29.85	5.43	32.20	5.84
27.55	5.02	29.90	5.44	32.25	5.84
27.60	5.03	29.95	5.45	32.30	5.85
27.65	5.04	30.00	5.46	32.35	5.86
27.70	5.05	30.05	5.47	32.40	5.87
27.75	5.05	30.10	5.48	32.45	5.88
27.80	5.06	30.15	5.48	32.50	5.89
27.85	5.07	30.20	5.49	32.55	5.89
27.90	5.08	30.25	5.50	32.60	5.90
27.95	5.09	30.30	5.51	32.65	5.91
28.00	5.10	30.35	5.52	32.70	5.92
28.05	5.11	30.40	5.53	32.75	5.93
28.10	5.12	30.45	5.54	32.80	5.94
28.15	5.13	30.50	5.55	32.85	5.94
28.20	5.14	30.55	5.56	32.90	5.95
28.25	5.15	30.60	5.57	32.95	5.96
28.30	5.16	30.65	5.58	33.00	5.97
28.35	5.17	30.70	5.59	33.05	5.98
28.40	5.18	30.75	5.59	33.10	5.99
28.45	5.19	30.80	5.60	33.15	5.99
28.50	5.20	30.85	5.61	33.20	6.00
28.55	5.21	30.90	5.62	33.25	6.01
28.60	5.22	30.95	5.63	33.30	6.02
28.65	5.23	31.00	5.64	33.35	6.03
28.70	5.24	31.05	5.64	33.40	6.04
28.75	5.24	31.10	5.65	33.45	6.04
28.80	5.25	31.15	5.66	33.50	6.05
28.85	5.26	31.20	5.67	33.55	6.06
28.90	5.27	31.25	5.68	33.60	6.07
28.95	5.27	31.30	5.69	33.65	6.08
29.00	5.28	31.35	5.69	33.70	6.09
29.05	5.29	31.40	5.70	33.75	6.09
29.10	5.30	31.45	5.71	33.80	6.10
29.15	5.30	31.50	5.72	33.85	6.11
29.20	5.31	31.55	5.73	33.90	6.12
29.25	5.32	31.60	5.74	33.95	6.13
29.30	5.33	31.65	5.74	34.00	6.14
29.35	5.34	31.70	5.75	34.05	6.14
29.40	5.35	31.75	5.76	34.10	6.15
29.45	5.36	31.80	5.77	34.15	6.16

% Acid	Correction	% Acid	Correction	% Acid	Correction
34.20	6.17	34.40	6.20	34.60	6.24
34.25	6.18	34.45	6.21	34.65	6.24
34.30	6.19	34.50	6.22	34.70	6.25
34.35	6.19	34.55	6.23		

Appendix B

GWBASIC and RPG Programs

```
C ********************************************************************************
C*          RPG CALCULATIONS SUBROUTINE THAT CAN BE USED TO CALCULATE
C*          THE LBS SOLIDS/GALLON FROM THE BRIX USING EQUATION 2-10
C ********************************************************************************
CSR          SUB1      BEGSR
CSR                    Z-ADD1          COUNT    20       INITIALIZE
CSR                    Z-ADD1.0000000  TAYLOR   157      VARIABLES
CSR          BRIX      ADD 330.872     F0       154
CSR                    Z-ADD1.0000      F1       158
CSR                    Z-ADD1.0000      F2       154
CSR                    Z-ADD1.0000      F3       154
CSR          F1        MULT F           F1        H       CALCULATE
CSR          F2        MULT 170435      F2                FACTORS
CSR          F1        DIV F2           F1        H        FOR
CSR          F1        DIV F3           F1        H       TAYLOR
CSR          F1        MULT F0          F1        H       SERIES
CSR                    Z-ADDF1          F4       158
CSR          RET       TAG
CSR                    SETOF                     10       TAYLOR
CSR          TAYLOR    ADD F4           TAYLOR            SERIES
CSR          F4        MULT F3          F4        H       SUMMATION
CSR          F4        MULT F1          F4        H
CSR          COUNT     ADD 1            COUNT     H
CSR          F3        MULT COUNT       F3        H
CSR          F4        DIV F3           F4        H
CSR          COUNT     COMP 20                            10
CSR     N10            GOTO RET
CSR                    SETOF                     10
CSR          TAYLOR    MULT .0437691    TAYLOR    H
CSR          TAYLOR    MULT BRIX        SPG      43H      FINAL SPG
CSR                    ENDSR                              VALUE
```

```
10 'PROGRAM TO ADJUST BRIX USING WATER OR HIGHER BRIX CONC.
20 INPUT "Brix on hand" ;BI
30 INPUT "Desired Brix" ;B
40 INPUT "Volume on hand" ;VC
50 IF BI>B THEN GOTO 200
55 INPUT "Brix of high Brix concentrate" ;BH
60 INPUT "Can the volume vary (Y/N)" ;Q$
70 IF Q$="Y" THEN GOTO 300
80 BB=BI
90 GOSUB 400
100 SPGI=SPG
110 BB=BH
120 GOSUB 400
130 SPGH=SPG
140 BB=B
150 GOSUB 400
160 VH=VC*(SPG-SPGI)/(SPGH-SPGI)
170 VH=CINT(VH)
180 PRINT "Volume of high Brix needed=" ;VH; "gallons"
190 END
200 INPUT "Can the volume vary (Y/N)" ;Q$
210 IF Q$="Y" THEN GOTO 300
220 BB=BI
230 GOSUB 400
240 SPGI=SPG
250 BB=B
260 GOSUB 400
270 VW=VC*(SPGI-SPG)/SPGI
275 VW=CINT(VW)
280 PRINT "Volume of water needed=" ;VW; "gallons"
290 END
300 BB=BI
310 GOSUB 400
320 SPGI=SPG
330 BB=BH
340 GOSUB 400
350 SPGH=SPG
360 BB=B
370 GOSUB 400
380 V=VC*(SPGI-SPG)/(SPG-SPGH)
385 V=CINT(V)
390 PRINT "Vol. of water or hi-Brix conc. needed=" ;V; "gallons"
395 END
400 SPG=.0437691*BB*EXP(BB+330.872)^2/170435!)
410 RETURN


10 'PROGRAM TO DETN CUTBACK JUICE NEEDED
20 INPUT "Brix of concentrate" ;BC
30 INPUT "Brix of single strength juice" ;BJ
```

```
40 INPUT "Desired final Brix" ;B
50 INPUT "Desired final Volume" ;V
60 BB=BC
70 GOSUB 200
80 SPGC=SPG
90 BB=BJ
100 GOSUB 200
110 SPGJ=SPG
120 BB=B
130 GOSUB 200
140 VJ=V*(SPGC-SPG)/(SPGC-SPGJ)
150 VJ=CINT(VJ)
160 VC=CINT(V-VJ)
170 PRINT "Volume of ";BC; "Brix conc. needed=";VC;"gallons"
180 PRINT "Volume of ";BJ;"Brix juice needed=";VJ;"gallons"
190 END
200 SPG=.0437691*BB*EXP((BB+330.872)^2/170435!)
210 RETURN


10 'THIS PROGRAM DETN'S THE AVERAGE BRIX OF A BLEND
20 DIM B(80):
25 'MAXIMUM NUMBER OF BLEND COMPONENTS IS 80 IN THIS
   PROGRAM. THE NUMBER CAN BE CHANGED IN STEPS 20-40.
30 DIM V(80)
40 DIM SPG(80)
50 N=N+1
60 PRINT "Brix of component ";N;
70 INPUT B(N)
75 IF B(N)=0 THEN GOTO 130
80 PRINT "Volume of component ";N;
90 INPUT V(N)
95 SPG(N)=.0437691*B(N)*EXP((B(N)+330.872)^2/170435!)
100 Z=Z+V(N)*SPG(N)
110 V=V+V(N)
120 GOTO 50
130 N=N-1
135 PRINT " #"," BRIX","VOL"
140 FOR I=1 TO N
150 PRINT I,B(I),V(I)
160 NEXT
170 S=Z/V
180 BF=S*10
190 B=BF
200 BF=S/(.0437691*EXP((B+330.872)^2/170435!))
210 IF ABS(BF-B)<.0001 THEN GOTO 230
220 GOTO 190
230 BF=CINT(BF*10)
240 BF=BF/10
250 PRINT "AVERAGE BRIX=";BF;TAB(21);"TOT VOL=";V
```

```
10 'THIS PROGRAM DETN'S THE VOL NEEDED FOR ONE COMPONENT
20 'IN ADJUSTING A BLEND TO A SPECIFIC BRIX
30 DIM B(80)
40 DIM V(80)
50 DIM SPG(80)
60 INPUT "Final desired Brix";B
70 INPUT "Brix of component used to adjust Brix";B1
80 N=N+1
90 PRINT "Brix of component";N
100 INPUT B(N)
110 IF B(N)=0 THEN GOTO 200
120 PRINT "Volume of component";N;
130 INPUT V(N)
140 BB=B(N)
150 GOSUB 320
160 SPG(N)=SPG
170 Z=Z+V(N)*SPG(N)
180 V=V+V(N)
190 GOTO 80
200 N=N-1
210 BB=B1
220 GOSUB 320
230 SPG1=SPG
240 BB=B
250 GOSUB 320
260 V1=CINT((V*SPG-Z)/(SPG1-SPG))
270 PRINT " #"," BRIX","VOL"
280 FOR I=1 TO N
290 PRINT I,B(I),V(I)
300 NEXT
305 V=V+V1
310 PRINT "ADD VOLUME OF";V1;"OF";B1;"BRIX COMPONENT"
311 PRINT "TO GET VOLUME OF";V;"OF";B;"BRIX PRODUCT"
315 END
320 SPG=.0437691*BB*EXP((BB+330.872)^2/170435!)
330 RETURN


10 'THIS PROGRAM CALCULATES THE UNCORRECTED BRIX NEEDED FROM THE EVAPORA-
TOR IN ORDER TO ACHIEVE A DESIRED GPL IN PROCESSING LEMON CONCENTRATE
20 INPUT "Uncorrected Brix of inbound lemon single strength juice";UBJ
30 'IF REFRACT CORRECTS FOR TEMP, DISREGARD STEPS 40-60
40 INPUT "Temperature for Brix correction to lemon SSJ";T
50 BB=UBJ
60 GOSUB 340
70 INPUT "mls of NaOH titrated with SSJ";MLS
80 'IF NaOH NORMALITY IS ALWAYS 0.3125N THEN DISREGARD STEPS 90-110
90 INPUT "Normality of NaOH";N
100 IF N=0 THEN N=.3125
110 MLS=MLS*N/.3125
```

```
120 INPUT "Weight of SSJ sample (10.4g = 10mls)";W
130 ACID=MLS*2/W
140 AC=.014+.192*ACID-.00035*ACID^2
150 BJ=UBJ+AC+TC
160 INPUT "Desired GPL";GPL
170 INPUT "Temperature of concentrate for Brix correction";T
180 SPGJ=.0437691*BJ*EXP((BJ+330.872)^2/170435!)
190 DENSITYJ=.524484*EXP((BJ+330.872)^2/170435!)
200 SPGC=GPL*SPGJ/(10*ACID*DENSITYJ)
210 BF=10*SPGC
220 B=BF
230 BF=SPGC/(.0437691*EXP((B+330.872)^2/170435!))
240 IF ABS(B-BF)>.0001 THEN GOTO 210
250 DENSITYC=.524484*EXP((BF+330.872)^2/170435!)
260 ACIDCON=GPL/(10*DENSITYC)
270 ACCON=.014+.192*ACIDCON-.00035*ACIDCON^2
280 BB=BF
290 GOSUB 340
300 EVAPBRIX=CINT((BF-ACCON-TC)*10)
310 EVAPBRIX=EVAPBRIX/10
320 PRINT "Uncorrected Brix for evaporator =";EVAPBRIX
330 END
340 A=BB^2*(.0001425-8.605E-06*T+7.138E-08*T^2)
350 AA=BB*(-.02009+.001378*T-1.857E-05*T^2)
360 AAA=-.7788+.017*T+.0011*T^2
370 TC=A+AA+AAA


10 'THIS PROGRAM CALCULATES THE GPL OF LEMON CONCENTRATES
20 INPUT "Uncorrected Brix";UB
30 'IF REFRACT CORRECTS FOR TEMP, DISREGARD STEPS 40-80
40 INPUT "Temperature for Brix correction";T
50 A=UB^2*(.0001425-8.605E-06*T+7.138E-08*T^2)
60 AA=UB*(-.02009+.001378*T-1.857E-05*T^2)
70 AAA=-.7788+.017*T+.0011*T^2
80 TC=A+AA+AAA
90 INPUT "mls NaOH titrated";MLS
100 'IF THE NORMALITY IS ALWAYS 0.3125N, THEN DISREGARD STEPS 110-130
110 INPUT "Normality of NaOH";N
120 IF N=0 THEN N=.3125
130 MLS=MLS*N/.3125
140 INPUT "Weight of sample";W
150 ACID=MLS*2/W
160 AC=.014+.192*ACID-.00035*ACID^2
170 BRIX=UB+AC+TC
180 DENSITY=.524484*EXP((BRIX+330.872)^2/170435!)
190 GPL=CINT(ACID*DENSITY*10)
200 PRINT "GPL =";GPL
210 INPUT "Desired GPL";DGPL
220 INPUT "Volume of high GPL concentrate";V
230 WATER=CINT(V*(GPL-DGPL)/GPL)
```

```
240 PRINT "Water needed = ";WATER
250 TOTAVOL=CINT(V+WATER)
260 PRINT "Total volume = ";TOTALVOL

10 'THIS PROGRAM DETN'S THE AVERAGE B/A RATIO OF A BLEND
20 DIM B(80)
30 DIM R(80)
40 DIM V(80)
50 DIM SPG(80)
60 N=N+1
70 PRINT "Brix of component ";N;
80 INPUT B(N)
90 IF B(N)=0 THEN GOTO 190
100 PRINT "Ratio of component ";N;
110 INPUT R(N)
120 PRINT "Volume of component ";N;
130 INPUT V(N)
140 SPG(N)=.0437691*B(N)*EXP((B(N)+330.872)^2/170435!)
150 Z=Z+V(N)*SPG(N)
160 ZZ=ZZ+V(N)*SPG(N)/R(N)
170 V=V+V(N)
180 GOTO 60
190 N=N-1
200 PRINT " #"," BRIX", "RATIO", "VOL"
210 FOR I=1 TO N
220 PRINT I,B(I),R(I),V(I)
230 NEXT
240 R=Z/ZZ
250 PRINT "AVERAGE RATIO= ";R;TAB(30);"TOT VOL=";V

10 'THIS PROGRAM DETN'S THE VOL NEEDED FOR ONE COMPONENT
20 'IN ADJUSTING A BLEND TO A SPECIFIC B/A RATIO
30 DIM B(80)
40 DIM R(80)
50 DIM V(80)
60 DIM SPG(80)
70 INPUT "Final desired Ratio";R
80 INPUT "Brix of component used to adjust ratio";B1
90 INPUT "Ratio of component used to adjust ratio";R1
100 N=N+1
110 PRINT "Brix of component ";N;
120 INPUT B(N)
130 IF B(N)=0 THEN GOTO 240
140 PRINT "Ratio of component ";N;
150 INPUT R(N)
160 PRINT "Volume of component ";N;
170 INPUT V(N)
180 BB=B(N)
190 GOSUB 370
200 SPG(N) =SPG
```

```
210 Z=Z+V(N)*SPG(N)*(1/R-1/R(N))
220 V=V+V(N)
230 GOTO 100
240 N=N-1
250 BB=B1
260 GOSUB 370
270 SPG1=SPG
280 V1=CINT(Z/(SPG1*(1/R1-1/R)))
290 PRINT " #"," BRIX","RATIO","VOL"
300 FOR I=1 TO N
310 PRINT I,B(I),R(I),V(I)
320 NEXT
330 V=V+V1
340 PRINT "ADD VOLUME OF";V1;"OF";R1;"RATIO COMPONENT"
350 PRINT "TO GET VOLUME OF";V;"OF";R;"RATIO PRODUCT"
360 END
370 SPG=.0437691*BB*EXP((BB+330.872)^2/170435!)
380 RETURN


10 'THIS PROGRAM CALCULATES THE BRIX, ACID, AND RATIO FROM LABORATORY DATA
20 INPUT "Uncorrected Brix";UB
30 'IF AN AUTO TEMP CORRECTING REFRACT IS USED, DISREGARD STEPS 40-80
   AND ANY REFERENCES TO "TC"
40 INPUT "Temperature for Brix correction";T
50 A=UB^2*(.0001425-8.605E-06*T+7.138E-08*T^2)
60 AA=UB*(-.02009+.001378*T-1.857E-05*T^2)
70 AAA=-.7788+.017*T+.0011*T^2)
80 TC=A+AA+AAA
90 INPUT "mls of NaOH titrated";MLS
100 'IF THE NORMALITY OF THE NaOH IS ALWAYS 0.1562, DISREGARD STEPS 110-130
110 INPUT "Normality of NaOH";N
120 IF N=0 THEN N=.1562
130 MLS=MLS*N/.1562
140 INPUT "Weight of juice sample (10.5g = 10 mls SSJ)";W
150 ACID=MLS/W
160 AC=.014+.192*ACID-.00035*ACID^2
169 'THE FOLLOWING FIXES THE CORRECT DECIMAL PLACES
170 BRIX=CINT((UB+AC+TC)*10)
180 BRIX=BRIX/10
190 ACID=CINT(ACID*100)
200 ACID=ACID/100
210 RATIO=CINT((BRIX/ACID)*10)
220 RATIO=RATIO/10
230 PRINT BRIX;"Brix"
240 PRINT ACID;"% acid"
250 PRINT RATIO;"B/A ratio"
260 INPUT "Desired Brix";DB
270 INPUT "Volume of high Brix juice";V
280 SPGI=.0437691*BRIX*EXP((BRIX+330.872)^2/170435!)
290 SPGF=.0437691*DB*EXP((DB+330.872)^2/170435!)
300 WATER=CINT(V*(SPGI-SPGF)/SPGF)
```

```
310 PRINT "Water needed =";WATER
320 TOTALVOL=CINT(V+WATER)
330 PRINT "Total final volume =";TOTALVOL


10 'THIS PROGRAM CALCULATES THE AVERAGE BRIX AND RATIO OF A BLEND
20 DIM B(80)
30 DIM R(80)
40 DIM V(80)
50 DIM SPG(80)
60 N=N+1
70 PRINT "Brix of component ";N;
80 INPUT B(N)
90 IF B(N)=0 THEN GOTO 190
100 PRINT "Ratio of component ";N;
110 INPUT R(N)
120 PRINT "Volume of component ";N;
130 INPUT V(N)
140 SPG(N)=.0437691*B(N)*EXP((B(N)+330.872)^2/170435!)
150 Z=Z+V(N)*SPG(N)
160 ZZ=ZZ+V(N)*SPG(N)/R(N)
170 V=V+V(N)
180 GOTO 50
190 N=N-1
200 PRINT " #"," BRIX","RATIO","VOL"
210 FOR I=1 TO N
220 PRINT I,B(I),R(I),V(I)
230 NEXT
240 SPG=Z/V
250 RATIO=CINT(10*Z/ZZ)
260 RATIO=RATIO/10
270 BF=SPG*10
280 B=BF
290 BF=SPG/(.0437691*EXP((B+330.872)^2/170435!))
300 IF ABS(BF-B)<.0001 THEN GOTO 320
310 GOTO 280
320 BRIX=CINT(BF*10)
330 BRIX=BRIX/10
340 PRINT "AVG BRIX=";BRIX;SPC(3);"AVG RATIO=";RATIO;SPC(3);"TOT VOL=";V
350 INPUT "Desired Brix";DB
360 DSPG=.0437691*DB*EXP((DB+330.872)^2/170435!)
370 WATER=CINT(V*(SPG-DSPG)/DSPG)
380 PRINT WATER;" of water needed."
390 VT=CINT(WATER+V)
400 PRINT "Total volume =";VT
410 INPUT "Add more components (Y/N)";Q$
420 IF Q$="Y" THEN GOTO 60


10 'THIS PROGRAM CALCULATES FRUIT SAMPLE TEST RESULTS FOR ORANGES,
   GRAPEFRUIT, AND TANGERINES
20 INPUT "net weight of fruit";WF
30 INPUT "net with of juice";WJ
```

```
40 INPUT "Uncorrected Brix";UB
50 'IF REFRACT CORRECTS FOR TEMP DISREGARD STEPS 60-100
60 INPUT "Temperature for Brix correction";T
70 A=UB^2*(.0001425-8.605001E-06*T+7.138E-08*T^2)
80 AA=UB*(-.02009+.001378*T-1.857E-05*T^2)
90 AAA=-.7788+.017*T+.0011*T^2
100 TC=A+AA+AAA
110 INPUT "mls NaOH titrated";MLS
120 'IF YOU ALWAYS USE A NORMALITY OF 0.1562 DISREGARD STEPS
130-150
130 INPUT "Normality of NaOH";N
140 IF N=0 THEN N=.1562
150 MLS=MLS*N/.1562
160 'IF 10 MLS OF JUICE ARE USED DISREGARD STEPS 170-200
170 INPUT "Weight of juice sample (10.5g = 10mls)";W
180 IF W=0 THEN W=10.5
190 ACID=CINT((MLS/W)*100)/100
200 AC=.014+.192*ACID-.00035*ACID^2
210 BRIX=CINT((UB+AC+TC)*10)/10
220 RATIO=CINT((BRIX/ACID)*10)/10
230 SOL=CINT((20*WJ*.85*BRIX/WF)*10)/10
240 DENSITY=4.37691*EXP((BRIX+330.872)^2/170435!)
250 GAL=CINT((2000*WJ*.85/(WF*DENSITY))*10)/10
260 PEEL=CINT(2000*(WF-WJ)/WF)
270 PRINT BRIX;" Brix",ACID;" % acid",RATIO;" B/A ratio"
280 PRINT SOL;" lbs sol/ton", GAL;" gal/ton",PEEL;" lbs peel/ton"
290 INPUT "net weight of fruit in load";NW
300 TGAL=CINT(GAL*NW/2000)
310 TSOL=CINT(SOL*NW/2000)
320 TPEEL=CINT(PEEL*NW/2000)
330 PRINT TSOL;" lbs solid",TGAL;" gallons",TPEEL;" lbs peel"

10 'THIS PROGRAM CALCULATES LEMON OR LIME FRUIT SAMPLE TEST RESULTS
20 INPUT "net weight of fruit";WF
30 INPUT "net weight of juice";WJ
40 INPUT "Uncorrected Brix";UB
50 'IF REFRACT CORRECTS FOR TEMP DISREGARD STEPS 60-100
60 INPUT "Temperature for Brix correction";T
70 A=UB^2*(.0001425-8.605001E-06*T+7.138E-08*T^2)
80 AA=UB*(-.02009+.001378*T-1.857E-05*T^2)
90 AAA=-.7788+.017*T+.0011*T^2
100 TC=A+AA+AAA
110 INPUT "mls NaOH titrated";MLS
120 'IF YOU ALWAYS USE A NORMALITY OF 0.3125 DISREGARD STEPS 130-150
130 INPUT "Normality of NaOH";N
140 IF N=0 THEN N=.3125
150 MLS=MLS*N/.3125
160 'IF 10 MLS OF JUICE ARE USED DISREGARD STEPS 170-200
170 INPUT "Weight of juice sample (10.5g = 10mls)";W
180 IF W=0 THEN W=10.5
190 ACID=CINT((MLS*2/W)*100)/100
```

```
200 AC=.014+.192*ACID-.00035*ACID^2
210 BRIX=CINT((UB+AC+TC)*10)/10
220 DENSITY=4.37691*EXP((BRIX+330.872)^2/170435!)
230 APT=CINT((20*WJ*ACID*.78/WF)*10)/10
240 GPT=CINT((2000*WJ*.78/(WF*DENSITY))*10)/10
250 PEEL=CINT(2000*(WF-WJ)/WF)
260 PRINT
270 PRINT BRIX;"Brix";SPC(3);ACID;"% acid";SPC(3);APT;"lbs
acid/ton";SPC(3);GPT;"gal/ton"
280 PRINT
290 PRINT TAB(20);PEEL;"lbs peel/ton"
295 PRINT
300 INPUT "net weight of load";NW
310 TAPT=CINT(APT*NW/2000)
320 TGPT=CINT(GPT*NW/2000)
330 TPEEL=CINT(PEEL*NW/2000)
340 PRINT
350 PRINT TAPT;"lbs acid";SPC(3);TGPT;"gallons";SPC(3);TPEEL;"lbs peel"
360 PRINT


10 'THIS PROGRAM DETN'S THE AVG OIL AND OIL ADJUSTMENT OF A BLEND
20 DIM B(80)
25 DIM O(80)
30 DIM V(80)
35 DIM SPG(80)
50 N=N+1
60 PRINT "Brix of component ";N;
70 INPUT B(N)
75 IF B(N)=0 THEN GOTO 130
76 PRINT "% oil of component ";N;
77 INPUT O(N)
78 O(N)=CINT(1000*O(N))/1000
80 PRINT "Volume of component ";N;
90 INPUT V(N)
95 SPG(N)=.0437691*B(N)*EXP((B(N)+330.872)^2/170435!)
100 Z=Z+V(N)*SPG(N)
105 ZZ=ZZ+V(N)*SPG(N)*O(N)
110 V=V+V(N)
120 GOTO 50
130 N=N-1
135 PRINT " #"," BRIX"," OIL","VOL"
140 FOR I=1 TO N
150 PRINT I,B(I),O(I),V(I)
160 NEXT
170 S=Z/V
175 O=CINT(1000*ZZ/Z)/1000
180 BF=S*10
190 B=BF
200 BF=S/(.0437691*EXP((B+330.872)^2/170435!))
210 IF ABS(BF-B)<.0001 THEN GOTO 230
220 GOTO 190
```

```
230 BF=CINT(BF*10)/10
250 PRINT "AVG BRIX =";BF;SPC(3);"AVG OIL=";O;SPC(3);"TOT VOL=";V
255 INPUT "Do you want to add oil (Y/N)";Q$
256 IF Q$="N" THEN END
260 INPUT "Desired final oil level";OF
261 INPUT "Is volume in drums (52-gallon)(Y/N)";Q$
265 IF Q$="Y" THEN DD=52 ELSE DD=1
270 OA=CINT((OF-O)*36.19*S*V*DD)
280 PRINT "Oil needed=";OA
290 INPUT "mls oil enhancer needed per specification";OE
295 OI=OF-OE/(V*S*36.19*DD)
300 IF OA>=OE THEN PRINT "mls oil enhancer needed=";OA: END
310 INPUT "Can the total volume of the blend change (Y/N)";Q$
330 INPUT "Number of blend component you wish to exchange";NN
340 FOR J=1 TO N
350 ZZZ=ZZZ+SPG(J)*V(J)*(OI-O(J))
360 NEXT
370 ZZZ=ZZZ-SPG(NN)*V(NN)*(OI-O(NN))
375 O1=CINT((OI+ZZZ/(SPG(NN)*V(NN)))*1000)/1000
380 IF Q$="N" THEN PRINT "% oil needed in component";NN;"=";O1
390 V1=CINT(ZZZ/(O(NN)*SPG(NN)*(1-OI/O(NN))))
400 IF Q$="Y" THEN PRINT "Volume needed in component";NN;"=";V1
410 ZZZ=0
420 GOTO 310


10 'THIS PROGRAM DETN'S THE AVG PULP AND PULP ADJUSTMENT OF A BLEND
20 DIM B(80)
25 DIM P(80)
30 DIM V(80)
35 DIM SPG(80)
50 N=N+1
60 PRINT "Brix of component ";N;
70 INPUT B(N)
75 IF B(N)=0 THEN GOTO 130
76 PRINT "% pulp of component ";N;
77 INPUT P(N)
78 P(N)=CINT(1000*O(N))/1000
80 PRINT "Volume of component ";N;
90 INPUT V(N)
95 SPG(N)=.0437691*B(N)*EXP((B(N)+330.872)^2/170435!)
100 Z2=Z2+V(N)*SPG(N)
105 Z1=Z1+V(N)*SPG(N)*P(N)
120 GOTO 50
130 N=N-1
135 PRINT " #"," BRIX"," PULP","VOL"
140 FOR I=1 TO N
150 PRINT I,B(I),P(I),V(I)
160 NEXT
170 PAVG=Z1/Z2
250 PRINT "AVG PULP=";PAVG;SPC(3);"TOT VOL=";V
310 INPUT "Can the total volume of the blend change (Y/N)";Q$
```

```
330 INPUT "Number of blend component you wish to exchange";NN
340 FOR J=1 TO N
350 Z3=Z3+SPG(J)*V(J)*(PAVG-P(J))
360 NEXT
370 Z3=Z3-SPG(NN)*V(NN)*(PAVG-P(NN))
375 PNN=CINT((PAVG+Z3/(SPG(NN)*V(NN)))*1000)/1000
380 IF Q$="N" THEN PRINT "% pulp needed in component";NN;"=";PNN
390 VNN=CINT(Z3/(P(NN)*SPG(NN)*(1-PAVG/P(NN))))
400 IF Q$="Y" THEN PRINT "Volume needed in component";NN;"=";VNN
410 Z3=0
420 GOTO 310


10 'THIS PROGRAM DETN'S THE AVERAGE LIMONIN AND LIMONIN ADJUSTMENT
OF A BLEND
20 DIM B(80)
25 DIM L(80)
30 DIM V(80)
35 DIM D(80)
50 N=N+1
60 PRINT "Brix of component ";N;
70 INPUT B(N)
75 IF B(N)=0 THEN GOTO 130
76 PRINT "limonin of component ";N;
77 INPUT L(N)
78 P(N)=CINT(1000*O(N))/1000
80 PRINT "Volume of component ";N;
90 INPUT V(N)
95 D(N)=4.37691*EXP((B(N)+330.872)^2/170435!)
100 Z2=Z2+V(N)*D(N)
105 Z1=Z1+V(N)*D(N)*L(N)
120 GOTO 50
130 N=N-1
135 PRINT " #"," BRIX"," LIM ","VOL"
140 FOR I=1 TO N
150 PRINT I,B(I),LI,V(I)
160 NEXT
170 LAVG=Z1/Z2
250 PRINT "AVG LIMONIN=";LAVG;SPC(3);"TOT VOL=";V
310 INPUT "Can the total volume of the blend change (Y/N)";Q$
330 INPUT "Number of blend component you wish to exchange";NN
340 FOR J=1 TO N
350 Z3=Z3+D(J)*V(J)*(LAVG-L(J))
360 NEXT
370 Z3=Z3-D(NN)*V(NN)*(LAVG-L(NN))
375 LNN=CINT((LAVG+Z3/(D(NN)*V(NN)))*1000)/1000
380 IF Q$="N" THEN PRINT "limonin needed in component";NN;"=";LNN
390 VNN=CINT(Z3/(L(NN)*D(NN)*(1-LAVG/L(NN))))
400 IF Q$="Y" THEN PRINT "Volume needed in component";NN;"=";VNN
410 Z3=0
420 GOTO 310
```

Appendix C

HP-41C Programs

The following HP-41C or HP-41CV programs can best be used by assigning the programs to convenient keys using the ASN function. Then the program will activate when the calculator is in the USER mode. The programs are designed for the calculator to be alloted 53 storage registers using the SIZE function (SIZE 053). Included with each program is a brief explanation of how to use the program. It is assumed that the user is familer with the HP-41C program language. Most programs are self-explanatory. You enter the data when prompted. Consult the text and relevant flow charts for program logic and the objectives of the programs.

PROGRAM TO ADJUST BRIX USING WATER OR HIGHER BRIX CONCENTRATE

This program will calculate the parameters needed to adjust a full tank of juice to the proper Brix by adding either water or higher Brix concentrate. If no entry is made for the needed Brix, the program defaults to 11.8°Brix. If the prompt does not apply, enter nothing.

1. LBL HILOBX	16. PROMPT	31. RCL 46
2. 11.8	17. STO 47	32. −
3. BRIX NEEDED?	18. RCL 44	33. /
4. TONE 5	19. XEQ 02	34. RCL 47
5. PROMPT	20. STO 44	35. X
6. STO 44	21. RCL 45	36. STO 44
7. HI BRIX?	22. XEQ 02	37. FIX 0
8. PROMPT	23. STO 45	38. LBL 03
9. STO 45	24. RCL 46	39. GAL HI=
10. 0	25. XEQ 02	40. ARCL 44
11. LOW BRIX?	26. STO 46	41. TONE 5
12. PROMPT	27. RCL 44	42. T0NE 7
13. STO 46	28. RCL 46	43. TONE 4
14. TOT GAL?	29. −	44. TONE 6
15. LBL 01	30. RCL 45	45. PROMPT

430

46. RCL 47
47. RCL 44
48. −
49. STO 45
50. RCL 46
51. X=0?
52. GTO 04
53. GAL LO=
54. ARCL 45
55. PROMPT
56. GTO 03
57. LBL 04
58. WATER=
59. ARCL 45
60. PROMPT
61. GTO 03
62. LBL 02
63. STO 00
64. 330.872
65. +
66. X^2
67. 170435
68. /
69. E^x
70. .0437691
71. X
72. RCL 00
73. X
74. END

PROGRAM THAT CALCULATES THE BRIX, ACID, AND RATIO FROM LAB DATA

This program will calculate the Brix, % acid, and B/A ratio from laboratory results. No entry when you are prompted for the temperature causes the program to default to 20°C or to apply no temperature correction to the Brix. If nothing is entered for the normality, the default is $0.1562N$. This can be changed in the program to $0.3125N$ if desired. If no entry is made when you are prompted for the sample weight, a weight of 10.5 g (10 ml of SSJ) will be assumed.

1. LBL BAR
2. UNCOR BRIX?
3. TONE 5
4. PROMPT
5. STO 44
6. 0
7. TEMP-DEG C?
8. PROMPT
9. X=0?
10. GTO 01
11. STO 48
12. X^2
13. STO 49
14. 7.138 E-8
15. X
16. RCL 48
17. 8.605 E-6
18. X
19. −
20. 1.425 E-4
21. +
22. RCL 44
23. X^2
24. X
25. RCL 49
26. 1.857 E-5
27. X
28. RCL 48
29. 1.378 E-3
30. X
31. −
32. CHS
33. .02009
34. −
35. RCL 44
36. X
37. +
38. RCL 49
39. .0011
40. X
41. RCL 48
42. .017
43. X
44. +
45. .7788
46. −
47. +
48. ST + 44
49. LBL 01
50. MLS NaOH?
51. PROMPT
52. STO 48
53. .1562
54. NaOH NORMAL?
55. PROMPT
56. RCL 48
57. X
58. .1562
59. /
60. STO 48
61. 10.5
62. WT JUC SAMP?
63. PROMPT
64. RCL 48
65. /
66. 1/X
67. STO 45
68. 11.7
69. X>Y?
70. GTO 03
71. X<>?
72. X^2

73. 1.9537 E-4
74. X
75. RCL 45
76. .184836815
77. X
78. −
79. .085917569
80. −
81. CHS
82. ST + 44
83. GTO 02
84. LBL 03
85. X < > Y?
86. .19

87. X
88. .01267
89. +
90. ST + 44
91. LBL 02
92. FIX 1
93. BRIX=
94. ARCL 44
95. TONE 6
96. TONE 7
97. TONE 4
98. TONE 5
99. PROMPT
100. FIX 2

101. % ACID=
102. ARCL 45
103. PROMPT
104. RCL 44
105. RCL 45
106. /
107. STO 43
108. FIX 1
109. RATIO=
110. ARCL 43
111. PROMPT
112. GOTO 02
113. END

SET OF PROGRAMS THAT ALLOWS ONE TO ENTER, DELETE, CORRECT, AND VIEW BLEND COMPONENTS AND CALCULATE THE AVERAGE BRIX AND RATIO

Each program below should be assigned a USER key. The BLD G?D program sets the volume mode of the blend. Gallons mode assumes all volumes are in gallons. Gal/drums mode assumes that all volumes less than 52 represent the number of 52-gallon drums, and all volumes 52 or over are gallons. For example, an entry of 45 would mean 45 drums of concentrate, whereas 58 would be taken as 58 gallons of concentrate. This allows the blending from bulk tanks and drums simultaneously without conversion of one quantity to the volume units of the other. Each component is numbered and appears in the prompts and displays. In the view program, if you enter nothing when prompted for the number of the blend component desired, the default is the first component. The calculator then will automatically step through the entered blend components for comparison to a blend manifest. You can add, delete, correct, or view blend components at any time by using the separate assigned USER keys. The blend calculation takes a few minutes. The calculation program also will prompt for the desired Brix and will calculate the water needed in the blend, as well as the resulting final volume of the blend.

Entering Blend
Component Data
1. LBL BLD ENT
2. FS?C 01
3. GTO 03
4. FS? 03
5. GTO 04
6. CLRG
7. LBL 03
8. 1
9. ST + 41
10. LBL 04

11. ^{T}(BLANK)
12. ASTO 51
13. FX?02
14. GTO 02
15. FIX O
16. GAL/DRUMS?
17. LBL 01
18. TONE 3
19. ARCL 51
20. ARCL 41
21. PROMPT
22. BRIX?

23. ARCL 51
24. ARCL 41
25. PROMPT
26. 1 E2
27. /
28. +
29. RATIO?
30. ARCL 51
31. ARCL 41
32. PROMPT
33. 1 E5
34. /

35. +
36. FS?C 03
37. RTN
38. STO IND 41
39. GTO 03
40. LBL 02
41. FIX O
42. GALLONS?
43. GTO 01
44. END

*Sets Gallons or
Drum/Gallons Mode*

1. LBL BLD G?D
2. FS? 02
3. GTO 01
4. SF 02
5. GALLONS MODE
6. TONE 9
7. PROMPT
8. GTO BLD G?D
9. LBL 01
1O. CF 02
11. GAL/DRM MODE
12. TONE 3
13. PROMPT
14. GTO BLD G?D
15. END

*Add or Delete Blend
Component Data*

1. LBL BLD ADD
2. 1
3. ST − 41
4. SF 01
5. GTO BLD ENT
6. LBL BLD DEL
7. ITEM NO?
8. PROMPT
9. STO 42
10. STO 43
11. LBL 01
12. 1
13. ST + 43
14. RCL IND 43
15. X=0?

16. GTO 02
17. STO IND 42
18. 1
19. ST + 42
20. GTO 01
21. LBL 02
22. 1
23. ST − 41
24. 0
25. STO IND 42
26. STO 42
27. SF 01
28. GTO BLD VU
29. END

*Correcting Blend
Component Data*

1. LBL BLD COR
2. ITEM NO?
3. PROMPT
4. RCL 41
5. STO 42
6. X<>Y
7. STO 41
8. SF 03
9. XEQ BLD ENT
10. STO IND 41
11. RCL 41
12. STO 43
13. RCL 42
14. STO 41
15. RCL 43
16. STO 42
17. SF 01
18. 1
19. ST − 42
20. GTO BLD VU
21. END

*Viewing Blend
Components*

1. LBL BLD VU
2. FS?C 01
3. GTO 01
4. 1
5. START NO?

6. TONE 3
7. PROMPT
8. 1
9. −
10. STO 42
11. LBL 01
12. 1
13. ST + 42
14. RCL IND 42
15. INT
16. STO 43
17. X=0?
18. GTO 02
19. RCL IND 42
20. FRC
21. 1 E3
22. X
23. INT
24. 10
25. /
26. STO 44
27. RCL IND 42
28. 1 E3
29. X
30. FRC
31. 1 E2
32. X
33. STO 45
34. $^{T}-$
35. ASTO 46
36. ^{T},
37. ASTO 47
38. CLA
39. FIX 0
40. ARCL 42
41. ARCL 46
42. ARCL 43
43. ARCL 51
44. FIX 1
45. ARCL 44
46. ARCL 47
47. ARCL 45
48. TONE 7
49. AVIEW
50. PSE
51. GTO 01

52. LBL 02
53. NO MORE DATA
54. PROMPT
55. GTO BLD VU
56. END

*Calculation of Avg Brix
and Ratio*

1. LBL BLD CAL
2. 0
3. STO 42
4. STO 44
5. STO 46
6. STO 47
7. CF 01
8. LBL 01
9. 1
10. ST + 42
11. RCL 41
12. 1
13. −
14. RCL 42
15. X > Y?
16. GTO 04
17. RCL IND 42
18. INT
19. FS? 02
20. GTO 03
21. 52
22. X < > Y
23. X < = Y?
24. XEQ 02
25. LBL 03
26. STO 43
27. ST + 44
28. RCL IND 42
29. FRC
30. 1 E3
31. X
32. INT
33. 10
34. /
35. STO 00
36. 330.872
37. +
38. X^2

39. 170435
40. /
41. E^x
42. .0437691
43. X
44. RCL 00
45. X
46. STO 45
47. RCL 43
48. X
49. ST + 46
50. RCL IND 42
51. 1 E3
52. X
53. FRC
54. 1 E2
55. X
56. RCL 43
57. /
58. 1/X
59. RCL 45
60. X
61. ST + 47
62. GTO 01
63. LBL 04
64. 0
65. STO 42
66. RCL 46
67. RCL 44
68. /
69. STO 48
70. LBL 05
71. RCL 00
72. 330.872
73. +
74. X^2
75. 170435
76. /
77. E^x
78. .0437691
79. X
80. RCL 48
81. /
82. 1/X
83. STO 49
84. RCL 00

85. −
86. ABS
87. .1
88. X > Y?
89. GTO 06
90. RCL 49
91. STO 00
92. GTO 05
93. LBL 06
94. FIX 1
95. RCL 49
96. FIX 0
97. CLA
98. ARCL 44
99. ARCL 51
100. FIX 1
101. ARCL 49
102. ARCL 51
103. FS?C 05
104. GTO 07
105. RCL 47
106. 1/X
107. RCL 44
108. X
109. RCL 48
110. X
111. STO 47
112. LBL 07
113. ARCL 47
114. BEEP
115. PROMPT
116. 0
117. BRIX NEEDED?
118. PROMPT
119. X=0?
120. 60
121. STO 00
122. 330.872
123. +
124. X^2
125. 170435
126. /
127. E^x
128. .0437691
129. X
130. RCL 00

131. X	145. RCL 45	159. FRC
132. STO 50	146. RCL 44	160. 1 E2
133. RCL 48	147. +	161. X
134. –	148. STO 45	162. X=0?
135. CHS	149. TOT GAL=	163. GTO 08
136. RCL 50	150. ARCL 45	164. RCL 43
137. /	151. PROMPT	165. X
138. RCL 44	152. SF 02	166. GTO 03
139. X	153. GTO 06	167. LBL 08
140. STO 45	154. LBL 02	168. RCL 43
141. WATER=	155. STO 43	169. 52
142. FIX 0	156. RCL IND 42	170. X
143. ARCL 45	157. 1 E6	171. END
144. PROMPT	158. X	

FRUIT SAMPLE CALCULATIONS

This program can be used to calculate fruit sample data for oranges, tangerines, grape-fruit, or other citrus fruit except lemons and limes. The default temperature for the temperature correction to the Brix is, again, 20°C or no temperature correction. The default normality is 0.1562, and the default sample weight is 10.5 g (10 ml SSJ).

1. LBL ORG FRT	24. X	47. RCL 48
2. WT OF FRUIT?	25. –	48. .017
3. TONE 9	26. 1.425 E-4	49. X
4. PROMPT	27. +	50. +
5. STO 42	28. RCL 44	51. .7788
6. WT OF JUICE?	29. X^2	52. –
7. PROMPT	30. X	53. +
8. STO 50	31. RCL 49	54. ST + 44
9. UNCOR. BRIX?	32. 1.857 E-5	55. LBL 03
10. PROMPT	33. X	56. MLS NaOH?
11. STO 44	34. RCL 48	57. PROMPT
12. 0	35. 1.378 E-3	58. STO 48
13. TEMP-DEG C?	36. X	59. 10.5
14. PROMPT	37. –	60. WT SAMPLE?
15. X=0?	38. CHS	61. PROMPT
16. GTO 03	39. .02009	62. RCL 48
17. STO 48	40. –	63. /
18. X^2	41. RCL 44	64. 1/X
19. STO 49	42. X	65. STO 45
20. 7.138 E-8	43. +	66. .19
21. X	44. RCL 49	67. X
22. RCL 48	45. .0011	68. .01267
23. 8.605 E-6	46. X	69. +

70. ST + 44
71. LBL 02
72. FIX 1
73. BRIX=
74. ARCL 44
75. TONE 6
76. TONE 6
77. TONE 7
78. TONE 9
79. PROMPT
80. FIX 2
81. % ACID=
82. ARCL 45
83. PROMPT
84. RCL 44
85. RCL 45
86. /
87. STO 43
88. FIX 1
89. RATIO=
90. ARCL 43
91. PROMPT
92. XEQ 01
93. 100
94. X
95. 1/X
96. RCL 50
97. X
98. RCL 42
99. /
100. 2000
101. X
102. .85
103. X
104. STO 46
105. FIX 2
106. GAL/TN=
107. ARCL 46
108. PROMPT
109. XEQ 01
110. RCL 44
111. X
112. RCL 46
113. X
114. STO 47
115. SOL/TN=
116. ARCL 47
117. PROMPT
118. RCL 42
119. RCL 50
120. −
121. RCL 42
122. /
123. 2 E3
124. X
125. FIX 0
126. STO 48
127. PEEL/TN=
128. ARCL 48
129. PROMPT
130. GTO 02
131. LBL 01
132. RCL 44
133. 330.872
134. +
135. X^2
136. 170435
137. /
138. E^x
139. .043769
140. X
141. END

CALCULATION OF AVERAGE OIL OR PULP

This program will calculate the average % oil or % pulp in a blend, as the two calculations are similar. The difference between oil and pulp is determined by the magnitude of their values. The Brix of each component must be entered, as well as the % oil or pulp. The volume prompt assumes that all units are the same (all drums or all gallons). When all the components have been entered, enter nothing, press R/S, and the weighted average % oil or pulp will appear.

1. LBL AVG O/P
2. 0
3. STO 42
4. STO 43
5. STO 44
6. LBL 01
7. 1
8. ST + 44
9. FIX 0
10. BRIX?
11. ARCL 44
12. TONE 9
13. PROMPT
14. X=0?
15. GTO 02
16. 1
17. X<>Y
18. X=Y?
19. GTO 04
20. XEQ 03
21. STO 45
22. LBL 04
23. % OIL/PLP?
24. ARCL 44
25. PROMPT
26. DRMS/GALS?
27. ARCL 44
28. PROMPT
29. RCL 45
30. X
31. ST + 42
32. X
33. ST + 43
34. GTO 01
35. LBL 02
36. FIX 3

37. RCL 43	46. ARCL 00	55. 170435
38. RCL 42	47. BEEP	56. /
39. /	48. PROMPT	57. E^x
40. .1	49. GTO AVG O/P	58. .0437691
41. X< =Y?	50. LBL 03	59. X
42. FIX 1	51. STO 00	60. RCL 00
43. X< >Y	52. 330.872	61. X
44. STO 00	53. +	62. END
45. WT AVG =	54. X^2	

LEMON/LIME FRUIT SAMPLE CALCULATIONS

This program calculates the fruit sample information for lemons and limes. The default for temperature is, again, 20°C; the default for the NaOH normality is 0.1562; and the default for the sample weight is 10.5 g.

1. LBL LEM FRT	30. X	59. .1562
2. WT OF FRUIT?	31. RCL 49	60. NaOH NORMAL?
3. TONE 8	32. 1.857 E-5	61. PROMPT
4. PROMPT	33. X	62. RCL 48
5. STO 47	34. RCL 48	63. X
6. WT OF JUICE?	35. 1.378 E-3	64. .1562
7. PROMPT	36. X	65. /
8. STO 46	37. −	66. STO 48
9. UNCOR. BRIX?	38. CHS	67. 10.5
10. PROMPT	39. .02009	68. WT SAMPLE?
11. STO 44	40. −	69. PROMPT
12. 0	41. RCL 44	70. RCL 48
13. TEMP-DEG C?	42. X	71. /
14. PROMPT	43. +	72. 1/X
15. X=0?	44. RCL 49	73. STO 45
16. GTO 01	45. .0011	74. 11.7
17. STO 48	46. X	75. X>Y?
18. X^2	47. RCL 48	76. GTO 03
19. STO 49	48. .017	77. X< >Y
20. 7.138 E-8	49. X	78. X^2
21. X	50. +	79. 1.9537 E-4
22. RCL 48	51. .7788	80. X
23. 8.605 E-6	52. −	81. RCL 45
24. X	53. +	82. .184836815
25. −	54. ST + 44	83. X
26. 1.425 E-4	55. LBL 01	84. −
27. +	56. MLS NaOH?	85. .085917569
28. RCL 44	57. PROMPT	86. −
29. X^2	58. STO 48	87. CHS

88. ST + 44	112. X	136. RCL 49
89. GTO 02	113. RCL 47	137. /
90. LBL 03	114. /	138. 1/X
91. X < > Y	115. 20	139. STO 48
92. .19	116. X	140. FIX 2
93. X	117. .78	141. GAL/TON=
94. .01267	118. X	142. ARCL 48
95. +	119. STO 49	143. PROMPT
96. ST + 44	120. LBL 05	144. RCL 47
97. LBL 02	121. FIX 2	145. RCL 46
98. FIX 1	122. ACID/TN=	146. −
99. BRIX=	123. ARCL 49	147. RCL 47
100. ARCL 44	124. PROMPT	148. /
101. TONE 7	125. RCL 44	149. 2000
102. TONE 9	126. 330.872	150. X
103. TONE 6	127. +	151. FIX 0
104. TONE 7	128. X^2	152. STO 50
105. PROMPT	129. 170435	153. PEEL/TN=
106. FIX 2	130. /	154. ARCL 50
107. % ACID=	131. E^x	155. PROMPT
108. ARCL 45	132. .0437691	156. GTO 02
109. PROMPT	133. X	157. END
110. RCL 45	134. RCL 45	
111. RCL 46	135. X	

CALCULATION OF UNCORRECTED BRIX NEEDED DURING EVAPORATION OF LEMON/LIME CONCENTRATES IN ORDER TO ACHIEVE A DESIRED GPL

This program will calculate the uncorrected Brix needed by evaporator operators in order to concentrate lemon or lime juices to a desired GPL level. The temperature, normality, and sample weight defaults are as before—20°C, 0.1562N, and 10.5 g, respectively. The default for the desired GPL is 400 GPL. If you want to subtract an acid correction from the Brix observed by the evaporator operator, then enter the temperature of the concentrate observed by the refractometer.

1. LBL LEM BRX	10. GTO 01	19. RCL 48
2. LEM SSJ BRX?	11. XEQ 10	20. X
3. TONE 6	12. LBL 01	21. .1562
4. PROMPT	13. MLS NaOH?	22. /
5. STO 44	14. PROMPT	23. STO 48
6. 0	15. STO 48	24. 10.5
7. TEMP-DEG C?	16. .1562	25. WT SAMPLE?
8. PROMPT	17. NaOH NORMAL?	26. PROMPT
9. X=0?	18. PROMPT	27. RCL 48

28. /
29. 1/X
30. STO 45
31. 11.7
32. X > Y?
33. GTO 03
34. X < > Y
35. X^2
36. 1.9537 E-4
37. X
38. RCL 45
39. .184836815
40. X
41. −
42. .085917569
43. −
44. CHS
45. ST + 44
46. GTO 02
47. LBL 03
48. X < > Y
49. .19
50. X
51. .01267
52. +
53. ST + 44
54. LBL 02
55. 400
56. DESIRED GPL?
57. PROMPT
58. 10
59. /
60. STO 47
61. RCL 44
62. STO 00
63. XEQ 05
64. STO 42
65. 100
66. X
67. 453.59237
68. X
69. 3785.306
70. /
71. RCL 45
72. X
73. RCL 47

74. /
75. 1/X
76. RCL 42
77. RCL 44
78. X
79. X
80. STO 43
81. 10
82. X
83. STO 00
84. LBL 07
85. XEQ 05
86. RCL 43
87. /
88. 1/X
89. STO 50
90. RCL 00
91. −
92. ABS
93. .1
94. X > Y?
95. GTO 06
96. RCL 50
97. STO 00
98. GTO 07
99. LBL 06
100. RCL 50
101. STO 00
102. XEQ 05
103. 100
104. X
105. 453.59237
106. X
107. 3785.306
108. /
109. RCL 47
110. /
111. 1/X
112. 11.7
113. X > Y?
114. GTO 08
115. X < > Y
116. STO 46
117. X^2
118. 1.9537 E-4
119. X

120. RCL 46
121. .184836815
122. X
123. −
124. .085917569
125. −
126. CHS
127. GTO 04
128. LBL 08
129. X < > Y
130. .19
131. X
132. .01267
133. +
134. LBL 04
135. ST − 50
136. 0
137. CONC TEMP?
138. PROMPT
139. X = 0?
140. GTO 09
141. XEQ 10
142. LBL 09
143. FIX 1
144. ST − 50
145. EVAP BRIX =
146. ARCL 50
147. TONE 7
148. TONE 9
149. TONE 8
150. TONE 9
151. PROMPT
152. GTO LEM BRX
153. LBL 05
154. RCL 00
155. 330.872
156. +
157. X^2
158. 170435
159. /
160. E^x
161. .0437691
162. X
163. RTN
164. LBL 10
165. STO 48

166. X^2	179. RCL 49	192. RCL 49
167. STO 49	180. 1.857 E-3	193. .0011
168. 7.138 E-8	181. X	194. X
169. X	182. RCL 48	195. RCL 48
170. RCL 48	183. 1.378 E-3	196. .017
171. 8.605 E-6	184. X	197. X
172. X	185. −	198. +
173. −	186. CHS	199. .7788
174. 1.425 E-4	187. .02009	200. −
175. +	188. −	201. +
176. RCL 44	189. RCL 44	202. ST + 44
177. X^2	190. X	203. END
178. X	191. +	

CALCULATION OF LEMON/LIME GPL

This program calculates the GPL from Brix and acid measurements. The temperature, normality, sample weight, and desired GPL defaults are 20°C, 0.3125N, 10.5 g, and 400 GPL, respectively. Any volumes can be used, but you must use the same units throughout the calculation. The program also calculates the water needed to adjust high GPL concentrates or juices.

1. LBL LEM GPL	24. X	47. +
2. UNCOR. BRIX?	25. RCL 49	48. ST + 44
3. TONE 8	26. 1.857 E-3	49. LBL 01
4. PROMPT	27. X	50. MLS NaOH?
5. STO 44	28. RCL 48	51. PROMPT
6. 0	29. 1.378 E-3	52. STO 48
7. TEMP-DEG C?	30. X	53. .3125
8. PROMPT	31. −	54. NaOH NORMAL?
9. X=0?	32. CHS	55. PROMPT
10. GTO 01	33. .02009	56. RCL 48
11. STO 48	34. −	57. X
12. X^2	35. RCL 44	58. .1562
13. STO 49	36. X	59. /
14. 7.138 E-8	37. +	60. STO 48
15. X	38. RCL 49	61. 10.5
16. RCL 48	39. .0011	62. WT SAMPLE?
17. 8.605 E-6	40. X	63. PROMPT
18. X	41. RCL 48	64. RCL 48
19. −	42. .017	65. /
20. 1.425 E-4	43. X	66. 1/X
21. +	44. +	67. STO 45
22. RCL 44	45. .7788	68. 11.7
23. X^2	46. −	69. X>Y?

70. GTO 02
71. X < > Y
72. X^2
73. 1.9537 E-4
74. X
75. RCL 45
76. .184836815
77. X
78. −
79. .085917569
80. −
81. CHS
82. ST + 44
83. GTO 03
84. LBL 02
85. X < > Y
86. .19
87. X

88. .01267
89. +
90. ST + 44
91. LBL 03
92. RCL 44
93. 330.872
94. +
95. X^2
96. 170435
97. /
98. E^x
99. .0437691
100. X
101. 100
102. X
103. 453.59237
104. X
105. 3785.306

106. /
107. RCL 45
108. X
109. 10
110. X
111. FIX 0
112. STO 42
113. LEM GPL=
114. ARCL 42
115. TONE 6
116. TONE 4
117. TONE 3
118. TONE 6
119. PROMPT
120. GTO LEM GPL
121. END

CALCULATION OF WATER NEEDED TO DILUTE JUICE TO A DESIRED BRIX

This program is convenient when dilution to a desired Brix is all that is wanted. The default desired Brix is 11.8°Brix. Any units of volume can be used as long as they are consistent.

1. LBL ORG DIL
2. HIGH BRIX?
3. TONE 7
4. PROMPT
5. XEQ 01
6. STO 48
7. 11.8
8. NEEDED BRIX?
9. PROMPT
10. XEQ 01
11. STO 43
12. RCL 48
13. −
14. CHS
15. RCL 43
16. /
17. VOL HI BRIX?

18. PROMPT
19. STO 47
20. X
21. STO 44
22. WATER=
23. ARCL 44
24. TONE 6
25. TONE 4
26. TONE 4
27. TONE 6
28. PROMPT
29. RCL 47
30. RCL 44
31. +
32. STO 45
33. TOT VOL=
34. ARCL 45

35. PROMPT
36. GTO ORG DIL
37. LBL 01
38. STO 00
39. 330.872
40. +
41. X^2
42. 170435
43. /
44. E^x
45. .0437691
46. X
47. RCL 00
48. X
49. END

GENERATION OF VOLUME TABLE FOR BULK TANKS

This program will generate data to construct a volume table for a particular bulk tank. After entering the needed tank dimensions (see text for greater details), you enter the distances from the hatch to the juice level, and the equivalent volume of the tank will appear. The hatch depth is the distance from the lower lip of the hatch to the edge of the tank.

1. LBL TANKS	39. RCL 46	77. /
2. TANK SIZE	40. RCL 45	78. X ≠ 0?
3. AVIEW	41. +	79. X < > Y
4. PSE	42. X > Y?	80. RCL 44
5. DIAMETER-IN?	43. GTO 02	81. X^2
6. TONE 2	44. RCL 50	82. PI
7. PROMPT	45. RCL 46	83. X
8. 2	46. −	84. STO 00
9. /	47. RCL 45	85. RCL 46
10. STO 44	48. −	86. X
11. SLOPE HT-IN?	49. RCL 48	87. 231
12. PROMPT	50. X	88. /
13. STO 45	51. RCL 47	89. +
14. CYLINDER HT?	52. X ≠ 0?	90. RCL 00
15. PROMPT	53. /	91. RCL 45
16. STO 46	54. X ≠ 0?	92. X
17. HATCH HT-IN?	55. X < > Y	93. 2
18. PROMPT	56. RCL 44	94. /
19. STO 47	57. −	95. 231
20. HATCH DEPTH?	58. CHS	96. /
21. PROMPT	59. ENTER	97. +
22. STO 48	60. ENTER	98. STO 00
23. JUICE LEVEL?	61. X	99. GTO 05
24. PROMPT	62. X	100. LBL 01
25. LBL 07	63. RCL 44	101. 2
26. STO 49	64. X^2	102. /
27. RCL 45	65. RCL 44	103. RCL 50
28. RCL 46	66. X	104. X > Y?
29. +	67. −	105. GTO 03
30. RCL 47	68. CHS	106. −1
31. +	69. RCL 47	107. STO 51
32. −	70. X	108. GTO 04
33. CHS	71. PI	109. LBL 03
34. STO 50	72. X	110. 1
35. RCL 45	73. 693	111. STO 51
36. X > Y?	74. /	112. LBL 04
37. GTO 01	75. RCL 48	113. RCL 50
38. X < > Y	76. X ≠ 0?	114. RCL 45

115. /
116. STO 43
117. 2
118. X
119. 1
120. −
121. CHS
122. RCL 51
123. X
124. RAD
125. ASIN
126. DEG
127. RCL 51
128. X
129. CHS
130. 1
131. RCL 43
132. −
133. RCL 43
134. X
135. SQRT
136. RCL 43
137. 2
138. X
139. 1
140. −

141. X
142. 2
143. X
144. +
145. PI
146. 2
147. /
148. +
149. RCL 44
150. X^2
151. X
152. RCL 50
153. X
154. 462
155. /
156. STO 00
157. GTO 05
158. LBL 02
159. RCL 44
160. X^2
161. PI
162. X
163. STO 00
164. RCL 45
165. X
166. 2

167. /
168. RCL 50
169. RCL 45
170. −
171. RCL 00
172. X
173. +
174. 231
175. /
176. STO 00
177. LBL 05
178. RCL 00
179. .5
180. X< =Y?
181. GTO 06
182. 0
183. STO 00
184. LBL 06
185. FIX 0
186. GALS=
187. ARCL 00
188. TONE 9
189. PROMPT
190. GTO 07
191. END

Appendix D

Answers to Select Questions and Problems

Chapter 2

Questions

3. No.
5. Temperature, acid level, lab technician or techniques, refractometer calibration, wavelength or color of light source, sample concentration.

Problems

3.

	0.76% acid	3.83% acid	24.76% acid	37.86% acid
EQ1-2	0.16	0.74	4.55	6.78
EQ1-3	0.16	0.75	4.55	6.75
TBL-1	0.15	0.75	4.53	not on table

4. 11.5°Brix, 10.6 B/A ratio.
7. 92 gallons of 64.5°Brix concentrate.
9. 659 gallons of 11.9°Brix cutback juice.

Chapter 3

Questions

2. The mitochondria of the juice cell.
7. Yes.

Problems

4. 4.82% acid
7. 60.6°Brix
10. 245 gallons of the 398 GPL concentrate, 3146 total gallons.

Chapter 4

Questions

3. Taste is caused by electrical disturbances on the tongue, usually by ionic or polar compounds or compounds most associated with electrical charge.

Problems

2. 12.7 using Equation 3-1.
3. 15 drums.

Chapter 5

Questions

4. Because the smaller and irregular size of the lemon fruit causes more fruit to be lost during conveyance, and the nonspherical shape results in less efficient juice extraction.

Problems

3. Yes. 104.8% efficiency is within 100–110% efficiency.
5. 53.6 sol/ton, 67.5 gal/ton, 1314.7 lb peel/ton.

Chapter 6

Questions

1. Citrus oils and aromas and their components.
8. Because the bromine reacts preferentially with the d-limonene.
10. No. A volatile condenser should be used as shown in Fig. 6-1. Straight tube condensers may not be capable of condensing the highly volatile alcohol/d-limonene mixture.
17. None.

Problems

4. 20,000 ml oil flavor enhancer.
7. Yes. It gives the 0.010% oil needed.

Chapter 7

Questions

2. Increased.
3. The yield would be low, also.
9. Pulp wash may be used in drinks or beverages, or in other food products.

Problems

4. 4.7% pulp.
5. 1.8% pulp.

Chapter 8

Questions

3. The deesterification of pectin by pectinase enzymes that form carboxyl groups, which, in turn, precipitate with divalent cations, such as calcium, removing juice cloud from single-strength juices or forming gels in concentrates.

Problems

5. 5.48×10^{-6} equivalents$/g_{soluble\,solids}$ — sec, compared to the method in the text which gives 4.12×10^{-5} equivalents$/$ml-min., which would be stable cloud.

Chapter 9

Questions

2. California orange juice is usually deeper in color than Florida orange juice because of California's drier climate.

Chapter 10

Questions

4. USDA: no standards. Florida: Grade A, 600 ppm naringin and 5.0 ppm limonin; Grade B, 750 ppm naringin and 7.0 ppm limonin.
7. Test the juice for limonin, heat, and retest until the heating no longer increases the limonin level.

Problems

2. 20.3 ppm limonin.
3. 7.2 ppm limonin.
6. 435 ppm naringin.

Chapter 11

Questions

4. Advantages: It emphasizes nutritional value and looks professional. Disadvantages: Because only a few of the mandatory nutrients are declared, it may give the impression of a lack of nutrition; it takes space on the label; it requires monitoring for compliance with the law.
6. Actually it is water because the definition of a nutrient in the first paragraphs of the chapter states that nutrients are those components that engender growth. Carbohydrates is the answer sought here, however.
10. 120% vitamin C and 8% thiamin.

Problems

4. • 1 g protein and less than 2% USRDA.
   • 20 g carbohydrates.
   • 120% USRDA vitamin C, which includes a 20% increase. (Again remember that naturally occurring nutrients can occur at 80% of the declared value.)
   • 8% USRDA thiamin which includes a 20% increase (for the same reason as for vitamin C).
   • 6% USRDA calcium. (Because this is added, it must be declared at the actual level.)
5. Ascorbic acid = 48.5 mg/100 ml; dehydroascorbic acid = 6.3 mg/100 ml. Because both have vitamin C potency, the total vitamin C level is the combined 54.8 mg/100 ml. To find what would be declared on the label, multiply 54.8 by 1.77 (177 ml/serving versus 54.8 mg per 100 ml) and divide by 60 mg (the USRDA value) to get 162% USRDA, or 160% USRDA rounded off according to the labeling rules.

Chapter 12

Questions and Problems

4. Extractors.
5. 58 sec^{-1}.

Chapter 13

Questions

3. The delayed bitterness and lighter color of the navel varieties.
5. Blood oranges have the best flavor, and mandarins have the best color.
7. Seedy and dark pigmented.

Chapter 14

Questions

4. Chewing tobacco in processing areas is prohibited according to Title 21 *Code of Federal Regulations* section 110.10 paragraph (b)(6).
9. No, but the FDA can be notified, which may result in a special inspection and subsequent action.
12. Quality of the products.

Chapter 15

Questions

9. This minimum plate count is unknown, perhaps in the tens of thousands or higher.
12. By a pure random correlation; or, it does not correlate.
14. Several months.

Problems

1. Too many to count (TMTC) or too numerous to count (TNTC), as over 300 colonies appeared on the plate.
4. 3.2 ppm.

Chapter 16

Questions

2. Food, water, light, sex odor, and temperature.
4. Because the peel oils and acidity of the juice are toxic or unpalatable to most insects.
6. Keep the plant clean, seal openings in the plant, keep doors closed, use air screens, and fog the plant with appropriate insecticides.

Chapter 17

Questions

2. Quality control personnel.
3. None.

Chapter 18

Questions

4. No black flakes are generally permitted.
8. During storage.
13. Acid hydrogens tie up the water molecules in concentrates, forcing the potassium and citrate to crystallize.

Problems

1. $2.35 - 2.27$ meq/100 ml $= 0.08$, less than the 0.10 that produces off flavors. Therefore, the shelf life has not expired, but it is about to do so shortly.
3. $Ksp\ (0°C) = 0.014\ M^2$

Chapter 19

Questions

4. Water addition (adding water to pasteurized orange juice), carbohydrate addition (adding sugars to FCOJ), cover-up (adding colors), and blending of unauthorized juices or juice products (adding grapefruit juice to orange juice).
6. Oxygen isotope analysis. Other methods are used because the oxygen isotope analysis is very costly and complex.
9. The determination of the isocitric acid, which can detect any type of dilution. As there is no commercial source of isocitric acid, it cannot be added to obscure the results.
14. Members of the industry themselves.

Problems

2.

ml Titrated	% Reducing Sugars	Action
5.21	4.59	none
6.03	3.97	diluted with sucrose
3.12	7.67	diluted with fructose

4. No. Early-season juice is authentic and may contain this much acid.

Chapter 20

Questions

2. "Nature's glue" holds membranes and tissues together.
9. Polyvalent cations.
11. The drain weight.
15. Juice drinks are cheaper to make, with a lower price to the consumer, and they provide a market for low-quality bitter juices, pulp washes, or other cheap juice-like products.
17. 30% protein, 10–15% hemicellulose, 5% pectin.

Problems

3. 688 pounds of citric acid, 13,134 gallons of water, 13,838 pounds of sugar, and 5797 gallons of juice.
5. 0 pounds of citric acid, 5,645 gallons of water, and 5,492 pounds of sugar.

Chapter 21

Questions

1. Protein, fiber, fat, and nitrogen-free extract.
6. To remove the slimy texture of the peel and aid in the expression of moisture from the peel.
8. Peel oil.

Problems

1. 25.0.
5. 6.3% alcohol; no.

Chapter 22

Questions

3. So they will not provide harborages for rodents and insects.
6. It produces no off odors.
9. 0.005% v/v.

Problems

2. 7.5 using Equation 22-2 or 22-4.
4. 1121 ppm COD and 728 ppm BOD.

Chapter 23

Questions

3. Significant figures determine the number of decimal places that legitimately can be used but do not indicate the actual error range. Error analysis does not indicate the precision of measurement, only the range of error.
4. By specification.
9. Least squares analysis is a form of regression analysis that uses calculus to mathematically determine the best fit of a particular curve to particular data.

Problems

2. 29.5, 711.5, 0.1552, 0.0000002.
3. VR = 15.25. No, there is a significant difference.

Chapter 24

Questions

2. Neither should be done. Production and quality control should be completely separate.
5. No. When an employee's personal problems affect his or her performance on the job, they become the business of the company, and the company has a right to act accordingly, especially to protect its products and other employees.

Chapter 25

Questions

3. To check customer complaints, to provide samples for potential customers, and for research purposes.
5. To continue quality control operations in the event of refractometer breakdown.

Problems

1. 84 samples and 9 acceptable rejects. One every 6 minutes.
3. 55,692 gallons

Index

Printed in the United States
by Baker & Taylor Publisher Services